Operations Research Using Excel

Operations Research Using Excel

A Case Study Approach

Vikas Singla

CRC Press
Taylor & Francis Group
Boca Raton London New York

CRC Press is an imprint of the
Taylor & Francis Group, an **informa** business

First edition published 2022
by CRC Press
6000 Broken Sound Parkway NW, Suite 300, Boca Raton, FL 33487-2742

and by CRC Press
2 Park Square, Milton Park, Abingdon, Oxon, OX14 4RN

First edition published by CRC Press 2022

CRC Press is an imprint of Taylor & Francis Group, LLC

ISBN: 9780367646431 (hbk)
ISBN: 9781032081076 (pbk)
ISBN: 9781003212966 (ebk)

DOI: 10.1201/9781003212966

Typeset in Times
by codeMantra

Contents

Preface..xi
Author..xiii

Chapter 1 Operations Research: An Introduction........................1

1.1 Introduction........................1
1.1.1 Genesis of OR........................2
1.2 Decision-Making Process in OR........................3
1.2.1 Problem Formulation........................3
1.2.2 Problem Analysis........................4
1.3 Model Development........................4
1.3.1 Case: Adidas AG........................5
1.4 Model Solution........................8
1.5 Unconstrained Optimization........................9
1.5.1 Case 1: University Press........................9
1.5.1.1 Problem Formulation........................9
1.5.1.2 Model Development........................9
1.5.1.3 Model Solution........................11
1.5.2 Case 2: Adidas AG........................12
1.5.2.1 Problem Formulation........................12
1.5.2.2 Model Development........................12
1.5.2.3 Model Solution........................13
1.6 Summary........................15
1.7 Case: Suzuki Motor Corporation........................16
1.8 Glossary........................17
1.9 Model Questions........................17

Chapter 2 Linear Programming........................19

2.1 Introduction........................19
2.2 Meaning of Linear Programming (LP)........................19
2.3 Assumptions........................20
2.4 Applications of LP........................23
2.4.1 Marketing Research........................23
2.4.2 Media Selection........................25
2.4.3 Financial Planning........................27
2.4.4 Product-Mix Problem........................29
2.4.5 Vendor Selection........................30
2.4.6 Make or Buy Problem........................33
2.4.7 Diet Problem........................35
2.4.8 Blending Problem........................37
2.4.9 Workforce Assignment........................39

2.5 Graphical Method ... 41
2.5.1 Illustration of Maximization: Adidas AG Retail Stores ... 41
2.5.2 Illustration of Minimization: Rose's Luxury Restaurant ... 49
2.6 Summary ... 56
2.7 Case 1: Federal Mogul Corporation ... 57
2.8 Case 2: Toyota Motors ... 58
2.9 Glossary ... 58
2.10 Model Questions ... 59

Chapter 3 Linear Programming: Simplex Method ... 67

3.1 Introduction ... 67
3.2 Illustration of Maximization ... 67
3.2.1 Case: Woodland Biomass Power (US) ... 67
3.2.1.1 Simplex Method ... 68
3.2.2 Case: Adidas AG Retail Stores ... 77
3.2.2.1 Simplex Method ... 78
3.3 An Illustration of Minimization ... 82
3.3.1 Case: Federal-Mogul ... 82
3.3.1.1 Simplex Method ... 84
3.3.2 Case: Rose's Luxury Restaurant ... 96
3.3.2.1 Simplex Method ... 97
3.4 Illustration of Maximization Problem with Greater Than Equal to Constraints ... 101
3.4.1 Simplex Method ... 102
3.5 Special Versions of LPP Solved by Simplex Method ... 107
3.5.1 Degeneracy ... 107
3.5.2 Unbounded ... 110
3.5.3 Infeasibility ... 111
3.5.4 Multiple Optimal Solutions ... 113
3.6 Summary ... 117
3.7 Case Study: Johnson Controls ... 118
3.8 Glossary ... 120
3.9 Model Questions ... 120

Chapter 4 Sensitivity Analysis and Duality Theory ... 125

4.1 Introduction ... 125
4.2 Fundamental Nature of Sensitivity Analysis ... 126
4.3 Applying Sensitivity Analysis ... 129
4.3.1 Change in Right Hand Side (RHS) Values of Constraint Functions (b_i) ... 129
4.3.2 Allowable Range of RHS Values (b_i) ... 132
4.3.3 Change in Objective Function Coefficient (Non-basic Variable) ... 134

4.3.4 Change in Objective Function Coefficient (Basic Variable) ... 137
4.4 Duality ... 144
4.4.1 Construction of Dual Problem ... 144
4.4.2 Relationship between Primal and Dual Problem ... 146
4.4.3 Dual Problem of Standard LPP ... 148
4.4.4 Dual Problem of Non-standard LPP ... 155
4.5 Solved Illustrations ... 164
4.5.1 Illustration of Sensitivity Analysis ... 164
4.5.2 Illustration of Duality ... 171
4.6 Summary ... 174
4.7 Case Study: Modern Foods India Limited ... 174
4.8 Glossary ... 176
4.9 Model Questions ... 176

Chapter 5 Network Model I: Transportation Model ... 179

5.1 Introduction ... 179
5.2 Structure of Transportation Model ... 180
5.3 Assumptions of Transportation Problems ... 182
5.4 Transportation Problem ... 182
5.4.1 Case: Musashi Auto Parts Michigan, Inc. ... 182
5.5 Transportation Solution Methods ... 185
5.5.1 Formulation of Model ... 185
5.5.2 Initial Solution ... 186
5.5.2.1 Least Cost Method ... 186
5.5.2.2 North-West Corner Method ... 188
5.5.2.3 Vogel's Approximation Method (VAM) ... 189
5.5.3 Optimality Test ... 191
5.6 Unbalanced Transportation Model ... 198
5.6.1 Scenario 1: Supply is More than Demand ... 198
5.6.2 Scenario 2: Demand is More than Supply ... 201
5.7 Degeneracy ... 206
5.7.1 Scenario 1: With a_{11} as New Basic Variable ... 207
5.7.2 Scenario 2: With a_{23} as New Basic Variable ... 207
5.8 Maximization ... 209
5.8.1 Linear Programming Formulation ... 215
5.9 Unacceptable Routes ... 216
5.10 Transshipment Problem: Theory ... 221
5.10.1 Case: Food Corporation of India ... 221
5.11 Summary ... 226
5.12 Case Study 1: Norland Plastics Co. ... 227
5.13 Case Study 2: Honda Motors ... 227
5.14 Glossary ... 228
5.15 Model Questions ... 229

Chapter 6 Network Model II: Assignment Model 233

6.1 Introduction 233
6.2 Assignment Problem: Construction of Model 234
6.2.1 Case: MarketOne International LLP 234
6.3 Assumptions 236
6.4 Comparison with Transportation Model 237
6.5 Hungarian Algorithm 238
6.6 Variations of Assignment Problem 244
6.6.1 Unbalanced Assignment Problem 244
6.6.1.1 Case 1: Number of Employees is More than the Number of Jobs 245
6.6.1.2 Case 2: Number of Employees is Less than Number of Jobs 250
6.6.2 Maximization Problem 253
6.6.3 Unacceptable Assignment 257
6.7 Crew Assignment 260
6.7.1 Case: Star Airlines 260
6.7.1.1 Case 1: Assignment of Different Types of Airplanes to Different Routes 260
6.7.1.2 Case 2: Assignment of Crew Members to Routes 262
6.8 Summary 270
6.9 Case 1: Assigning Workers to Processes 271
6.10 Case 2: Assigning Swings to Kids 272
6.11 Glossary 272
6.12 Model Questions 273

Chapter 7 Network Model III: Travelling Salesman, Vehicle Routing and Shortest Path Problem 281

7.1 Introduction 281
7.2 Travelling Salesman Problem 281
7.2.1 Branch and Bound Method 286
7.3 Vehicle Routing Problem 295
7.3.1 Clark–Wright Savings Algorithm 298
7.4 Shortest Path Problem: Dijkstra's Algorithm 302
7.5 Summary 305
7.6 Glossary 306
7.7 Case Study: JTEKT Corporation 307
7.8 Model Questions 310

Chapter 8 Project Scheduling: PERT and CPM 315

8.1 Introduction 315
8.2 Network Planning 316
8.2.1 Rules for Construction of Network Diagrams 317

8.3 Critical Path Method ... 321
 8.3.1 Case: Mattel Inc. ... 321
8.4 Program Evaluation and Review Technique ... 331
 8.4.1 Case: East Fork Roofing ... 331
8.5 Crashing: Time – Cost Trade-Offs ... 338
 8.5.1 An Illustration of Crashing ... 340
8.6 Resource Planning ... 346
 8.6.1 Resource Limited Scheduling ... 347
 8.6.2 Resource Levelling ... 351
8.7 Summary ... 354
8.8 Glossary ... 354
8.9 Case Study: Polyplastics Industries India Pvt. Ltd. ... 355
8.10 Model Questions ... 357

Chapter 9 Game Theory ... 363

9.1 Introduction ... 363
9.2 Characteristics of Game Theory ... 364
9.3 Elements of Game Theory ... 365
9.4 Solving Games: 5G Technology ... 365
 9.4.1 Explanation of Payoff Table ... 366
9.5 Games with Saddle Point ... 367
 9.5.1 Principle of Dominance ... 369
9.6 Games with Mixed Strategies ... 371
9.7 Graphical Method ... 373
 9.7.1 For 2*m Games ... 373
 9.7.2 Algebraic Method for 2*m Games ... 375
 9.7.3 For m*2 Games ... 376
 9.7.4 Algebraic Method for m*2 Games ... 378
9.8 Linear Programming Formulation ... 379
9.9 Summary ... 381
9.10 Glossary ... 382
9.11 Case Study: Vendor–Retailer Relationship ... 383
9.12 Model Questions ... 384

Index ... 391

Preface

The thought of the author behind writing this book is to cater to the need of understanding the subject of Operations Research (OR) of both under and postgraduate students in the disciplines of management, commerce and engineering. The endeavour is to combine practical and theoretical approaches in presenting the subject. Every chapter would begin with an industrial case representing a relevant problem requiring one of the techniques of OR for its resolution. The theoretical aspects of each chapter would provide students with a step-wise and easy approach to solve the mentioned case problem. After understanding and solving the case problem, each concept would be complimented by few solved examples. The approach fulfilled the need of students to be proficient in both practical and mathematical aspects of the subject. Other similar books predominantly focus on one of the aspects. Thus, this book intends to fill the gap. This book would emanate from more than ten years of teaching this course. The book is proposed to have nine chapters discussing various techniques of OR.

One of the techniques that deal especially with problems of resource allocation has been discussed in chapters of *Linear Programming.* The chapter explains the process of creating models in which a company faces constraints such as production hours, capacity, working hours, financial, etc. in order to fulfil the objectives of either maximizing profits or minimizing costs. These problems behave in a linear fashion, i.e. a unit change in constraints would have a unit effect on the objective. The solution of these problems has been illustrated by using an algebraic method known as the simplex method and a graphical method. As companies face multiple constraints at different levels, solutions through graphical methods become difficult, thus making usage of algebraic methods a necessity. The extensive application of Solver application in Excel software is explained in the book to solve complex questions of linear programming.

To understand different scenarios under which the same problem occurs *Sensitivity analysis and Duality* is examined. This chapter discusses two aspects. First, what would be the optimal limits within which the same problem is feasible? For instance, linear programming would only tell the feasible solution but not the range within which a changed scenario is viable or feasible. This aspect has been dealt with in sensitivity analysis. The concept of duality has significant application in implementation and interpretation of sensitivity analysis. It helps in identifying optimal solution for a primal problem.

Transportation problem discussed is a special kind of linear programming problem that is used to determine the number of units to be distributed from limited supply centres to known demand centres. In the majority of cases, there are more than one supply and demand centres. This makes the problem of allocating number of units from one supply centre to a particular demand centre peculiar. With the increase in the number of supply and demand centres, the problem becomes more complex. Typically, the objective of a transportation model is to find the minimum cost required to fulfil demand of various destinations when material is sourced from various suppliers. This model has been discussed with appropriate cases and solved examples.

Assignment model is a special form of transportation problem discussed in Chapter 5. The objective of the assignment model is to allocate or assign jobs to machines or jobs to people. The most common application of the assignment model is assigning various jobs to various limited employees. This objective can be fulfilled based on certain criteria. Various workers can be assigned to various jobs based on the criteria of least time taken to perform the job or least cost incurred to the company. The assignment model has a special application in location decision-making. Which department should be assigned to which building in a University? Should grocery store be located within a supermarket? An apparel store in a mall is located in which part of the mall? Assignment model can be helpful in solving such location problems.

Out of various variants, the chapter on Travelling Salesman, Vehicle Routing and Shortest Path Problem has discussed three special, practical and complex network flow models. The traveling salesman problem is considered to be very vital in both theoretical and industrial applications, though solving it becomes increasingly complex with the increase in the number of nodes. The problem deals with a salesman or a vehicle starting from origin, visiting every other city (node) only once and then coming back to source. The objective of Vehicle Routing problem is to allocate a vehicle to which cities so that savings in terms of distance travelled or time taken is maximum. The Clark-Wright savings algorithm has been discussed to solve vehicle routing problems. Finally, the shortest path problem was discussed using Dijkstra's algorithm to find the shortest path out of multiple paths from a source to reach to a destination without the need to come back to origin.

Project Scheduling: PERT and CPM deals with the planning of projects. A project involves a number of aspects and its successful completion requires a scientific and systematic approach. PERT and CPM are two such approaches which answer questions such as how many individuals are required to perform a particular activity? Can they be used for the performance of other activities? How much time does each activity take? What is the cost required to perform each activity? These two methods are network-oriented techniques used primarily to determine project completion time.

Game theory is very relevant in deciding management strategy in the light of competition. For instance, which advertisement would be successful when two competing products are trying to launch similar products? Which politician would win an election? The chapter covers the concepts of pure and mixed strategy games. Which player should follow a strategy that results in the maximum number of game needs, the construction of game matrix and finding value of the game? This was explained by illustrating the concepts of saddle point, dominance method, algebraic and graphical method.

Finally, any suggestions regarding corrections, modifications including addition or deletion would be highly appreciated. Such suggestions would be very helpful in formation of a helpful and successful book.

Vikas Singla

Author

Vikas Singla is currently holding the position of Assistant Professor in the School of Management Studies at Punjabi University, Patiala, India. By qualification, he has done Industrial Engineering, MBA and Ph.D. After gathering fruitful experience from the manufacturing industry, he has been in the teaching profession for the last 15 years. He teaches Business Administration and handles subjects in the area of business statistics, research methodology and operations management. He has been associated with the ministry of human resource development in formulation modules in the subject of operations management and advance quantitative techniques for both under and postgraduate students. His specific research areas include lean manufacturing techniques. He has contributed to these fields academically by publishing research papers in various national and international journals.

1 Operations Research
An Introduction

1.1 INTRODUCTION

Two kids are competing to build a house out of plastic building blocks. Both would most likely use a different, but not completely different, method of combining blocks in order to build a house, despite using similar resources and trying to produce similar output. In a more complex example, car manufacturing involves more than ten thousand parts grouped into hundreds of components to build a single car. Two competing manufacturers would be using similar inputs to produce similar output of a car. However, one house or a particular car is better than the competitor. It could be better on one or more of various key performance indicators such as quality, aesthetics, durability, creation time, etc. The question is why there is a difference in output when similar inputs are used. Here comes the importance of understanding of operations. In the simplest terms, *operations* can be defined as the process of doing work. This process is an interaction or a combination of various activities arranged in a certain way so as to convert resources into formation of a desired product or service (Figure 1.1). A kid or a car manufacturer applying operations in a more efficient and effective manner would be able to provide more productive outputs and be able to defeat competitor.

Historically, to make a process more effective, the work has been categorized into more specialized arenas. F.W. Taylor, considered as the father of scientific management, very successfully segregated managerial work from operations work. Specialization of work by breaking down into as minute entity as possible and segregating them into various disciplines such as finance, marketing, sales, personnel management, etc. resulted in studying of operations as a separate discipline. Henry Ford through its Model T established such specialization to its highest degree when the company by using similar resources as its competitors built a cheaper version of car by modifying the production process to mass production. No competitor was

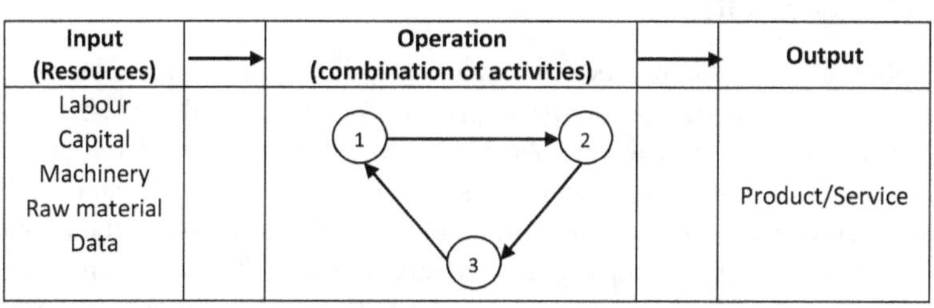

FIGURE 1.1 Operation as a conversion process.

DOI: 10.1201/9781003212966-1

able to match the efficiency of Ford cars even by using the same resources, similar processes and comparable output. Thus, operation is further defined as the method of *managing processes.*

However, specialization of disciplines does not imply isolation or working in separation. By definition, operations deal with processes or how a work is being done. Processes are involved in every discipline of an organization. Accounting, finance, product design, production, marketing, quality control, etc. and many more functions work by carrying out certain activities. Each function performs activities according to a certain process to fulfil its role and then interlinks with other functions. Marketing department through its feedback surveys analyses consumer preferences and places them with production department, which in order to produce more or new items, asks for more budget from finance to procure materials and hire more people. Human resource manager steps in to apply processes of hiring. Purchase department starts processes of procuring more material and machines and so forth. This entails that in addition to internal processes of every function, they should gel with each other effectively. This makes managing of processes or study of operations management as encompassing every function of an organization and not as a separate entity. Also as discussed, a process is a way of doing a work and work is being performed by every organization. A small carpenter, a retail store, manufacturing companies, service firms, and defence equipment producers all indulge in processes making application of operations management in order to manage their processes essential. Scale and complexity of companies in the present hypercompetitive business scenario force them to produce better whether in terms of quality, price, delivery time or any other feature. The availability of limited resources which are being fought over by various players further puts pressure to use valuable and restricted resources to produce more valuable products. Proper planning, coordination of various functions and managing of overall operations have been found very effective in achieving such goal.

Solving diverse and complex process problems requires a systematic and mathematical approach by using quantitative analysis. A variety of monikers such as management science, decision science, and operations research (OR) have been used to describe body of knowledge dedicated to study and research of mathematical structured methods used in decision-making. The above discussion focussed on explaining the importance of operations in a variety of functions of different organizations. Thus, this book used Operations Research (OR) as its title.

1.1.1 Genesis of OR

Promulgation of scientific methods of management in the early 20th century encouraged the establishment of quantitative approach to decision-making. This approach became increasingly useful during World War II when the armed forces strongly applied quantitative methods to help in strategic problems. For instance, British forces were considered to first utilize methods of OR to understand the application of radar technology in intercepting enemy aircrafts. Due to the all-encompassing nature of OR as discussed in solving process problems of different functions, such application of OR methods was found to be successful by forming teams consisting of experts from different disciplines such as mathematics, engineering, supply chain, etc. Effective results obtained from OR methods provided an impetus to its flourish into other fields such as agriculture, manufacturing and services.

From various definitions available pertaining to OR, management science or decision sciences, a common understanding that emanates is that the OR discipline is a scientific method providing decision makers quantitative tools for solving problems. The key points in these definitions are: first, the usage of the term scientific method, which implies a systematic procedure of solving problems, and second, the use of mathematical quantitative tools to provide feasible results that can be repeated in the case of recurrence of similar problem. A thorough understanding of quantitative methods drastically improves decision-making effectiveness. For example, a logistics company intending to reduce its transportation cost incurred in supplying materials from different warehouses to various retailers by adopting an optimum approach would make an effective decision by using a quantitative approach. Qualitative analysis would ask for adoption of the shortest possible route. However, a quantitative well-structured approach would provide comparison of costs incurred on various approaches and would allow manager to make an informed decision. However, such an approach requires clear and concise formulation of problem, which is the first step in the OR process. Details have been discussed in the subsequent section.

1.2 DECISION-MAKING PROCESS IN OR

The entire OR process applied for decision-making by using quantitative tools can be divided into two categories (Figure 1.2). The first category discusses the essential steps in formulating a problem accurately, and the second category describes analysing the problem. To establish the sound conceptual understanding of decision-making process using OR methods, one should be conversant of steps involved in such endeavour.

1.2.1 Problem Formulation

This step includes the following sequence:

1. **Problem should be well defined**: The first and foremost step is to fully understand and formulate the problem. A problem occurs when there is a discrepancy between the actual and desired outcomes. For example, undergraduate students in business management in their year of specialization face a problem of deciding which discipline to pursue. If there had been no choice, then problem would not occur, but the number of choices causes an individual to make a decision.

Problem formulation						Problem analysis		
Defining the problem	→	Identification of number of choices	→	Selection of criteria	→	Alternative evaluation	→	Selection of alternative

FIGURE 1.2 OR process.

2. **Identification of a number of choices**: The student can specialize in marketing, finance, decision sciences, personnel management, and supply chain analytics, among others. A plethora of choices make decision-making problematic.
3. **Selection of criteria**: Decision has to be made by evaluating all alternatives based on certain observable and measurable criteria. For example, one criterion could be job offers that a discipline provides. A single criterion would make selection of alternatives easy. A discipline that offers maximum job opportunities would be selected. But seldom is such case. A candidate might also evaluate each discipline in terms of salary package, interest, education background, peer pressure, etc. Multiple criteria make problem difficult but more realistic. In the beginning of chapter, a kid constructing blocks can create a house by using a variety of set of pieces and providing different alternatives. If only a particular piece fits, then it would be easy; however, if selection has to be made from different pieces, then the kid has to adopt a selection criteria which could be single or multiple.

1.2.2 Problem Analysis

For a scientific solution to a problem, the OR process applies model development for analysing various selected alternatives and then provides the most optimal solution by solving the model quantitatively. Problem analysis primarily does two things:

1. **Alternative evaluation**: All the alternatives, for instance, all disciplines on offer, have to be evaluated on identified criterion/criteria. No one alternative provides the best result. A discipline may offer the best job opportunities, but an individual lacks interest or does not have the acumen to understand it. A frequent problem faced in businesses is how to maximize sales without incurring heavy expenditure on costs. Should new stores be opened to make the product range more accessible? Should more sales people be employed? Should prices be reduced to lure more customers? OR methods provide quantitative tools to evaluate such alternatives by considering all criteria.
2. **Selection of alternatives**: Importantly, no best alternative can be achieved, so instead of achieving the maximum or minimum objective, the *optimum* level of result is achieved. An ideal situation would be increased in maximum sales without incurring any additional cost that is implausible. Therefore, decisions are optimized rather than maximized or minimized. Achieving of optimum result would make selection of an alternative, which is referred to as making a decision.

1.3 MODEL DEVELOPMENT

Development of model is important for executing problem analysis. The main purpose is to evaluate different alternatives. It has been emphasized that OR discipline focusses on the application of quantitative tools to help in decision-making by adopting a systematic approach to problem solving. This approach uses mathematical models that, through mathematical expressions, equations and symbols, represent real problem or situations. Models simulate real situation. They are used to examine an optimal solution in as real environment as possible. Models are a very effective way to analyse and make inferences about real problems as they are constructed as an archetype of problem at hand. For instance, to examine the effect of variation in total units to be supplied on cost of transportation, a model could be constructed involving cause-and-effect variables. In this case, units that are to be transported would be causal variables, and cost of transportation is the outcome or effect. By varying or experimenting with causal variables, we can analyse the effect on cost. This process involves less time and cost, and also is less risky than doing experimentation practically. From centre A to B, if cost of transporting one unit is $10 per unit, then a mathematical model of situation can be represented by equation: Cost (C) = 10x; where x is the number of units. For 1 unit, cost would be $10; for 2 units, it would be $20; and so on. Decision maker has to decide which conclusion suits best.

However, model conclusion provides the best results only if they represent the real situation accurately. That is why problem identification and formulation are so important. In the above example, cost could be the function of other variables and not only the number of units such as the size of truck or the number of possible routes, etc. Thus, more closely a model represents the problem, more conclusive results would be.

The following case explains the development of a model by highlighting various elements used in its formulation.

1.3.1 Case: Adidas AG

Adidas AG is the second largest sportswear manufacturer in the world. It is primarily a male brand catering to the needs of burgeoning young population with disposable income. The company is looking to open retail stores in a major capital city. The retailer has shortlisted two locations: a famous shopping mall (location A) located at the outskirts of a city and a popular shopping plaza (location B) located at the heart of the city. Adidas wanted to estimate the number of units that could be sold from two geographic areas, which would maximize sales. Profit generated per unit from location A was estimated to be $80, whereas from location B was $40.

Profit was found to be a direct function of the number of units sold, which in turn depended on the accessibility of store measured in terms of distance, population density of each geographic area and lastly median household income. From location analysis, a retailer considered his shop to be attractive in the range of 14 miles. Location A customers needed to travel an average distance of 9 miles, whereas customers of location B travelled 4 miles to access offerings of the retail shop. Second, population density in terms of the number of households was estimated to be 2,000 and 7,000 for A and B, respectively. Population density represents prospective clients or potential demand from a stores' trade area. A household can buy more than one unit, so the retailer decided to cater to maximum demand of 10,000 households. Finally, demographic data indicated that the maximum average household income of two demand areas was $20,000. Median household income would decide the extent to which customers would be willing to spend on products. It was found that the mall (location A) population had a median income of \$5,000 and the shopping plaza (location B) population had \$10,000. Under these circumstances, the retailer wants to estimate the maximum profit that could be earned by selling specific units to customers of A and B.

The following elements can be deduced from the above illustration:

- **Decision variables**: are those variables that directly influence the achievement of an objective. In this case, the objective is to maximize the profit of a sportswear retailer, which is dependent on the number of units being sold at stores at two different locations. Therefore, the numbers of units being sold – symbolized by x_1 and x_2 representing sales at two locations A and B, respectively – becomes *decision variables.*
- **Objective function**: the store manager wants to estimate the number of units of sportswear being sold at each store, which would maximize profits of the entire chain. Thus, in this case, the objective is *profit maximization.* In some cases, such as costs, time to travel, distance, etc., the objective function could be minimization. However, the objectives of minimization and maximization are sometimes misnomers because no cost can be minimized to zero or profits can be increased to the highest value because every productive activity is performed under certain constraints such as limited resources, competition, etc. Thus, more appropriate form of understanding the objective is that the solution achieved is optimal rather than minimum or maximum. The objective function is denoted by 'Z'.
- **Constraints**: Constraints are the limitations under which any productive activity is performed. In this case, three constraints are illustrated. The first constraint is accessibility. The average effective trade area of both stores was estimated to be a maximum of 14 miles. This implies that a store's potential customers would be within the reach of this distance. The second constraint is population density. It was estimated that the store could fulfil the demands of an area with a population of 10,000. More than that would cause inefficiencies in the operations of the store. Finally, income has a role in influencing purchase intentions. In this case, it was mentioned that the Adidas store can offer products to customers whose median income is maximum \$20,000. Thus, the objective has to be achieved by considering these two limiting factors.
- **Coefficients**: Coefficients used in both constraint and objective function equations associated with each decision variable indicate the level of activity per unit. In terms of accessibility, the case illustrates that store can cater to population residing within 9 and 4 miles of locations A and B, respectively. Similarly, other associations between decision variables and coefficients of constraints are formulated. In other population density constraints, coefficients were 2,000 and 7,000, and for constraint of median income, they were \$5,000 and \$10,000. For the objective function of maximization, \$80 per unit profit from A and \$40 from B are coefficients.

- **Inequality**: Inequalities such as less than (<), less than equal to (≤), greater than (>) or greater than equal to (≥) represent the relation between level of activity and maximum or minimum resources available corresponding to that activity. For instance, the above illustration indicates that maximum accessibility is 14 miles; maximum population density that can be served is 10,000 with a maximum median income of $20,000.

These elements used in model formulation are denoted by certain symbols. These generally used symbols are as follows:

x_j = number of decision variable (for j = 1, 2,…., n)
Z = objective function of maximization or minimization
a_{ij} = coefficients of decision variables with regard to constraint equations indicating the amount of resource i that is available for a particular decision variable *j*
c_j = coefficients with regard to each decision variable in the objective function indicating either profit or cost for a particular decision variable
b_i = constraints indicating the maximum or minimum amount of resource available to carry out a particular activity i (for i = 1, 2,…., n).

Thus, a general form of linear programming problem would be formulated as follows:
Objective function of maximization or minimization

$$Z = c_1x_1 + c_2x_2 + \cdots\cdots + c_nx_n$$

Subject to: $a_{11}x_1 + a_{12}x_2 + \cdots\cdots + a_{1j}x_j \leq b_1$

$$a_{21}x_1 + a_{22}x_2 + \cdots\cdots + a_{2j}x_j \leq b_2$$

$$\cdots\cdots + \cdots\cdots + \cdots\cdots + \cdots\cdots \leq \cdots\cdots$$

$$a_{i1}x_1 + a_{12}x_2 + \cdots\cdots + a_{ij}x_j \leq b_i$$

For the above case and by using explained symbols, the mathematical representation of the model is as follows:

The objective is to maximize profits that are dependent on the number of units sold at each store. Therefore, if

x_1: represents the number of units sold at a shopping mall in thousands (location A)
x_2: represents the number of units sold at a shopping plaza in thousands (location B)
Then, the objective function would be:

$$\textbf{Maximize } Z = 80x_1 + 40x_2$$

However, the achievement of an objective is constrained by certain limitations. Therefore,

Constraint 1: **Accessibility**: Data in the case indicates that, on an average, a client would travel 9 miles to a shopping mall to buy one unit, so buying x_1 units by the entire estimated population in the given area would be $9x_1$. Similarly, considering location B would give the following equation:

$$9x_1 + 4x_2 \leq 14$$

Constraint 2: **Population density**: As explained, one unit can be accessed by any of 2,000 households in area A, so x_1 units can be accessed by $2{,}000x_1$ or $2x_1$. Similarly, considering for location B would give the following equation:

$$2x_1 + 7x_2 \leq 10$$

Constraint 3: **Median income**: Profit earned by a store depends on money being spent on purchase of Adidas products, which in turn is a function of income. This gives the following equation:

$$5x_1 + 10x_2 \leq 20$$

Finally, it is important to mention that the number of units sold could not be negative. Therefore, very importantly, $x_1, x_2, x_3 \geq 0$.

1.4 MODEL SOLUTION

This step involves the application of algorithms to derive a solution. It is important to understand two aspects of solution. As discussed above, a problem formulation is done in a way to fulfil either the objective of maximization or minimization. Therefore, a solution that represents the best of this objective function is considered most acceptable. Such kind of solution is termed as *optimal solution*. For the above illustration, the algorithm should be capable of identifying the number of units sold that would maximize profit. Second, the solution has to be *feasible* as well, which means the solution should not violate any of the constraints (Figure 1.3). Profit can be increased by increasing the accessible geographic area but it cannot be more than 9 and 4 miles for stores located at A and B respectively. therefore, though a solution could provide maximum provide, it might not be feasible if it requires more than designated accessible area. Thus, the best possible solution is not the maximum or minimum but the most optimum. This concept is termed as *optimization*.

Model Development	→	**Model Solution**
Decision variables Objective function Constraints Coefficients Inequalities		Optimal Feasible

FIGURE 1.3 Problem analysis.

1.5 UNCONSTRAINED OPTIMIZATION

The purpose of quantitative tools applied in OR is to find an optimum solution to a relevant problem. Two types of optimization are discussed in this text: one is *unconstrained*, and the other is *constrained.* Unconstrained optimization implies that the problem solution has no resource constraints. The mathematical expression consists of an objective function and a decision variable with no constraint equations. The following cases illustrate the solution of unconstrained optimization by examining the lot size to be ordered which would minimize total cost of inventory.

1.5.1 Case 1: University Press

1.5.1.1 Problem Formulation

East South University conducts examinations of approximately 2 lakh students enrolled in its numerous affiliated colleges. The examinations are conducted twice a year. To print exams and answer sheets, the University has a centralized printing press at its premises. As the number of students enrolled and the examination system remains almost constant, the demand for paper required for printing also remains constant. This also results in ordering a fixed quantity of paper every time an order is placed. The University has long-term contracts with suppliers of paper, making them very reliable. They provide a fixed batch size of paper required in fixed and certain time period. Finally, the press never goes out of stock as order gets replenished when stock reaches zero level. The press has to incur cost of placing an order mostly twice a year and cost for maintenance of paper in the warehouse. The press buys paper in 1,500 pound rolls for printing. Annual demand is 2,500 rolls. The cost per roll is $800, and annual holding cost is 15% of the cost. Each order costs $50. How many rolls should press order at a time so that the total cost of keeping that many rolls is minimized?

1.5.1.2 Model Development

Solving this requires understanding of variables used in mathematical expression required to describe the problem.

Total cost is a function of only ordering and holding costs. Notations used are: D: Annual demand; H: Holding cost per unit; S: Ordering or setup cost per order; and Q: Batch size.

Holding Cost

On average, at any particular unit, Q/2 units are being held up in inventory between the start and end of cycle (time period t). Figure 1.4 shows that as the number of units in inventory increases, the annual holding cost increases. If the cost of holding one unit is H, then:

Annual holding cost = Average inventory level * Holding cost per unit

$$=(Q/2)*H$$

Ordering Cost

Suppose the annual demand for an item is 1,000 units and the manager orders 100 units per order. Thus, he/she has to place an order 10 times. Thus, (D/Q) represents

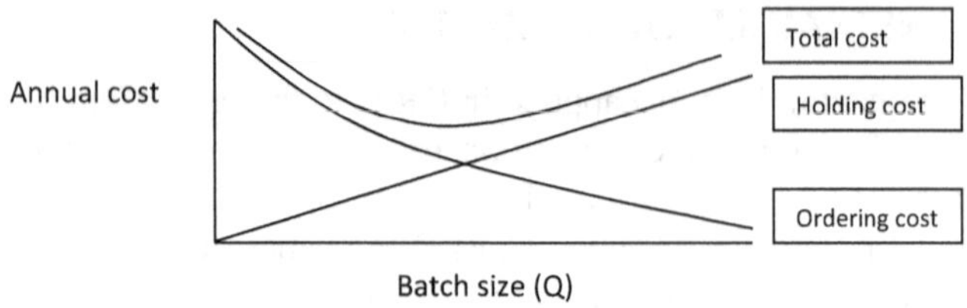

FIGURE 1.4 Cost curves in EOQ.

the number of orders. As shown in Figure 1.4, as the number of units stored in inventory increases, the annual ordering cost decreases. If S is set up or ordering cost per order, then:

Annual ordering cost = Number of orders per year * Ordering or set up cost per order

$$= (D/Q) * S$$

Figure 1.4 deduces a very important relationship between holding and ordering costs. The graph clearly shows that as the number of units stored increases, holding cost keeps on increasing and ordering cost decreases. This is quite understandable, as with bigger batch size, the frequency of repeating orders automatically decreases. Thus, ordering and holding costs have an inverse relationship with each other. Importantly, the intersection of the holding and ordering costs curve indicates the number of units that should be ordered per batch.

At this intersection point, the most *optimum* total annual cost can be computed. In addition, the intersection point if extended further to touch the total cost curve, it would be touching at the lowest point of curve that is minimum point. Thus, a solution by using an algorithm involves finding the lowest point for finding out the objective of minimization. Similarly, in the case of maximization, it would be to find the highest point. These points are called 'extreme points' of the curve. At the extreme point, change in objective function with change in decision variable is zero implying a flat rather than increasing or decreasing curve. Mathematically, it can be explained as change in 'y' with change in 'x' to be zero. An optimal solution would be said to be achieved at this point. For unconstrained optimization, it is achieved as explained below.

As total cost is a combination of two types of cost:

$$\text{Total cost} = \text{Annual holding cost} + \text{Annual ordering cost}$$

$$TC = (Q/2) * H + (D/Q) * S$$

So, we know:
Annual demand D = 2,500;
Holding cost, H = 0.15 * 800 = $12
Ordering cost, S = $50

Objective function (Z) is to minimize total cost:

$$\text{Minimize TC} = (Q/2) * 12 + (2{,}500/Q) * 50$$

Thus, in unconstrained optimization, the objective function Z is represented as a function of one decision variable. As discussed, minimization would be achieved when the slope changes, i.e. $\Delta y/\Delta x = 0$.

1.5.1.3 Model Solution

1. *First derivative of TC equation*:

$$f'(Q) = \{12/2\} * \{\Delta(Q)/\Delta(Q)\} + (-)\{2{,}500 * 50\} * Q^{-2}$$

$$= 6 - 125{,}000/Q^2$$

2. *Solve for decision variable Q by putting the first derivative to zero*:

$$6 - 125{,}000/Q^2 = 0$$

$$Q = \sqrt{125{,}000/6} = 144.34 \text{ or rounded off to 145 units.}$$

3. *Is this the lowest point on curve? For that, take the second derivative*:

$$f''(Q) = 250{,}000/Q^3 > 0.$$

   It would always be greater than zero, implying a calculated Q of 145 would be the minimum point on the curve (Figure 1.5). It would yield a minimum total cost.

4. *Find total cost by putting a value of Q = 145 units:*

$$\text{TC} = (145/2) * 12 + (2{,}500/145) * 50$$

$$= \$1732.19$$

Examining the graph (Figure 1.5) provides a very good understanding of the solution. The objective function was to find the minimum quantity that should be ordered as a batch size that would minimize the total inventory cost. The minimum point on the graph is the lowest point where it becomes flat, implying slope to be zero or any change in 'x', i.e. quantity ordered would result in no change in 'y'. The graph shows that if Q = 10, then the total cost would be 12,000, and if it is increased to 20 units, then total cost decreases to 8,000, implying a slope of (12,000–8,000)/(20–10) = 400. However, this slope is zero at the lowest point, indicating the minimum value of Q.

To summarize the rules of model solution, i.e. to find maximum or minimum for unconstrained optimization are:

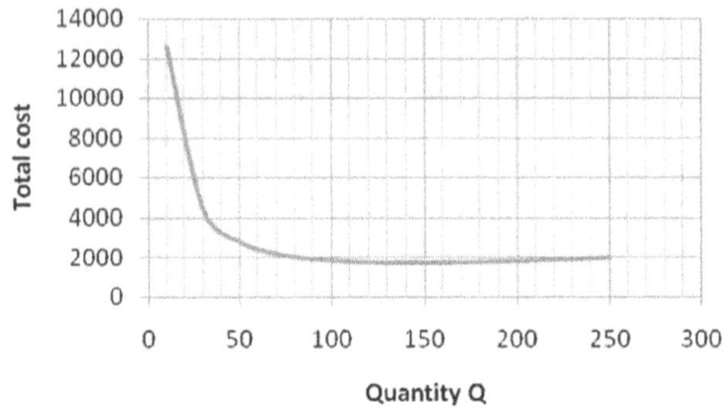

FIGURE 1.5 Total cost curve.

- The slope has to be zero, i.e. (change in y)/(change in x) = 0 represented in the form of the first derivative as $f'(x) = 0$.
- Take the second derivative ($f''(x)$) in order to ascertain whether the identified extreme point is minimum or maximum.
  - If $f''(x) > 0$, then extreme point is minimum.
  - If $f''(x) < 0$, then extreme point is maximum.
  - If $f''(x) = 0$, then it is inconclusive.

1.5.2 Case 2: Adidas AG

1.5.2.1 Problem Formulation

Adidas AG is the second largest sportswear manufacturer in the world. The company is looking to open one of its retail stores in a major capital city. The retailer has short-listed two locations: a famous shopping mall (location A) which is located at the outskirts of city and other a popular shopping plaza (location B) located at heart of city. The fixed cost of location A is $7,500 and of location B is $7,000 per month. Variable costs at A and B are expected to be $4 and $3 per unit, respectively. Total cost is a combination of fixed and variable costs. It is a function of the number of units sold. Even if no units are sold, the total cost would be equal to the fixed cost. However, with an increase in sales, total cost would increase by variable cost incurred to sell that unit as the retailer has to pay for utilities and extra labour. The quantity of units produced depends on the demand for shoes. Demand varies with the change in price of shoes. At initial price with no increase, demand was found to be 4,500 units per month at A and 5,200 at B. However, with every increase in price by one unit, the decrease in demand was estimated to be by 50 units at A and by 40 at B. Given this information, the company intends to find the price of shoe at both locations, which would yield maximum profit.

1.5.2.2 Model Development

The following notations are used to form a mathematical expression for given problem:

Number of units sold = x
Price of one unit = p
Fixed cost = FC = \$7,500 for A and \$7,000 for B
Variable cost = VC = \$4 for A and \$3 for B

$$\text{Total cost} = TC = FC + VC$$

$$\text{Profit} = P = \text{revenue} - \text{cost}$$

VC is a function of the number of units sold. So for A it would be 4x and for B 3x.
So, the cost function for location A = f (no. of units x) = 7,500 + 4x
Cost function for location B = f(no. of units x) = 7,000 + 3x
Demand function for location A = f(price per unit p) = 4,500 – 50p
Demand function for location B = f(price per unit p) = 5,200 – 40p
The number of units sold 'x' is represented by demand at a particular location. Therefore, the cost function is converted as follows:
So, the cost function
for location A: f(no. of units x) = 7,500 + 4x

$$f(p) = 7{,}500 + 4*(4{,}500 - 50p) = 25{,}500 - 200p$$

Similarly, for B:

$$f(p) = 7{,}000 + 3x = 7{,}000 + 3*(5{,}200 - 40p) = 22{,}600 - 120p$$

Revenue is a function of the price of product and demand for shoes.
So, the revenue function

$$\text{at location A} = (4{,}500 - 50p)*p = 4{,}500p - 50p^2$$

$$\text{at location B} = (5{,}200 - 40p)*p = 5{,}200p - 40p^2$$

The objective function (Z) is to maximize profit, which is formulated as follows:

$$\text{Maximize Profit }(A) = (4{,}500p - 50p^2) - (25{,}500 - 200p)$$

$$= -50p^2 + 4{,}300p - 25{,}500$$

$$\text{MaximizeProfit }(B) = (5{,}200p - 40p^2) - (22{,}600 - 120p)$$

$$= -40p^2 + 5{,}320p - 22{,}600$$

1.5.2.3 Model Solution

Maximization of profit for *location A*:

1. First derivative of profit equation:

$$f'(p) = f'(-50p^2 + 4{,}300p - 25{,}500)$$

$$= -100p + 4{,}300$$

2. Solve for decision variable Q by putting the first derivative to zero:

$$-100p + 4{,}300 = 0$$

   Thus, p = $43
3. Is this the highest point on curve? For that, take the second derivative:

$$f''(p) = -100 < 0.$$

   It would always be less than zero, implying calculated p would be the maximum point on the curve (Figure 1.6). It would yield maximum profit.
4. Find profit by putting a value of p = $43

$$\text{Profit (A)} = -50p^2 + 4{,}300p - 25{,}500$$

$$= -50*43*43 + 4{,}300*43 - 25{,}500$$

$$= \$66{,}950$$

Maximization of profit for *location B*:

1. The first derivative of profit equation:

$$f'(p) = f'(-40p^2 + 5{,}320p - 22{,}600)$$

$$= -80p + 5{,}320$$

2. Solve for decision variable Q by putting the first derivative to zero:

$$-80p + 5{,}320 = 0$$

   Thus, p = $66.5
3. Is this the highest point on curve? For that, take the second derivative:

$$f''(p) = -80 < 0.$$

   It would always be less than zero, implying calculated p would be the maximum point on the curve (Figure 1.7). It would yield maximum profit.
4. Find profit by putting a value of p = $66.5

$$\text{Profit (B)} = -40p^2 + 5{,}320p - 22{,}600$$

$$= -40*66.5*66.5 + 5{,}320*66.5 - 22{,}600$$

$$= \$154{,}290$$

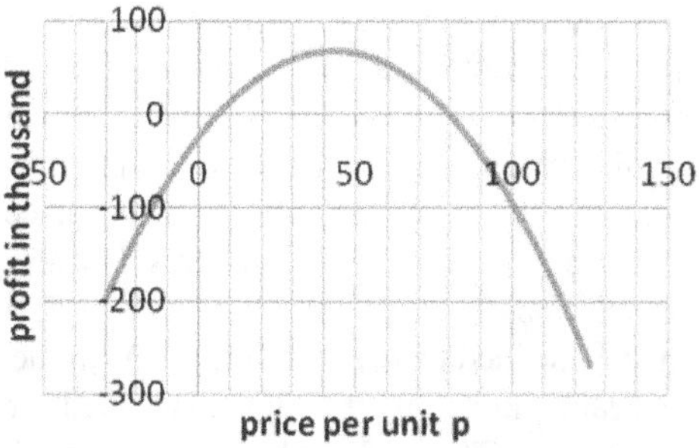

FIGURE 1.6 Profit function for location A.

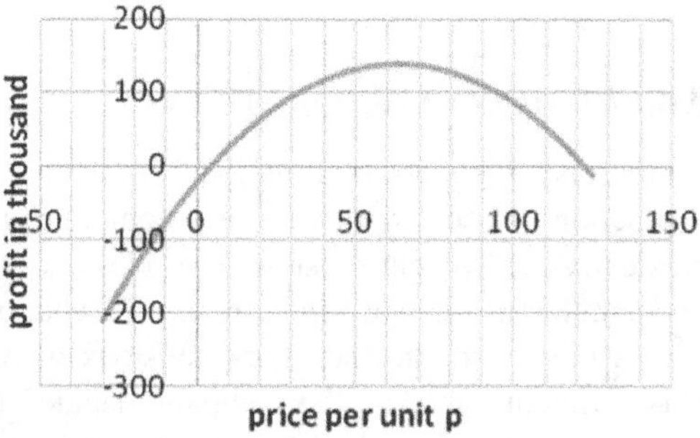

FIGURE 1.7 Profit function for location B.

So, for company to yield the maximum profit, location B proves to be more beneficial.

Graph for location A (Figure 1.6) shows the highest point at $43 per unit, giving a maximum profit of $66,950. At the highest point, the change in profit with change in price is constant, resulting in a slope of zero. After the highest point, the curve starts falling, indicating that if shoes are sold at a higher price than $43, profits will begin to fall. Similar interpretation can be made for location B (Figure 1.7).

The above illustrated cases explain quantitative solution of identified problems with no constraints on resources. However, in practical business situations, no problem is bereft of constraints. Chapter 2 would describe the powerful quantitative tool of linear programming in solving such problems. The case of Federal Mogul Corporation would be solved and used for explaining constrained optimization.

1.6 SUMMARY

This chapter discusses the genesis of OR as a discipline by describing its origin from a combination of various functions such as engineering, product development, operations management, etc. It emphasized the importance of OR methods in scientific decision-making through the application of mathematical procedures. The purpose of this chapter was to make the reader aware of the importance of the OR process in helping decision makers to make better decisions.

Problem identification was described as the most important step in the OR process. Accurate description of problem and identifying all possible alternative solutions was essential in decision-making. The development of mathematical models as a representative of real problems or situations would provide an effective way of prescribing feasible and optimum solution. These models solved by applying different OR tools would be discussed in subsequent chapters. However, their purpose was to find the best possible solution.

Problems are always constrained by certain limitations in the production process. Those without any constraints are termed as unconstrained, and their solutions as unconstrained optimal solutions. This chapter has briefly described solutions to such problems. The important understanding of these solutions was the systematic application of the decision-making process, which would be emulated further in the cases of constrained problems.

1.7 CASE: SUZUKI MOTOR CORPORATION

Suzuki Motor Corporation is a Japanese multinational company involved in designing, development, production and selling of various automobiles ranging from four-wheel vehicles such as cars, two wheelers like motorcycles and heavy vehicles like trucks, etc. It operates manufacturing plants in more than 20 countries with a number of suppliers from across the world. In one of its divisions involving manufacturing of emblems for its different vehicles, the company decided to increase capacity by installing another production line dedicated to the production of high-quality emblems for luxury cars. Such an addition demanded expansion or modification of facility. Expansion is a highly costly and time-consuming process, so modification of the existing layout was finalized.

Problem identification and formulation required collection of data such as dimensions of shop floor, each work centre, equipment and machinery involved. Relation between operations was judged by observing distance data between processes. Threadbare analysis was performed to check the correctness of data, and various alternative solutions were forwarded. Alternatives involved altering the distance between various workstations and calculating distance between them for each alternative. The aim of alternative layouts was to reduce distance between the processes for the whole facility. Layout alternatives were evaluated based on the criteria of cost of change and overall material flow. Mathematical models were developed to evaluate the alternatives. OR models, such as linear programming and transportation models, were applied to reach an accurate decision. By applying these models, the management was able to decide on an alternative which reduced material flow by 50%. In addition, the cost of change was found to be minimum, making the solution to be optimal and feasible.

When the mathematical results were implemented, they produce effective outcomes, as the company was saved from the expensive proposition of expansion. Rather, major modifications to the existing layout fulfilled the objective. This case is intended to highlight the importance of OR methods in solving real problems quantitatively.

1.8 GLOSSARY

- **Operation**: process of doing work.
- **Process**: combination of activities arranged in a technological sequence for conversion of input resources into desired output.
- **Problem formulation**: involves identification of the problem, selection of possible alternatives to the problem, and choosing criterion or criteria to examine those alternatives.
- **Problem solving**: involves analysing the alternatives through mathematical model development and solving the model by applying proper OR techniques to reach a feasible and optimum solution.
- **Model**: as real as possible representative of actual problem.
- **Objective function**: aim of problem is either to maximize or minimize.
- **Constraints**: limitations within which the best possible solution has to be achieved.
- **Optimal solution**: the best possible solution.
- **Feasible solution**: fulfils all the constraints and does not violate them.
- **Unconstrained optimization**: problems without constraints and can apply unlimited resources.

1.9 MODEL QUESTIONS

1.9.1 Suzuki at one of its production facilities produces two emblems for its cars: one with the symbol 'S', and the other with the name 'Suzuki'. One unit of 'S' requires 3 hours, and one unit of 'Suzuki' requires 2 hours for production. The maximum production time available to fulfil the demand of 25,000 units of A and B was 1,500 hours. The cost of production at the facility of 'S' and of 'Suzuki' was $2 per unit and $3 per unit, respectively. The fixed cost was $2,000 for both products as they are being produced at the same plant. Assuming 'x' units of 'S' and 'y' units of 'Suzuki' develop mathematical expression for:

i. Objective function of minimizing the cost of production
ii. Constraint equation for capacity and production time
iii. Total cost equation for both products
iv. If pl is the price of one unit of 'S' and p2 of 'Suzuki', what would be the revenue equation?
v. Profit function for both products

1.9.2 Fun blast manufactures dolls for kids. In addition to the existing variety of dolls, company decided to launch two new dolls, 'Blonde' and 'Brunette' for the upcoming festive season. A maximum of 15,000 blondes and 22,000 brunettes were to be produced. To manufacture one unit of blonde, a worker requires a maximum of 18 seconds and 24 seconds for a brunette. Maximum labour time available to fulfil demand was 5,000 minutes. Profit per unit of blonde and brunette was estimated to be $1 and $2, respectively. Develop mathematical expression for:

i. The objective function to maximize profit by assuming 'x' units of blonde and 'y' units of brunette.
ii. Capacity constraint
iii. Production time constraint

1. A distributor of refrigerators supplies to three major retailers (R1, R2 and R3) in the city. The maximum capacity of distributor per month was 20,000 units, whereas demand of each retailer R1, R2 and R3 was 5,000, 8,000 and 4,500 units, respectively. The transportation costs per unit from warehouse to retailers were \$9, \$10 and \$7. Write a mathematical expression for:
   i. objective of minimizing transportation cost and
   ii. constraint of demand of retailers

1.9.3 Transportation cost is a function of the number of units to be shipped from one supply centre to another demand centre and cost of transporting per unit. Fastway distribution centre intends to develop a mathematical model, which would model the real problem of estimating transportation cost incurred on supplying to two demand centres. If 'x' and 'y' are the number of units that can be shipped at \$4 and \$5 per unit, respectively, then form an objective function. Also if maximum units that Fastway can supply is 8,000 units and maximum demand of two centres is 3,000 and 4,000 units, what would be the constraint functions?

1.9.4 Cost is a function of the number of units manufactured. If zero units are produced, then the cost of production is \$90,000. It increases by \$400 for every unit produced. Demand is a function of the price of a product. If the price is not increased, then demand is 250 units. However, with every increase in the price of product, demand decreases by 5 units. Formulate:

i. Cost function
ii. Demand function
iii. Profit function
iv. How many units should be sold to break even?
v. What would be the price of one unit so as to maximize the profits?

1.9.5 Annual fixed costs for a convenience store were \$85,000. Variable costs were broken down as: direct materials \$3, direct labour \$1, overhead \$0.20 and cost of selling \$1.50. The average selling price per unit was \$18.00. Formulate:

i. Cost and profit function
ii. Number of units that should be sold to break even
iii. How many units need to be sold to earn a profit of \$1,000?

2 Linear Programming

2.1 INTRODUCTION

Chapter 1 discussed the importance of OR discipline as a scientific way of solving well-formulated problems, which help to make accurate decisions. The decision-making process by using OR was detailed as involving two steps. The first step deals with formulation of a well-defined problem. It included the identification of problem and alternative solutions. The second step was termed as problem analysis, and it involved model development and solution. Model simulates real problem as accurately as possible so that solution represents the solution accurately. Linear programming (LP) has been considered as an important and effective OR technique, which has the ability to simulate a complex problem through a mathematical model and provide solutions by applying algorithms. It is regarded as one of the major and most significant discoveries in the field of mathematics and problem solving of the last century. The significance of technique increases manifold as it can be used for model development and solution for simple or complex problems, for small or large industry and for any sector whether government or private. The importance gained acceptance with increase in computing power. Complex problems involving number of decision variables and constraint equations would be impossibly hard to solve manually. For instance, profit by selling a product could be a function of price only. However, most likely, there could be other variables such as price of competitor product, distribution costs, and other products in product line, which could estimate profit very difficult. Computers made such solutions easy and more precise. Thus, adoption and applicability of LP as an effective OR technique was made easy by advancements in computer.

2.2 MEANING OF LINEAR PROGRAMMING (LP)

The meaning of LP can best be understood through some illustrations.

- An emblem for a car during machining stage goes through four different machines. A company manufactures different types of emblems for various car manufacturers. During their machining process, all emblems require operation on the same four machines. The production manager estimates the time for each emblem at every stage and also the cost of producing them. The manager wants to allocate resources of machines in such a manner so as to reduce cost to the minimum.
- A laptop manufacturer launched a new product with advanced technology and wanted to promote its product during pre-summer season on various media channels. The manager wanted to spend limited promotion budget so as to maximize awareness about new product. Should more money

DOI: 10.1201/9781003212966-2

be allocated to Internet, print or TV media? Due to limited resources, it becomes very important to understand quantitatively the effect of allocating certain amount to a particular media channel on objective of awareness.

- A retail investor wants to maximize returns by investing in various stock market instruments such as shares, mutual funds, etc. The investor does not have unlimited money to invest so only certain amount has to be distributed among number of options.

All these problems face with similar concern of allocation of limited resources to maximize returns, profits, awareness, sales, etc. or minimize cost, delay, etc. LP is a very effective operations management technique to resolve such issues quantitatively. Such problems arise because of the availability of limited resources. No organization can avail a manpower, which is entirely competent, cannot have all the raw materials for production or have all the money to buy resources. Similarly, every organization wants to achieve maximum profits at minimum cost, but again, it is not possible because of various limitations of competition, government policies, etc. Thus, organizations have to operate with limited resources that need to be utilized effectively. This effective utilization necessitates proper allocation of resources. Here, the LP approach, which predominantly deals with problems of resource allocation by considering multiple constraints simultaneously, becomes very useful. The chapter 'Linear Programming' consists of two terms: 'Linear' and 'Programming'. Linearity indicates the proportional relationship between independent variables that affect the objective or purpose. This implies that if a variable that contributes in deciding purpose increases or decreases by a unit, then the objective also changes by the same amount. Programming implies the formulation of a model that simulates real situation. This model would be used to examine an optimal solution in the most realistic possible environment.

2.3 ASSUMPTIONS

As discussed in Chapter 1, the objective to solve a problem is to obtain a feasible and optimal solution. It was emphasized that in real business scenarios, no problem is without constraints. Their solution was termed as constrained optimization. In this chapter, we will discuss only such problems by understanding applications of LP and the graphical method of solving them.

A quantitative tool such as LP can only be applied if it fulfils certain assumptions. To understand these assumptions of LP, we have used the same case of Adidas AG as was discussed in Chapter 1. Without discussing in detail, we have reproduced the final model here.

$$\textbf{Maximize} \quad Z = 80x_1 + 40x_2$$

However, the achievement of objective is constrained by certain limitations. Therefore,

Constraint 1: Accessibility:

$$9x_1 + 4x_2 \leq 12$$

Constraint 2: Population density:

$$2x_1 + 7x_2 \quad \leq 10$$

Constraint 3: Median income:

$$5x_1 + 10x_2 \quad \leq 20$$

Finally, it is important to mention that the number of units sold could not be negative. Therefore, very importantly,

$$x_1, x_2, x_3 \geq 0$$

1. **Proportionality**: this assumption proposes that contribution of each decision variable is directly proportional to the level of activity indicated by coefficients of constraints (a_{ij}) and in objective function (c_j). If a decision variable is increased by one unit, then objective function Z would increase by c_j, whereas constraint would change by a_{ij}. In the illustrated case, the objective function is:

$$\textbf{Maximize} \quad Z = 80x_1 + 40x_2$$

where x_1 and x_2 represent the numbers of units. If suppose $x_1 = x_2 = 1$, then objective function Z would be

$$Z = 80*1 + 40*1$$

$$= \$120$$

If decision variables are increased by one unit, then change in Z would be proportional to its coefficients. Thus, if $x_1 = x_2 = 2$, i.e. decision variables are doubled from the previous scenario, then objective function would also be doubled:

$$Z = 80*2 + 40*2$$

$$= \$240$$

This assumption should also be used to check the feasibility of optimal solution achieved. This is accomplished by putting values of decision variables in constraint equations. In the case of constraints, a one unit change in decision variables will necessitate a proportional change by their coefficients, i.e. a_{ij}. Logically, decision variable values also indicate the amount of resources consumed. If the number of units sold from location A is one unit, then the process would have consumed similar units of resources proportional to coefficients. For instance, if the median income of a household is \$5,000, then purchasing one unit depletes some of the \$5,000 in available resources. If two units are purchased, then more of that resource is consumed. As a result, in order for a solution to be feasible, the purchase of not more than available resource can be made. In the case of manufacturing, if a machine can work for a maximum of 300 hours to produce a desired amount and one unit takes 3 hours to produce, then 3 more hours will be spent to manufacture another unit. Therefore, a solution would be feasible if it can produce a maximum of 100 units.

For instance, in the illustrated case, the first constraint is:
Constraint 1: Accessibility:

$$9x_1 + 4x_2 \leq 14$$

If for suppose $x_1 = x_2 = 1$, then the left-hand side of constraint 1 would be $9*1+4*1 = 13$ which is less than the right-hand side of 14. Thus, it can be inferred that the solution is feasible. Now applying proportionality assumption if $x_1 = x_2 = 2$, then the left-hand side of constraint would be $9*2+4*2 = 26$ which is more than the right-hand side making solution infeasible. Thus, this indicates two aspects: first change is proportional to change in decision variable, and second, feasibility of solution is examined regarding such every change.

2. **Additivity**: This assumption says that the achievement of an objective or constraint function is the combination of contributions of all decision variables. It is important to remember that a decision variable represents a productive activity, and that any change in production would result in modification of such activity. Thus, if production is a function of more than one activity, then the calculation of the objective function would be influenced by a combination of all such modifications. For instance, in the illustrated case, the function of achieving optimal profit is a function of the number of units sold at two different stores. Therefore, if x_1 and x_2 are the numbers of units sold, then total profit would be the sum of x_1 and x_2. Similarly, it can be seen in model formulation that constraint equations also use this assumption.
3. **Divisibility**: This assumption says that decision variables in a LP problem can take any value within found feasible region. The feasible region is bounded by constraint limitations. In this feasible region, decision variables may have both integer and non-integer values. For instance, if accessibility constraint is

$$9x_1 + 4x_2 \leq 14$$

Then, at $x_1 = 0$; $x_2 = 3.5$ which is a non-integer value.
at $x_2 = 0$; $x_1 = 1.55$ which is a non-integer value.
4. Certainty: According to this assumption, all coefficients in objective and constraint functions are assumed to be known constant. For instance, in the illustrated case, coefficients of objective function indicate profit per unit from two stores, i.e. \$80 and \$40. These are finite known constant values.

Similarly in constraint 1 of accessibility, store at location A has a trade area of 6 miles, and for store at location B, it is 4 miles. Similarly coefficients of other constraints indicating consumption of resources per unit are finite and known with certainty.

2.4 APPLICATIONS OF LP

2.4.1 Marketing Research

The function of marketing research involves understanding consumers' attitudes, preferences and ever changing needs. This process of research involves various activities including problem formulation, designing the method to solve such problems, data required and proper analysis to provide accurate information to the client so as to make an informed decision. The organizations specialized in marketing research study customer perspective and competitive forces that exist in similar market. One of the major functions of such agencies is to collect accurate data from the respondents. This involves a lot of risks as respondents may fit different profiles, some may not answer correctly and some may not know the subject of research. Data collection for a particular identified problem could be performed by various means such as through structured questionnaire, online subjective questions, etc. The selection of a proper method of collecting data poses another major problem for the researcher as each step involves cost. The following case illustrates this aspect of marketing research by developing a LPP to figure out a number of respondents required in the case of multiplicity of data collection instruments.

Logo of a company reflects characteristics of offerings, i.e. products or services that it offers. These characteristics in essence should match those of targeted customer. In hyper-competitive scenario where an individual is spoiled by choices, a strategy of communicating the fulfilment of customer needs through a logo becomes essentially important. This communication is made through elements of logo which are colour, font and design. The selection of a particular kind of colour or font or design is considered to be instrumental in matching company's intentions and customers' preferences. However, with a change in economic, demographic and competitive landscape, every company feels the need of modifying or changing one or all of these elements of its logo. This is performed to communicate and reflect aspirations of a new customer. This process of change in elements of logo is termed as rebranding.

Telco Limited has been a major player in providing telecom services such as voice and data for over 20 years, but increased competition from both domestic and foreign players as well as a shift of younger customer towards data communication from voice forced the company to embrace these changes. One of the strategies adopted by company to communicate adoption of such changes and provider of new services to younger customer was carried out by modifying elements of logo. After conducting a preliminary survey, it was decided that all three elements of logo, namely, colour, design and font would be modified in concurrence. To understand customers' perspective regarding changes in each of the element, the company decided to conduct three market surveys one for each of the element. The survey was conducted through three questionnaires one each for a particular logo element. A questionnaire consisted of a series of questions asking respondents to ascertain their emotions towards attributes represented by different elements of logo. A colour questionnaire was prepared containing 20 questions that has to be scored on a scale of 1–10. Thus, minimum score from one respondent could be 10 out of maximum of 2,000. Similarly, design and font questionnaire had 17 and 15 questions; thus, the maximum score could be 1,700 and 1,500, respectively. It was also decided that a minimum of 3,000 questionnaires would be filled. Due to higher importance of colour in logo as suggested by various studies, the importance to colour, design and font elements was given in the ratio of 5:3:4 implying that one colour questionnaire would be given importance of 5 units as compared to other questionnaires. The purpose of manager was to assess number of each type of respondents which would minimize cost of survey if cost of administering colour, design and font type of questionnaire is \$2.5, \$1.3 and \$1.5, respectively. To solve the problem, it is necessary to formulate a LPP.

Solution:

In formulating the LP model for the company, it is important to define the decision variables. These variables would reflect the number of respondents those should be subjected to questionnaires of colour, design and font. Thus, notation utilized for each decision variable is:

x_1 = number of respondents who were subjected to colour questionnaire,
x_2 = number of respondents who were subjected to design questionnaire, and
x_3 = number of respondents who were subjected to font questionnaire.

Administering one questionnaire consisting questions relating to colour attributes would incur a cost of \$2.5 so if there are x_1 respondents, then the total cost would be $2.5x_1$. Thus, using cost per interview data, the following minimization objective was formulated:

$$\textbf{Minimize} \quad Z = 2.5x_1 + 1.3x_2 + 1.5x_3$$

Constraint regarding minimum score implied that a respondent needs to respond to 20 questions in the case of colour questionnaire. Thus, the score of one respondent if he/she gives minimum i.e. score of ‘1’ to all questions would be at least 20. On similar lines, the maximum score for one question is earmarked as 10 so the total maximum score for all 20 questions would be 2,000. Hence, a respondents’ score would lie between 20 and 2,000. This understanding is used for the construction of constraint regarding colour questionnaire. One respondent scores minimum score of 10 so x_1 respondents would give a score of $10x_1$.

$$\text{Score constraint w.r.t. colour questionnaire}: \quad 20x_1 + 0x_2 + 0x_3 \leq 2{,}000$$

Similar understanding can be applied to design and font questionnaires resulting in the following constraints:

$$\text{Score constraint w.r.t. design questionnaire:} \quad 0x_1 + 17x_2 + 0x_3 \leq 1{,}700$$

$$\text{Score constraint w.r.t. font questionnaire:} \quad 0x_1 + 0x_2 + 15x_3 \leq 1{,}500$$

Finally, importance given to each type of questionnaire was in the ratio of 5:3:4 implying that colour questionnaire is given the maximum importance in relation to design and font. Thus, if one response to colour questionnaire is given importance of 5 units, then x_1 responses would entail importance of $5x_1$. The minimum number of responses required to understand customers' perspective towards change in colour, design and font elements was decided to be 3,000. Therefore, constraint would be:

$$\text{Minimum number of responses}: \quad 5x_1 + 3x_2 + 4x_3 \geq 3{,}000$$

The results of this LPP would help manager of Telco Ltd. in allocating a number of questionnaires to each category of respondent and collecting data regarding changes in perception due to modification in colour, design and font elements of logo at a minimum cost.

2.4.2 Media Selection

Serene Hill Resorts, Inc. has developed new resorts and banquet halls in picturesque surrounded by forests. The primary market for these resorts included rich and affluent individuals. To promote their offering, the resort manager hired an advertising company and asked them to make targeted market aware about their offerings. The budget allocated for promotion was decided to be not more than $25,000. Advertising company decided to allocate this amount on four different media channels, namely, electronic media involving TV and online channels, largest selling newspaper and radio. It was decided that TV ads would be limited to be shown only during evening prime time. During this time, due to the huge demand for advertisement slots, the rate was quite high which was $2,000 for an ad shown once. Maximum time the advertisement could be shown was only twice during these hours. Thus, the maximum available slots on TV in a month of 30 days for which the ads would be shown on different media channels were 60. In case of business daily, the cost of full page ad during weekdays was $500 per day. As it was decided to display ad only on weekdays, thus maximum times the ad can be promoted through newspaper in given month was 20. Due to high popularity of social media, a specific sum was also allocated for the promotion of resort facilities. Cost per ad on selected social media was $200, and a maximum of 40 slots were available to display the ad online. Finally, radio was found to be the cheapest mode of promotion channel with an ad slot costing only $75. The prime time available slots on radio were restricted to maximum of 35 for given month. The efficacy of promotion campaign was evaluated after a month. One of the measures of evaluating efficacy of campaign was potential customers reached during the campaign period. The advertisement manager set a target of at least reaching 70,000 customers with 2,500, 1,200, 2,000 and 500 customers being reached during one time of ad being shown on different media channels of TV, newspaper, online banner ad and radio, respectively. The objective of an advertisement campaign was to increase the awareness level among targeted market. The ad company measured awareness level of ads shown in different media by conducting a survey after the period during which ads were shown. Standardized questions were asked regarding ad recall and ad recognition which was established measures of ad awareness. These questions pertain to each type of media channel and respondents' score was taken on a scale of '0–100'. An average score of these two questions was recorded, which represented awareness level of an ad shown once. The evaluated scores were 70, 30, 55 and 20 for TV, newspaper, online and radio ad, respectively. The manager of company wanted to maximize this awareness level by finding number of times an ad should be shown on a particular media channel.

Solution:

The objective of manager was to maximize the awareness level among targeted customers. The achievement of this objective depends on certain decision variables, which need to be determined. Thus, the first step is to define these decision variables. Awareness level would increase if an individual watches a particular ad maximum number of times. As in this case there are four different types of media channels so each decision variable would represent the number of times an ad is seen on a particular media channel. Hence,

x_1 = number of times ad is shown on TV,

x_2 = number of times ad is shown on newspaper,

x_3 = number of times ad is shown online (banner ad), and

x_4 = number of times ad is displayed on radio.

As average awareness level regarding TV, newspaper, online and radio per ad exposure is given to be 70, 35, 55 and 20, respectively, so objective function would be

$$\textbf{Maximize} \quad Z = 70x_1 + 35x_2 + 55x_3 + 20x_4$$

The achievement of maximum awareness level depended on certain constraints. In this case, the purpose was constrained by three limitations, namely, budget allocated for promotion, media availability and maximum customer reach.

First, the resort management had limited budget of $25,000 which needs to be distributed among selected four media channels. The cost incurred depended on identified decision variables. This led to the formulation of following constraint:

$$2{,}000x_1 + 500x_2 + 200x_3 + 75x_4 \leq 25{,}000$$

Second, each type of media was available for limited slots. Thus, the second constraint was formulated for each type of media:

$$\text{Media availability for TV:} \quad 1x_1 + 0x_2 + 0x_3 + 0x_4 \leq 60$$

$$\text{Media availability for newspaper:} \quad 0x_1 + 1x_2 + 0x_3 + 0x_4 \leq 20$$

$$\text{Media availability for online:} \quad 0x_1 + 0x_2 + 1x_3 + 0x_4 \leq 40$$

$$\text{Media availability for radio:} \quad 0x_1 + 0x_2 + 0x_3 + 1x_4 \leq 35$$

Finally, manager wanted to reach as many customers as possible. It was estimated that promotion campaign would be considered successful if it reaches minimum of 70,000 customers. Thus, the last constraint would be:

$$\text{Customer reach}: \quad 2{,}500x_1 + 1{,}200x_2 + 2{,}000x_3 + 500x_4 \geq 70{,}000$$

Thus, the solution of the above-discussed four decision variable and six constraint LP model would provide maximum awareness level reached among targeted customers.

2.4.3 Financial Planning

LP models can also be employed to solve problems related to finance, investment, portfolio selection, etc. problems. Such an application arises from utility of LPP in resolving multi-criteria problems. When a decision to be made is based on more than single criteria, LPP can be used effectively. Organizations source its capital from more than one source such as loans, stocks, return on investments in fixed assets, etc. where each source asks for different costs and benefits making it a multi-criteria problem. Similarly, organizations have to pay certain fixed liabilities such as pensions, salaries, money for development, etc. All such decisions are made in different scenarios considering interest rates, time horizons, and the number of employees making them a complex decision making environment. In such cases, LPP plays an effective role in making informed and accurate decisions. Moreover, LPP models can also be constructed for individual problems such as portfolio selection. In such scenarios, an individual intends to select certain financial instruments to save income tax and avail maximum returns over short- or long-term period. One such scenario is discussed in the following section.

According to government guidelines, a female employee can invest up to 2 lakh rupees from total annual income in various saving instruments, which would be taxed at zero percent. Thus, every working female employee avails for this saving scheme to avail tax benefits. A female employee of a public University who receives assured annual income selected five instruments to invest her 2 lakh rupees which were a pension plan, a government security and three stocks (A, B and C) of different risk and return prediction. The selection criteria were to diversify funds in low- to high-risk category instruments. Stocks are generally considered as of high risk relative to long-term pension plans and government securities. Therefore, she selected three stocks to invest earmarked money for stock investment. Another and more important selection criteria were the returns that each instrument would provide. Past data from reliable sources indicated that a pension plan would provide long-term return of 5%, whereas low-risk government security would give a return of 8.8% per annum. The returns estimated by low-, medium- and high-category risk stocks were estimated to give a return of 9%, 11% and 13% over the next year, respectively. The investment is followed by certain guidelines which were:

- Pension plan should not have more than 10% of all other plans invested.
- Government security should be invested with at least $50,000.

- At least 20% of medium- and high-risk category stocks should be invested in low-risk category stock.
- Total amount be invested in stocks should not be more than \$20,000.

Formulate a LPP to estimate the amount invested in each of the instruments, which would maximize returns.

Solution:

The objective of maximization of returns after investing in different financial instruments depends on the amount of total money to be invested in each one of them. This results in defining of decision variables where each one of them pertains to the amount invested in a particular instrument. Thus,

x_1 = amount of money invested in pension scheme
x_2 = amount of money invested in government security
x_3 = amount of money invested in stock A
x_4 = amount of money invested in stock B
x_5 = amount of money invested in stock C

Understanding and formulation of decision variables help in creating objective function. In this case, the objective is to maximize returns. A unit of pension plan, government security, stocks A, B and C gives on an average an annual return of 5%, 8.8%, 9%, 11% and 13%, respectively. Thus, investment indicated by decision variables would lead to objective function of:

$$\textbf{Maximize} \quad Z = 0.05x_1 + 0.088x_2 + 0.09x_3 + 0.11x_4 + 0.13x_5$$

The investment is governed by certain limitations. First, a female employee has decided to make investment plans regarding only 2 lakh of her total income. Thus,

Total investment criteria:

$$x_1 + x_2 + x_3 + x_4 + x_5 = 200{,}000$$

Second, investment in pension plan to the maximum was decided to be not more than 10% of all other investments. Thus,

Pension plan investment:

$$x_1 \le 0.1(x_2 + x_3 + x_4 + x_5)$$

$$x_1 - 0.1x_2 - 0.1x_3 - 0.1x_4 - 0.1x_5 \le 0$$

$$\text{or} - x_1 + 0.1x_2 + 0.1x_3 + 0.1x_4 + 0.1x_5 \ge 0$$

Third, government security being considered as least risky so employee decided to invest a major chunk of investment in this financial instrument. It was decided that at least \$50,000 would go into selected government security. Therefore:

Government security investment:

$$x_2 \ge 50{,}000$$

Fourth, regarding investment in stock instruments a maximum of \$20,000 was considered to be invested in combination of all three stocks. Therefore:

Investment in stocks:

$$x_3 + x_4 + x_5 \leq 20{,}000$$

Also an employee being risk averse decides to limit investment in medium- and high-risk category stocks by considering minimum investment in low-risk stock (A) to be at least 20% of combined investment in B and C. Thus:

Investment in stock A:

$$x_3 \geq 0.20\left(x_4 + x_5\right)$$

$$x_3 - 0.20x_4 - 0.20x_5 \geq 0$$

Five variable and five constraint LPP problem can be solved to indicate the amount of investment in selected financial instruments.

2.4.4 Product-Mix Problem

Majority of production problems deal with allocating limited resources to manufacture more than one type of product. Every organization employing either a batch or mass system indulges in manufacturing more than one kind of product. A packaging machine manufacturer uses the same resources namely, employees with similar skill set, similar type of parts components, similar vendors to source material, etc. in the process of assembling both soap and biscuit packaging machines. For instance, in case of a piston machine shop, a particular machine is used to produce pistons for both tractors and cars. Thus, in the process of producing multiple types of products using similar resources, the manager must decide how much of resources such as production capacity, labour capacity, availability of raw material, etc. to be used to meet demand requirements of each product while keeping production cost at minimum level. LPP can be used effectively to solve such product-mix problems, allowing the manager to design and implement cost-efficient production schedule while fulfilling the maximum demand for each type of product.

A packaging machine manufacturer involves in producing biscuit packing machine. The machine requires numerous components, some of which are manufactured in-house and majority of them are ordered from various suppliers. Out of in-house components three P1, P2 & P3 are manufactured by using four workstations M1, M2, M3 & M4. The operation for manufacturing a component say P1 involves processing on all four machines in the sequence of M1---M2---M3 and M4. Similar sequence is adopted for all other components. However, a machine at a time can be involved in manufacturing only one component. This constraint leads to product-mix problem where manager has to decide under such production capacity constraint how many units of each component would be manufactured. The following data indicates production time constraint for each component at each machine. One unit of P1 takes 3 and 3.5 hours on M1 and M2. Similarly, one unit of P2 takes 4 and 4.5 hours on M2 and M4 and one unit of P3 takes 4 and 5 hours on M1 and M3, respectively. The maximum time available at each workstation is 150 hour profit earned by selling one unit of each type of product is \$5, \$6 and \$9. Formulate a LPP to find optimum number of units of three type of products that should be manufactured, which would give maximum profits.

Solution:

Formulation of a LP problem involves defining decision variables which would directly influence the objective function. In this case, the objective of obtaining maximum profits depends on managers' decision to produce a certain number of three components P1, P2 and P3. The notation for these three components would be:

x_1 = number of P1 components,

x_2 = number of P2 components, and

x_3 = number of P3 components.

Profit earned by producing one unit of P1 is \$5 so from x_1 units profit would be $5x_1$. Similar consideration for P2 and P3 would result in following objective function:

$$\textbf{Maximize} \quad Z = 5x_1 + 6x_2 + 9x_3$$

The production of each type of component depends on the available production hours on each workstation, which was 150 hours implying that M1, M2, M3 and M4 can maximum work for 150 hours within which maximum production needs to be done. Now workstation M1 is utilized to produce components P1 and P3, workstation M2 produces P1 and P2, M3 produces P3 and finally M4 produces only P2. Thus, time taken to produce one unit of each component at each workstation production constraints for each workstation would be:

$$\text{Constraint for M1}: \quad 3x_1 + 0x_2 + 4x_3 \le 150$$

$$\text{Constraint for M2}: \quad 3.5x_1 + 4x_2 + 0x_3 \le 150$$

$$\text{Constraint for M3}: \quad 0x_1 + 0x_2 + 5x_3 \le 150$$

$$\text{Constraint for M4}: \quad 0x_1 + 4.5x_2 + 0x_3 \le 150$$

Solving this LPP model consisting of three decision variables and four constraints would result in amount of each component that should be produced to earn maximum profits.

2.4.5 Vendor Selection

One of the major problems in production management is selection of suppliers or vendors by the manufacturer so as to receive ordered parts at the lowest possible cost and within a specified time. Such problems are catered under the section of vendor selection. The need for ordering parts from various vendors arises from different reasons. One reason is fluctuation in demand. Manufacturer does not want to lock limited resources in building fixed capability to produce more units. There could be periods when demand is sluggish so resources locked in fixed assets would go waste that ultimately results in losses. Second, manufacturer cannot produce everything by itself. For instance, a car manufacturer/assembler requires thousands of parts that go into production of car. Therefore, it would be more productive that manufacturer receives different parts from outside sources. In addition, manufacturer might not have capability to produce every type of part. Certain specialized components can only be produced by certain supplier. Finally, a manufacturer most of the time places an order for similar component with more than one vendor to address the issue of reliability. Thus, from vendors' perspective, how many units should be manufactured under given constraints which would maximize profits is discussed.

Car assembler who selects and places an order with the vendor has an objective of minimization of cost, whereas vendors who supply these parts would like to maximize their profits. Achievement of this objective is again governed by number of units being manufactured in stipulated time period. The manufacturing process for three types of seats (S1, S2 and S3) is the same for both vendors, with cutting, sewing and finishing taking place in the same order. Production time of each process for each type of seat per unit is shown in the following table. Also, maximum production time available for each process is given. This production time pertains to fulfilment of order of 540 seats. Under the given constraints of production time of each process and fulfilment of total demand, both vendors intended to estimate; number of seats of each type to be manufactured as well as the maximum profit that can be earned. Profit per unit of S1, S2 and S3 for vendor V1 was calculated to be \$9.6, \$11.0 and \$12.5, respectively. Similarly, profit for S1, S2 and S3 for V2 was \$12, \$13.5 and \$14.0.

	Vendor V1				**Vendor V2**			
	Time (Hours)			**Max. Production Time Available (Hours)**	**Time (Hours)**			**Max. Production Time Available (Hours)**
	S1	**S2**	**S3**		**S1**	**S2**	**S3**	
Cutting	1.5	1.3	2	300	3.7	3.4	2.4	380
Sewing	2	2.4	2.9	430	2.7	2.2	3.7	300
Finishing	4	5	3	620	5	6	7	575

Formulate LPP for both vendors. Are both solutions feasible? Which vendor generates more profit?

Solution:

The objective for both vendors is to maximize their profits by selling three different types of seats (S1, S2 and S3) to the car manufacturer. Maximization of profits is dependent on the number of units produced. The more number of seats of each type produced, the more would be the profit. This leads to formulation of decision variables.

x_1 = number of seats S1 manufactured by vendor V1,
x_2 = number of seats S2 manufactured by vendor V1,
x_3 = number of seats S3 manufactured by vendor V1,

x_4 = number of seats S1 manufactured by vendor V2,
x_5 = number of seats S2 manufactured by vendor V2, and
x_6 = number of seats S3 manufactured by vendor V2.

Understanding the decision variables is important in constructing objective function. For instance, for vendor V1 profit earned by selling one unit of S1 was $9.6, so selling x_1 units of S1 would generate a profit of 96 x_1. Thus, objective function for V1 would be:

$$\textbf{Maximize} \quad Z = 9.6x_1 + 11x_2 + 12.5x_3$$

Production of a certain number of units of each type of seat is constrained by the number of production hours. As each production of a seat requires three types of processes, namely cutting, sewing and finishing in that sequence so constraints would be constructed for each process. For instance, cutting of one unit of S1, S2 and S3 requires 1.5, 1.3 and 2 hours, respectively. Thus, for cutting of x_1, x_2 and x_3, units of these seats production time consumed would be $1.5x_1$, $1.3x_2$ and $2x_3$. The maximum production could be carried in maximum available time, which in the case of cutting process was 300 hours. Thus, constraint equation for cutting process of vendor V1 would be:

$$\text{Cutting process}: \quad 1.5x_1 + 1.3x_2 + 2x_3 \leq 300$$

On similar lines constraint equations for sewing and finishing process would be:

$$\text{Sewing process}: \quad 2x_1 + 2.4x_2 + 2.9x_3 \leq 430$$

$$\text{Finishing process}: \quad 4x_1 + 5x_2 + 3x_3 \leq 620$$

Finally, vendor V1 should utilize these resources to fulfil a maximum demand of total seats of 540 units. This implies that the resources at managers' disposal might be enough to fulfil either entire demand or less than 540 or even more than it but should not exceed the given limit. Thus, demand constraint would be:

$$\text{Maximum demand}: \quad 1x_1 + 1x_2 + 1x_3 \leq 540$$

The manager needs to find out whether the production process is capable of fulfilling the entire demand, and if so, what proportion of each type of seat would generate the maximum profit.

The results of V1 are compared to the results of vendor V2 using the same type of constraints and objective function. Thus, formulation of objective function and constraints for V2 would be:

$$\textbf{Maximize} \quad Z = 12x_4 + 13.5x_5 + 14x_6$$

$$\text{Cutting process}: \quad 3.7x_4 + 3.4x_5 + 2.4x_6 \leq 380$$

$$\text{Sewing process}: \quad 2.7x_4 + 2.2x_5 + 3.7x_6 \leq 300$$

$$\text{Finishing process}: \quad 5x_4 + 6x_5 + 7x_6 \leq 575$$

$$\text{Maximum demand}: \quad 1x_4 + 1x_5 + 1x_6 \leq 540$$

Solution of this three variable and four constraint LPP problem for each vendor would provide the results regarding the number of units of each type of seats to be produced and amount of profit earned. Results would indicate which vendor is more profitable and by how much.

2.4.6 Make or Buy Problem

LP models can be formulated to solve problems of how many units a manufacturer should produce and how many units to purchase from vendor. In production management, such decisions are referred to as make-or-buy decisions. Due to variation in demand, limited capacity and capability, etc., every manufacturer faces this problem. In such cases, LPP plays a constructive role in resolving issues of vendor selection and how many units to be ordered from which supplier. Such model formulation is illustrated in the following example.

Packaging machine manufacturing company produces two component parts (P1 and P2) used in assembly of its soap-packaging machine line. The weekly demand of P1 was estimated to be 10,000 units, whereas that of P2 to be 8,000 units. Capacity of this line to produce these two parts on weekly basis was 5,000 and 4,500 units, respectively, at a cost of $1.5 and $1.6 per unit. To fulfil the unmet demand, the company decided to purchase certain units of both parts from three suppliers (S1, S2 and S3). Thus, demand for each part is met from manufacturing in-house and purchasing from outside sources. S1 can supply a maximum of 3,000 units of P1 and 2,200 units of P2 at a cost of $1.7 and $1.9. Production capacity of S2 allows manufacturing a maximum of 2,000 units of P1 and 1,500 units of P2 at a cost of $1.8 and $1.2, respectively. S3 manufactures and supplies a maximum of 3,500 units of P1 at a cost of $1.4 per unit and a maximum of 1,700 units of P2 at a cost of $1.8 per unit. The company to fulfil its requirements orders parts from all four sources, i.e. its in-house manufacturing facility and three supplier Formulate LPP model to estimate: total number of units of P1 and P2 manufactured and purchased from three suppliers; total minimum cost, and finally the part which would incur less cost by following this strategy.

Solution:

The problem requires formulation of LPP model for production of part P1 and P2 by manufacturer and three supplier. As each part can be manufactured by all four players so different LPP models would be formulated for both parts. First, let us discuss LPP model for part P1. The purpose is to manufacture certain number of units of P1 at the lowest possible cost. As certain units are also sourced from various suppliers, managers' intention is twofold: first, how many units should be manufactured in-house and how many to be sourced from each supplier; second, what would be optimum or the lowest possible cost. This led us to define following decision variables:

x_{11} = number of units of part P1 produced by manufacturer.
x_{12} = number of units of part P1 produced by supplier S1,
x_{13} = number of units of part P1 produced by supplier S2, and
x_{14} = number of units of part P1 produced by supplier S3.

The objective function is to minimize cost, which includes cost of manufacturing per unit from manufacturer and supplier. Thus, objective function for part P1 would be:

$$\textbf{Minimize} \quad Z = 1.5x_{11} + 1.7x_{12} + 1.8x_{13} + 1.4x_{14}$$

The achievement of this objective depends on certain constraints. One of the constraint deals with limited demand of P1 which was 10,000 units weekly. Thus,

$$\text{Demand constraint:} \quad x_{11} + x_{12} + x_{13} + x_{14} = 10{,}000$$

Four other constraints dealt with limited capacity of manufacturer and each supplier. The capacity limits the capability of production. Thus,

$$\text{Capacity constraint of manufacturer:} \quad x_{11} + 0x_{12} + 0x_{13} + 0x_{14} \leq 5{,}000$$

$$\text{Capacity constraint of supplier S1:} \quad 0x_{11} + 1x_{12} + 0x_{13} + 0x_{14} \leq 3{,}000$$

$$\text{Capacity constraint of supplier S2:} \quad 0x_{11} + 0x_{12} + 1x_{13} + 0x_{14} \leq 2{,}000$$

$$\text{Capacity constraint of supplier S3:} \quad 0x_{11} + 0x_{12} + 0x_{13} + 0x_{14} \leq 3{,}500$$

Similarly model regarding minimum cost of production of part P2 can be formulated. For P2 decision variables could be defined on similar lines as:

x_{21} = number of units of part P2 produced by manufacturer,
x_{22} = number of units of part P2 produced by supplier S1,
x_{23} = number of units of part P2 produced by supplier S2, and
x_{24} = number of units of part P2 produced by supplier S3.

Objective function for part P2 would be:

$$\textbf{Minimize} \quad Z = 1.6x_{21} + 1.9x_{22} + 1.2x_{23} + 1.8x_{24}$$

The achievement of this objective depends on certain constraints.

$$\text{Demand constraint:} \quad x_{21} + x_{22} + x_{23} + x_{24} = 8{,}000$$

$$\text{Capacity constraint of manufacturer:} \quad x_{21} + 0x_{22} + 0x_{23} + 0x_{24} \leq 4{,}500$$

$$\text{Capacity constraint of supplier S1:} \quad 0x_{21} + 1x_{22} + 0x_{23} + 0x_{24} \leq 2{,}200$$

$$\text{Capacity constraint of supplier S2:} \quad 0x_{21} + 0x_{22} + 1x_{23} + 0x_{24} \leq 1{,}500$$

Capacity constraint of supplier S3: $0x_{21} + 0x_{22} + 0x_{23} + 0x_{24} \leq 1{,}700$

These two models related to P1 and P2 each involving four decision variables can be solved by using LPP. It would help in identifying number of units to be produced in-house and units to be sourced from each supplier at the lowest possible cost.

2.4.7 Diet Problem

The LP method of model formulation to solve multi criteria problems can also be used to find number of units of each diet an individual should take which would fulfil calorie intake at lowest possible cost. Or, from producers' perspective how many units of combination of various diets or foods should be offered which would maximize the profits. Thus, the purpose of maximization or minimization depends on the number of units of alternative diets. The application of LPP in solving such questions is illustrated in the following case.

A fast food restaurant intended to change consumers' perception from selling unhealthy foods to healthy foods. The manager from previous data estimated that most liked and sold items/diets were chicken burger along with regular fries and a regular shake. Most of customers tend to buy these items to one combined meal. The manager created a promotion campaign wherein calorie intake of these three diets was shown, and it was emphasized that such intake was within healthy limits. This emphasis was highlighted by displaying calorie in grams of each nutrient in each of the diet. Each diet or food item, namely burger, fries and shake, was indicated to have certain common nutrients, which were sodium, sugar, trans-fat, protein and fibre. Furthermore, as most of customers combine these three food items to form one combined meal so calorie intake is sum of calories from these food items i.e. maximum calorie intake of 2,000 g of sodium meant that by having one unit of chicken burger along with fries and shake would result in calorie intake of less than or equal to 2,000 g. Similarly, data regarding calorie of all nutrients for each food item was displayed. Maximum calorie intake data implies that combination of one unit of each type of diet would be within healthy limits.

The major purpose of showing such detailed data to customers was primarily to increase profits from enhanced sales of new items. Restaurant manager wanted to estimate number of units of each food item/diet that should be sold to earn maximum profit if per unit of burger, fries and shake give a profit of \$1.5, \$0.6 and \$0.8 respectively. The formulation of LPP for such problem is illustrated next. The following data indicates nutrient intake of these food items in grams per serving.

	Chicken Burger	Regular Fries	Regular Shake	Maximum Intake of Calories
Sodium	550	260	110	2,000
Sugar	4	0	28	25
Trans-fat	0.2	0.1	0.3	0.4
Protein	32	5	8	30
Fibre	5	5	0	7

Solution:

In the first step, decision variables that would decide objective function are defined as follows:

x_1 = number of units of chicken burger,

x_2 = number of units of regular fries, and

x_3 = number of units of regular shake.

Using the data regarding profit earned by selling one unit of each type of food item the objective function of maximization would be:

$$\textbf{Maximize} \quad Z = 1.5x_1 + 0.6x_2 + 0.8x_3$$

Health-conscious customers would indulge in buying these three food items if the intake is limited by amount of calories. Consuming the combination of three items i.e. burger, fries and shake would result in intake of calories of each nutrient. Thus, an individual would be consuming sodium, sugar, trans-fat, protein and fibre all together with consumption of one serving. The data indicates that each serving can have a certain maximum level of each type of nutrient, which would keep the diet within healthy limits. These limits are measured in terms of amount of calories consumed. Taking into consideration maximum limit of 2,000, 25, 0.4, 30 and 7g of sodium, sugar, trans-fat, protein and fibre following constraints were formulated:

$$\text{Constraint of sodium:} \quad 550x_1 + 260x_2 + 110x_3 \leq 2{,}000$$

$$\text{Constraint of sugar:} \quad 4x_1 + 0x_2 + 28x_3 \leq 25$$

$$\text{Constraint of trans-fat:} \quad 0.2x_1 + 0.1x_2 + 0.3x_3 \leq 0.4$$

$$\text{Constraint of protein:} \quad 32x_1 + 5x_2 + 8x_3 \leq 30$$

$$\text{Constraint of fibre:} \quad 5x_1 + 5x_2 + 0x_3 \leq 7$$

This problem of maximization of profits by consuming maximum number of units of a diet was formulated with three decision variables and five constraints by using the LPP approach.

Corollary 2.1

The above illustration indicates the number of units of each food item/diet which would maximize the profit. Suppose in the illustration if a dietician strictly restricts an individuals' calorie intake to be 2,200 calories from three diets then how the constraints would be modified. This implies calorie intake should not increase the limit of 2,200g when one unit of each diet is consumed in combination. Thus, another constraint would be:

$$\text{Maximum calorie intake:} \quad 1x_1 + 1x_2 + 1x_3 \leq 2{,}200$$

2.4.8 Blending Problem

The LPP model can be formulated for problems when the same items are combined in different proportion to produce different products. For instance, an alloy is created by adding different metals in different quantities, and if these quantities are varied, then a different metal would be formed. Thus, a problem of identifying amount of each metal to be added in such a proportion so that final demand of different products is met is resolved by using the LPP model. Blending problems mostly occur in oil industry where petrol and diesel are primary products, which are produced by adding similar chemicals in different proportions to crude oil. The constraint of limited supply of ingredients that go into production of final product limits the quantity produced. Thus, the formulation of LPP in such cases also indicates the maximum number of final products that could be manufactured with this limited availability of ingredients so that the optimum profit can be achieved. The following illustration explains the process of formulating LPP model with regard to an oil production company.

Petrol India Ltd. uses three types of chemicals (C1, C2 and C3) in different proportions to produce three different types of oil (A, B and C). Each oil takes a maximum limit of each type of chemical. For instance, a maximum of 60%, 20% and 15% of C1, C2 and C3 are required to produce a barrel of oil A. Similarly, a barrel of B requires a maximum of 55%, 20% and 30% of C1, C2 and C3. Finally, a maximum of 70%, 10% and 15% of C1, C2 and C3 are required to produce a barrel of C. The production of each type of oil is constrained by the availability of each type of ingredient. For a particular production schedule, a maximum of 4,000, 3,200 and 1,400 tonnes of C1, C2 and C3 are available. The purpose is to estimate number of barrels of each type of oil that can be produced by blending different chemicals so as to generate maximum profit if one barrel of A, B and C gives a profit of \$15, \$14 and \$13.5, respectively.

Solution:

The formulation of LPP requires defining decision variables. As in the case of blending problem, each ingredient, in this case, a chemical, is utilized in limited quantities to produce three different final products, so a decision variable would indicate the amount of a particular ingredient used in the production of a particular final product. Thus:

x_{11}: amount of chemical C1 required to produce one unit of oil A,
x_{12}: amount of chemical C1 required to produce one unit of oil B,
x_{13}: amount of chemical C1 required to produce one unit of oil C,
x_{21}: amount of chemical C2 required to produce one unit of oil A,
x_{22}: amount of chemical C2 required to produce one unit of oil B,
x_{23}: amount of chemical C2 required to produce one unit of oil C,
x_{31}: amount of chemical C3 required to produce one unit of oil A,
x_{32}: amount of chemical C3 required to produce one unit of oil B, and
x_{33}: amount of chemical C3 required to produce one unit of oil C

The objective function of maximum profit could be achieved by finding out number of units of each type of oil that is produced. Now per unit or barrel of oil A is produced by combining certain unit of three types of chemicals. Hence, the total number of units of oil would be $x_{11}+x_{21}+x_{31}$. Similar equation can be formulated for oil B and C resulting in the following objective function.

Maximize $Z = 15*(x_{11} + x_{21} + x_{31}) + 14*(x_{12} + x_{22} + x_{32}) + 13.5*(x_{13} + x_{23} + x_{33})$

The objective function is achieved under certain production limitations. In blending problem, there are two types of such constraints. One is the maximum proportion of each chemical that can be added to produce particular kind of final product as ignoring this limitation would change the constituents of oil. For instance, maximum 60% of C1 in oil A implies that out of total ingredients of chemical i.e. $(x_{11}+x_{21}+x_{31})$ not more than 60% is x_{11}. Thus,

$$\text{C1 in oil A:} \quad x_{11} \leq 0.6 * (x_{11} + x_{21} + x_{31})$$

$$\text{or} \quad 0.4x_{11} - 0.6x_{21} - 0.6x_{31} \leq 0$$

$$\text{C2 in oil A:} \quad x_{21} \leq 0.2 * (x_{11} + x_{21} + x_{31})$$

$$\text{or} \quad -0.2x_{11} + 0.8x_{21} - 0.2x_{31} \leq 0$$

$$\text{C3 in oil A:} \quad x_{31} \leq 0.15 * (x_{11} + x_{21} + x_{31})$$

$$\text{or} \quad -0.15x_{11} - 0.15x_{21} + 0.85x_{31} \leq 0$$

$$\text{C1 in oil B:} \quad x_{12} \leq 0.55 * (x_{12} + x_{22} + x_{32})$$

$$\text{or} \quad 0.45x_{12} - 0.55x_{22} - 0.55x_{32} \leq 0$$

$$\text{C2 in oil B:} \quad x_{22} \leq 0.20 * (x_{12} + x_{22} + x_{32})$$

$$\text{or} \quad -0.20x_{12} + 0.80x_{22} - 0.20x_{32} \leq 0$$

$$\text{C3 in oil B:} \quad x_{32} \leq 0.30 * (x_{12} + x_{22} + x_{32})$$

$$\text{or} \quad -0.30x_{12} - 0.30x_{22} + 0.70x_{32} \leq 0$$

$$\text{C1 in oil C:} \quad x_{13} \leq 0.70 * (x_{13} + x_{23} + x_{33})$$

$$\text{or} \quad 0.30x_{13} - 0.70x_{23} - 0.70x_{33} \leq 0$$

$$\text{C2 in oil C:} \quad x_{23} \leq 0.10 * (x_{13} + x_{23} + x_{33})$$

$$\text{or} \quad -0.10x_{13} + 0.90x_{23} - 0.10x_{33} \leq 0$$

$$\text{C3 in oil C:} \quad x_{33} \leq 0.15 * (x_{13} + x_{23} + x_{33})$$

$$\text{or} \quad -0.15x_{13} - 0.15x_{23} + 0.85x_{33} \leq 0$$

Finally, constraints regarding the supply of each chemical would be formulated. The production of each type of oil is constrained by limited availability of each ingredient, i.e. C1, C2 and C3. A maximum of 4,000, 3,200 and 1,400 tonnes of them are available per production schedule. Therefore, supply constraints would be:

Supply of chemical C1:

$$x_{11} + x_{12} + x_{13} \leq 4{,}000$$

Supply of chemical C2:

$$x_{21} + x_{22} + x_{23} \leq 3{,}200$$

Supply of chemical C3:

$$x_{31} + x_{32} + x_{33} \leq 1{,}400$$

Thus, this blending consists of 12 constraints and 9 decision variables. It could be clearly interpreted that number of constraints and decision variables increases with increase in the number of ingredients that go into production of final product.

2.4.9 Workforce Assignment

Every organization tasks its managers to assign employees to specific jobs. In mass production systems where each employee is trained to perform efficiently mostly only kind of task such assignment is a simpler task. However, in service industries operating under batch production system such as banks, retail stores, hotels, etc., employees can perform variety of tasks. In such cases, assignment of workforce to specific jobs depends on the selection of objective, which could be performing job in minimum possible time, satisfying customers so as to generate maximum sales or identifying sales area for an employee where he or she can target maximum customer. The following example illustrates one of such problems wherein a manager of retail store needs to assign sales area to different employees.

Western fashion has three stores (S1, S2 and S3) located at different areas in the city. Each store caters to three geographic sectors (Sector 1, Sector 2 and Sector 3). Thus, an individual residing in Sector 1 can access any of the three Western fashion stores. According to location decision-making theory, the selection of store by an individual is directly influenced by accessibility, i.e., distance between store and customer. These distances between customers of each sector and each store are given in kilometres in the following table. For instance, a distance of 5 miles between S1 and sector 1 implies a customer residing in sector 1 has to travel an average distance of 5 miles to access S1. Similarly, other distances can be interpreted. Location theory also implies that with increase in distance store attractiveness decreases and customers tend to shop from stores closer to their residing place rather than travel long distances. Thus, data in table also indicates the maximum distance for each store within which it can hope to extract a sale or area within which the store is considered attractive.

	Distance between Store and Sector (Miles)			Maximum Area Within which Store is Considered Attractive (Miles)
	Sector 1	Sector 2	Sector 3	
Store1 (S1)	5	3.5	1.2	12
Store2 (S2)	4.7	3	5	12
Store3 (S3)	2.9	5.2	2.7	14

Taking distances into consideration, manager needs to assign 40 salesmen to three geographic areas in the ratio of 1:2:2. This implies that for every employee assigned to Sector 1 two employees would be assigned to Sectors 2 and 3. Manager has estimated from past data that average sales per month from Sectors 1, 2 and 3 with regard to all three stores was 6,000, 5,500 and 5,000 units, respectively. In this case, sales has been considered as directly dependent on optimal assignment of workforce with distance as limiting variable so formulate a LPP estimating number of employees per geographic area so as to maximize sales.

Solution:

The objective of achieving maximum sales in a given case is considered to be directly dependent on optimum allocation of workforce for each geographic area. Thus, decision variables would be:

x_1: number of salesmen allocated to sector 1,

x_2: number of salesmen allocated to sector 2, and

x_3: number of salesmen allocated to sector 3.

Objective function denoted by Z would be:

$$\textbf{Maximize} \quad Z = 6{,}000x_1 + 5{,}500x_2 + 5{,}000x_3$$

Each store generates sales depending on the distance between store and customers of three geographic areas. These sales are limited by each stores' trade area, i.e. distance within which store offerings are considered attractive. Thus,

Distance constraint of S1:

$$5.0x_1 + 3.5x_2 + 1.2x_3 \leq 12$$

Distance constraint of S2:

$$4.7x_1 + 3.0x_2 + 5.0x_3 \leq 12$$

Distance constraint of S3:

$$2.9x_1 + 5.2x_2 + 2.7x_3 \leq 14$$

Finally, store manager has a maximum of 40 salesmen at his disposal, which needs to be distributed among three areas in the ratio of 1:2:2. Thus,

Maximum salesmen:

$$1x_1 + 2x_2 + 2x_3 \leq 40$$

Thus, the LPP model results in formulation of three decision variable and four constraint problem with achievement of maximization objective.

2.5 GRAPHICAL METHOD

Graphical method is limited by the fact that problems of only two decision variables can be resolved. The method is a pictorial representation of the constraints within which objective function has to be achieved. Lines are drawn for each constraint, and then their bounded area is identified. If a constraint has less than or equal to inequality, then area below the line and towards the origin is selected, whereas in the case of greater than or equal to inequality area, beyond the line and away from origin is selected. After drawing all lines and inferring their areas, a common area is selected which would indicate the area of feasible solution. Intersecting points would provide a solution to an objective function. However, these intersection points also known as corners of the common area could be more than one. Each of these corners would provide an optimal solution, but only one of them would give a feasible solution. Feasibility implies non-violation of all constraints simultaneously, i.e. all inequalities in the constraint equations are satisfied.

The process of achieving an optimal and feasible solution using the graphical method is depicted in the following illustrations, one on maximization and the other on minimization.

2.5.1 Illustration of Maximization: Adidas AG Retail Stores

Case of Adidas AG illustrated in first chapter would be used to explain graphical process for achieving a solution. Model has been reproduced here again:

$$\textbf{Maximize Profit} \quad Z = 80x_1 + 40x_2$$

$$\text{Constraint 1:} \quad 9x_1 + 4x_2 \leq 14$$

$$\text{Constraint 2:} \quad 2x_1 + 7x_2 \leq 10$$

$$\text{Constraint 3:} \quad 5x_1 + 10x_2 \leq 20$$

$$x_1, x_2, x_3 \geq 0$$

Solution involves following steps:

Step 1*:* As it is observed from the above equations that power of decision variables is one making model a linear one. Also as explained earlier, it fulfils other assumptions as well. Thus, when constraint equations are drawn on a graph, they would be represented by a straight line. Every straight line

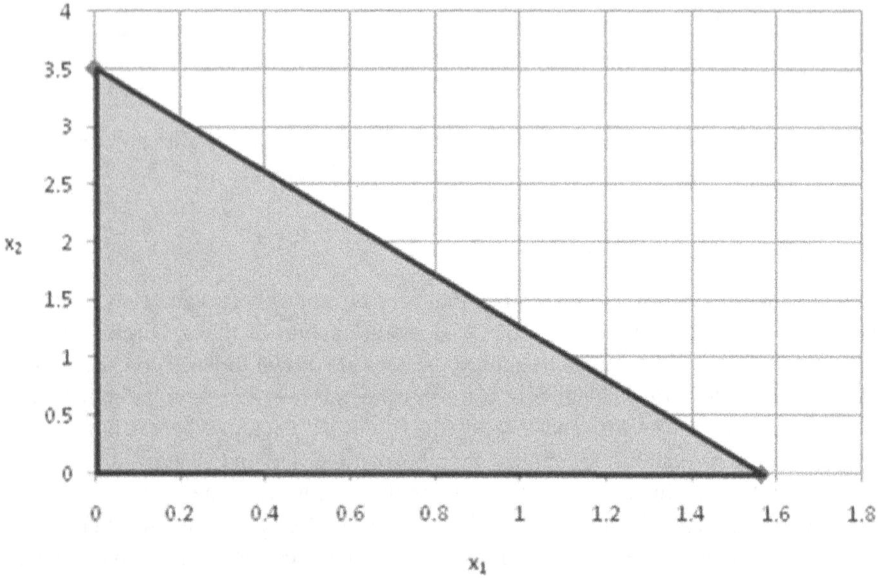

FIGURE 2.1 Line graph of constraint 1.

equation requires two extreme points. They are calculated by putting one decision variable at a time equal to zero and finding value for other. For instance, in the case of constraint 1, putting $x_1 = 0$ would give $x_2 = 3$ and putting $x_2 = 0$ would give $x_1 = 2$. Conceptually, it implies that when no units are being sold from location A then all sales would come from location B and vice versa. Line graph of constraint 1 is shown in Figure 2.1.

Two important understandings are derived from graphical representations.

- First, linear equation provides a straight line which in turn indicates that slope of line is constant at all points on the line. This further infers that solution of objective function Z for different points on the line would be the same. For example, at $x_1 = 0.2$; x_2 is approximately equal to 3 which would give Z as 136. At $x_1 = 0.4$; x_2 is equal to 2.6 which again gives Z as 136. Value of Z is the same in both cases. Similarly, at any other point on the line, results would be the same, implying almost infinite optimal solutions.
- Second, as constraint has inequality of less than equal to (≤), all values below this line would provide feasible solutions. For example, putting $x_1 = 0.4$ and $x_2 = 1$ left-hand side of constraint 1 would be 7.6, which is less than right-hand side thus not violating the constraint making solution feasible. However, at $x_1 = 1.4$, x_2 is approximately equal to 0.4 which would make left-hand side of constraint 14.2, which is greater than right-hand side violating the constraint making solution infeasible. However, this point does give an optimal value of Z. Thus, very importantly, *every optimal solution is not feasible.*

Similarly, lines of other two constraints can be drawn. For constraint 2, extreme points would be (0, 1.43) and (5, 0), and line graph is shown in Figure 2.2.

For constraint 3, extreme points would be (0,2) and (4,0). This line is shown in Figure 2.3.

However, a feasible solution requires all constraints to be satisfied simultaneously. Therefore, all line graphs needs to be superimposed. The real scenario would be represented by graph shown in Figure 2.3.

As discussed, as all the constraints have less than or equal to inequality, so feasible area would be below the line representing these inequalities but not below zero. Common area would indicate *feasible area* for all three constraints. All values in

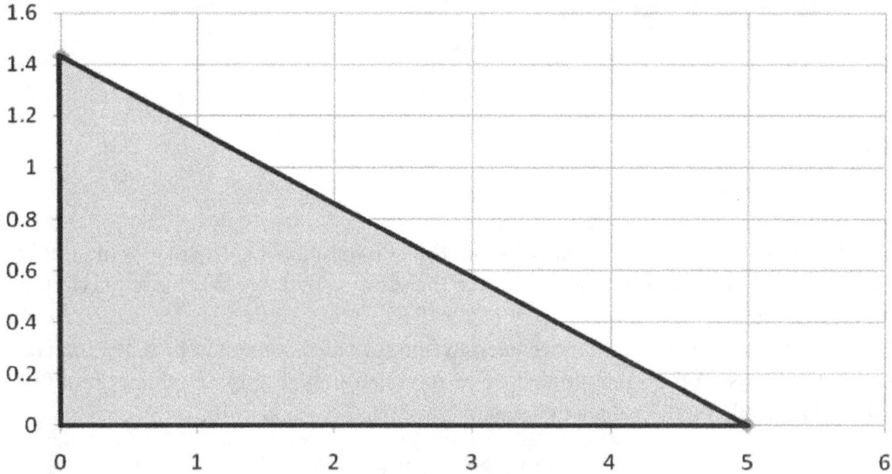

FIGURE 2.2 Line graph of constraint 2.

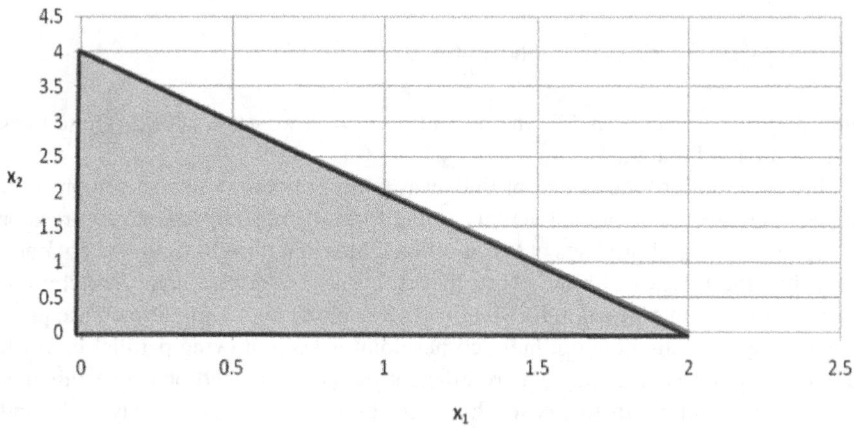

FIGURE 2.3 Line graph of constraint 3.

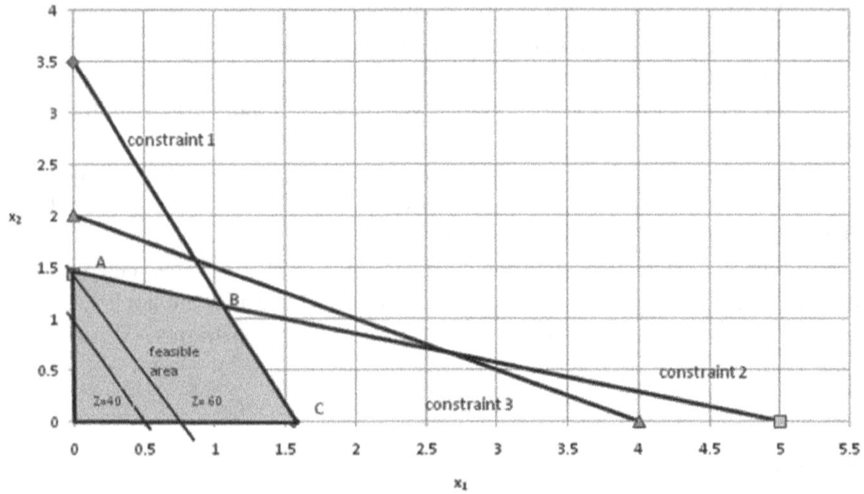

FIGURE 2.4 Feasible region.

this area would provide a feasible solution. But is it optimal? Optimality would imply that under given constraints what would be the solution that would achieve maximum objective function.

For finding optimal solution, we need to find for values of x_1 and x_2. These can be estimated by using the same procedure as in constraints. Let us say that at profit of $40, what would be the values of decision variables? So at Z = 40;

$$40 = 80x_1 + 40x_2$$

Putting $x_1 = 0$, we get x_2 as 1, and by putting $x_2 = 0$, we get $x_1 = 0.5$.

Similarly at Z = 60;

$$60 = 80x_1 + 40x_2$$

Putting $x_1 = 0$, we get x_2 as 1.5, and by putting $x_2 = 0$, we get $x_1 = 0.75$. These lines are indicated in Figure 2.4.

Therefore, upward movement of line away from zero shows increase in profit. It implies increase in operational activity in the form of consumption of resources or in this case selling of products at two locations. Maximum profit would be obtained when this line touches extreme point, that is, corner of feasible area. Stretching it beyond corner points would make solution infeasible though it might increase profits. One way of finding out optimum corner point is keep drawing parallel lines till an extreme point is reached. A more efficient method is to find objective function at all extreme points. In this case, there are three such points, namely A, B and C. Therefore, there could be three possible optimal solutions. One that would give maximum profit would be selected.

Solution of point A by looking at graph is very simple which is $x_1 = 0$ and $x_2 = 1.43$.

$$Z\,(A) = 80*0 + 40*1.43 = 57.2$$

Similarly for C where $x_1 = 1.56$ and $x_2 = 0$.

$$Z\,(C) = 80*1.56 + 40*0 = 124.8$$

However, point B is intersection of constraints 1 and 3. Two linear equations after converting them into equalities can easily be solved as shown below:

$$\text{Constraint 1:}\quad 9x_1 + 4x_2 \le 14$$

$$\text{Constraint 3:}\quad 5x_1 + 10x_2 \le 20$$

Two simultaneous equations can easily be solved by using algebra. However, if there are number of such interacting points, then we can use excel.

Excel Solution

For two decision variables, place their coefficients in two rows. As shown in the following snapshot (Figure 2.5), B2:C2 row is devoted to constraint 1 with coefficients 9 and 4. Similarly, constraint 3 is written in cells B3:C3. Right-hand side resources are written in E2 and E3 cells, respectively. Now select two empty rows G2 and G3 simultaneously for calculating values of x_1 and x_2. Write the formula =MMULT(MINVERSE(B2:C3), E2:E3) and press ctrl+shift+enter together. This would give answer for x_1 and x_2 in cells G2 and G3, respectively.

It gives $x_1 = 1.05$ and $x_2 = 1.13$.

$$Z\,(B) = 80*1.05 + 40*1.13 = 129.2$$

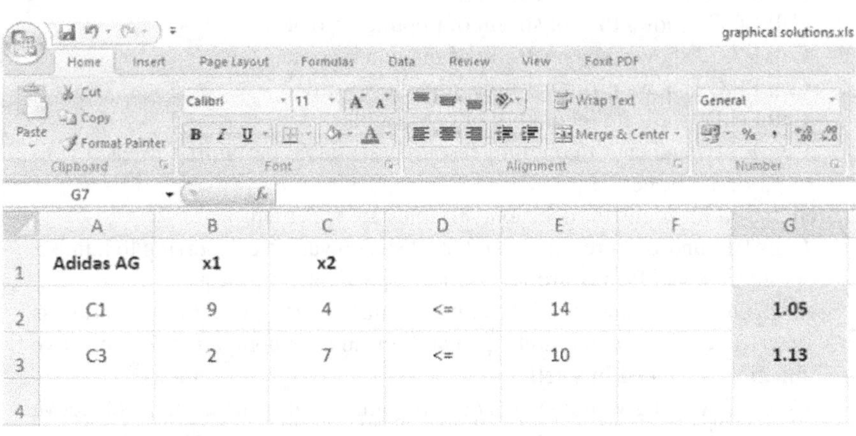

FIGURE 2.5 Solving of simultaneous equations by Excel.

At point B, objective function is maximum implying its optimality.

Therefore, maximum profit earned by selling 1.05*1,000 = 1,050 units from location A and by selling 1.13*1,000 = 1,130 units at location B would be 129.2*1,000 = $12,920.

To check for feasibility, the values of decision variables are put in all constraints and check whether these are satisfied or violated.

$$\text{Constraint 1:} \quad 9x_1 + 4x_2 \quad \leq 14$$

$$9*1.05 + 4*1.13 = 13.94$$

It is less than right-hand side indicating non-violation of inequality.

$$\text{Constraint 2:} \quad 2x_1 + 7x_2 \quad \leq 10$$

$$2*1.05 + 7*1.13 = 10$$

It is just equal to right-hand side indicating non-violation of inequality.

$$\text{Constraint 3:} \quad 5x_1 + 10x_2 \quad \leq 20$$

$$5*1 + 10*1.5 = 16.55$$

It is less than right-hand side indicating maximum available resource is not fully utilized. This idle resource available is called as *slack variable.*

As all the constraints are satisfied, thus optimal solution is also feasible.

Formulation of model in Excel worksheet:

Any LP model when formulated in Excel requires following steps. These steps are illustrated in Figure 2.6 shown as follows.

Step 1: Constraints data are entered first in the worksheet.

Cell B5:C5 shows the constraint of accessibility

Cell B6:C6 shows the constraint of population density

Cell B7:C7 shows the constraint of median income

Cell B9:C9 shows profit per unit from each ward

Cell D5:D7 indicate inequality of constraints

Cell E5:E7 indicate maximum resource available pertaining to each constraint respectively

Step 2: Cell locations for decision variables are specified.

Cells B12 and B13 are earmarked for decision variables referring to number of unit sold from wards A and B, respectively.

Step 3: Objective function data either of maximization or minimization is entered below the constraints data and formula for computing the objective function is created in a cell.

Cell B14 would give final value of maximum profit earned, i.e. the objective function. For its calculation, a formula has to be created which is the

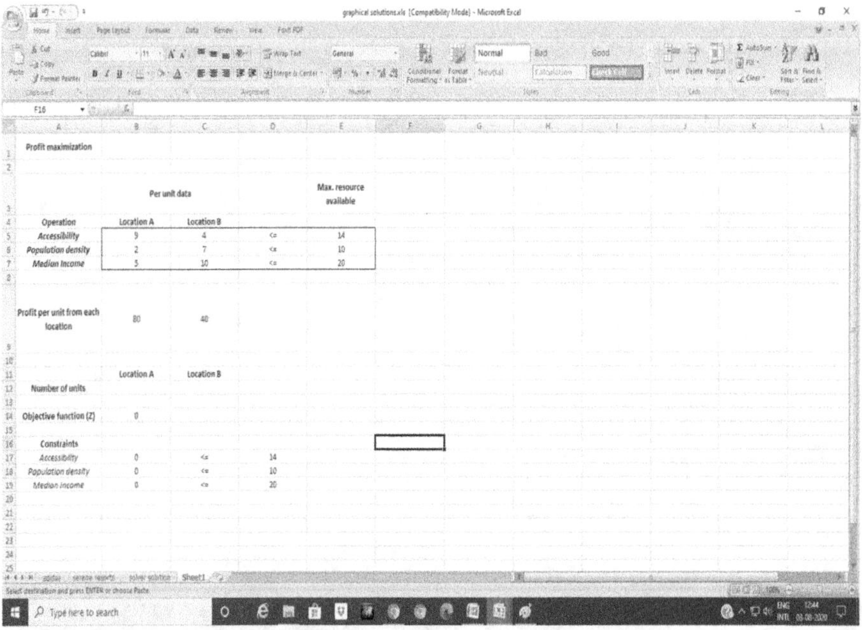

FIGURE 2.6 Excel Model formulation.

sum of product of profit per unit from each ward and number of units sold from that ward. Thus:

Cell B14: = B9*B12+C9*C12

Step 4: Formula for computing the left side of constraint equation is formulated. In excel sheet, select a cell for entering the formula of each constraint. Thus:

Cell B17: =B5*B12+C5*C12

Cell B18: = B6*B12+C6*C12

Cell B19: = B7*B12+C7*C12

Step 5: Similarly, formula for computing the right side of constraint equations are created in the same row in which the formulas are created in step 5. Thus, in excel worksheet (Figure 2.6):

Cell D17: = E5

Cell D18: = E6

Cell D19: = E7

The zero values in cells are formulas, which would be seen in formula bar.

Excel Solution

In Microsoft Excel, after entering entire LP data in the worksheet, the following steps would lead to a solution.

Step 1: Select Data menu in the toolbar.

Step 2: In Data menu, select Solver application.

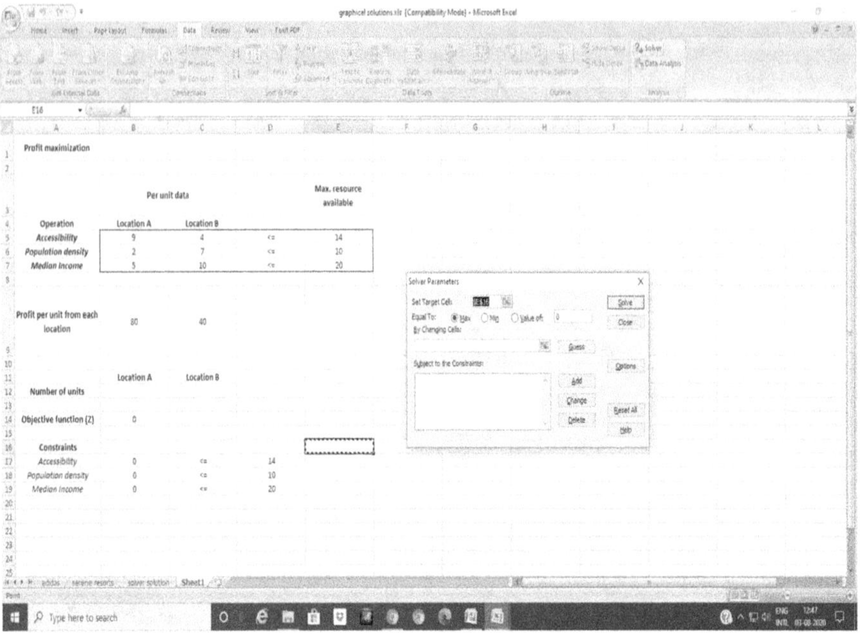

FIGURE 2.7 Solver.

Step 3: Open Solver application (Figure 2.7).
In Solver parameters dialog box
Enter B14 in set target cell.
Select purpose of maximization.
Enter B12:C12 in By changing cells box.
To enter constraint equations click on Add button.

Step 4: When Add constraint dialog box opens it would have three boxes: firstly cell reference; secondly inequalities of ≤ and lastly constraint box.

To enter accessibility constraint enter B17 in cell reference box; enter inequality of ≤ and D17 in constraint box. Then click on Add for adding more constraints.
For population density constraint enter B18; inequality ≤ and D18 as constraint.
Finally, for median income enter B19; ≤ and D19.
After entering all constraints click OK.

Step 5: Now choose Options. Select Assumptions of non-negativity and model to be linear. Click OK.

Step 6: When Solver parameters dialog box appear, click on Solve.

Step 7: Finally when final solution appears on the worksheet, select Keep Solver Solution and click OK.

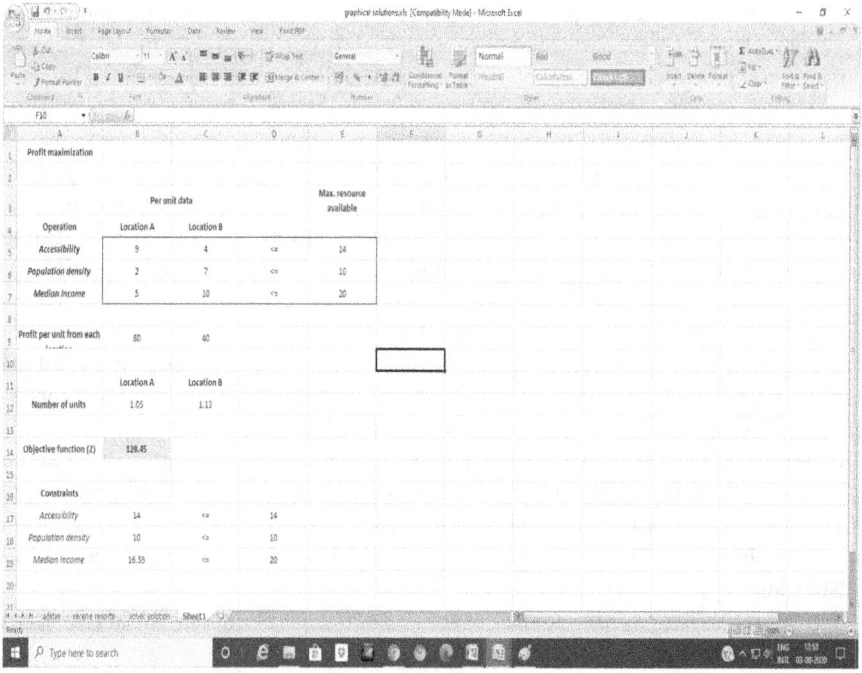

FIGURE 2.8 Excel solution.

The solution is shown in Figure 2.8.

Figure 2.8 gives final result with values of decision variables in cells B12 and C12. As the number of units sold is in thousands so sales from location A shown in cell B12 would be 1.05*1,000 = 1,050 units and sales from location B would be 1.13*1,000 = 1,130 units.

$$\text{Maximum profit would be } 129.2 * 1{,}000 = \$129{,}200.$$

Feasibility of solution is indicated by fulfilment of constraints. Right side of constraint equations are shown in cells B17, B18 and B19 for three constraints. All of them are fulfilled as right side values are either less or equal to left side.

2.5.2 Illustration of Minimization: Rose's Luxury Restaurant

The US hotel industry experienced strong growth from a period of 2009–2018 as indicated by increase in gross bookings from $116 billion to $185 billion (https://www2.deloitte.com/content/dam/Deloitte/us/Documents/consumer-business/us-consumer-2019-us-travel-and-hospitality-outlook.pdf). This had a positive spillover effect on associated sectors such as transportation and restaurants. Such unprecedented growth driven by vigorous economy and high consumer spending encouraged many restaurants to expand their operations. Rose's luxury in Washington DC intended to tap into favourable market conditions and decided for expansion of its restaurant. The restaurant provides eclectic menu to its clients. The restaurant serves predominantly two types of customers. Two member couples and four member families were its prime customers who enjoyed their services especially during dinner time. To cater to both types of customers, restaurant manager set two types of seating arrangements. One type of arrangement (i) catering to couples constituted two chairs and a round table. Such arrangement occupied maximum of 16 ft^2 of space. Other arrangement (ii) catering to four member families constituted four chairs and a rectangular table. It occupied maximum of 30 ft^2 of space. Various numbers of both types of seating arrangements had to be arranged in maximum of 1,000 ft^2 of space. It was estimated by the manager from past data that combined minimum demand of A type of seating arrangement in one cycle was 20 and that of B was 15. The kitchen of restaurant can prepare maximum of 40 meals to cater to demand of both types of customers if all the seating arrangements are taken. It was estimated that during peak evening hours to serve one arrangement of A type incurs a cost of $10, whereas for B type, the cost is $15. Manager wants to find the optimal number of seating arrangements, which would minimize cost. A higher number of seats arranged would cause congestion, which would increase the cost, while a lower number of seats would result in extra vacant spaces, implying improper utilization of space, which would increase the cost again.

Solution:

The objective is to minimize the cost of arranging a certain number of two types of seating arrangements A and B. Suppose,

x_1 represents the number of A type of seating arrangements and

x_2 represents the number of B type of seating arrangements

Also given is cost of serving one arrangement of both A and B types which is $10 and $15, respectively.

Thus, objective function would be

$$\textbf{Minimize} \quad Z = 10x_1 + 15x_2$$

Subject to the following constraints:

Constraint 1 (Floor space): One A type of arrangement takes 16 ft^2 whereas one B type takes 30 ft^2 out of maximum available 1,000 ft^2. So manager has to organize all A and B types in limited space of 1,000 ft^2.

Thus $16x_1 + 30x_2 \leq 1{,}000$

Constraint 2 (Demand of A): Minimum demand of A type is given as 20 tables and for B is 15 tables. This implies that given space of restaurant can at least organize 20 and 15 tables of A and B types, respectively.

Thus, $1x_1 + 0x_2 \geq 20$

Constraint 3 (Demand of B) $0x_1 + 1x_2 \geq 15$

Constraint 4 (Capacity): Maximum of 40 meals can be produced to cater to full demand of both A and B types.

Thus, $1x_1 + 1x_2 \leq 40$

Solution steps are given below:

Step 1: Coordinates of four lines to be drawn with regard to four constraints are calculated and shown in the following table.

Decision variables	x_1	x_2
Constraints		
Constraint 1	0	33.33
	62.5	0
Constraint 2	0	0
	20	0
Constraint 3	0	15
	0	0
Constraint 4	0	40
	40	0

Step 2: The graph is shown in Figure 2.9 giving coordinates of four constraints and common area.

Area bounded by constraint 1 is from the line and till origin as inequality is less than or equal to. Constraints 2 and 3 are with greater than or equal to inequalities so their solution area is away from origin. However, the area covered by these lines is unbounded, which would not give a feasible solution. However, we are looking for a common area bounded by four constraints simultaneously. Constraint 4 with less than inequality helps in providing a common feasible area shown by a filled area on

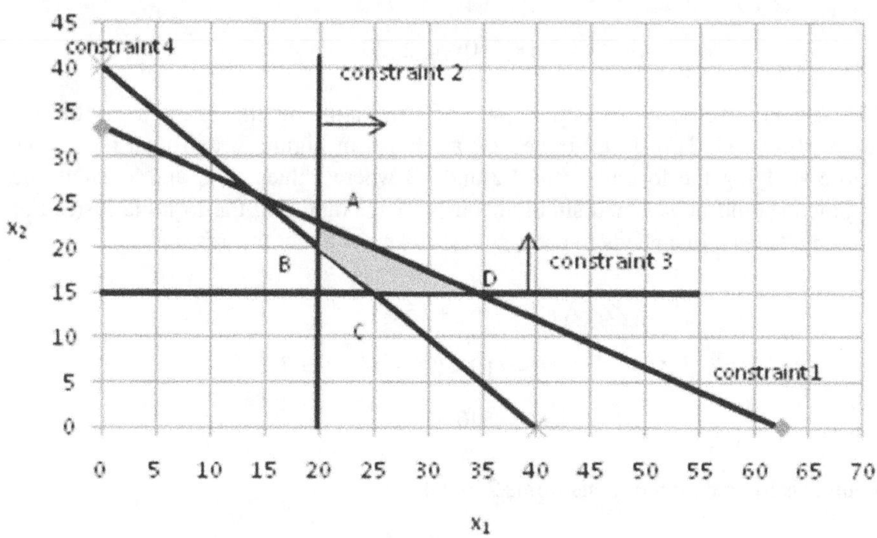

FIGURE 2.9 Graphical solution of minimization problem.

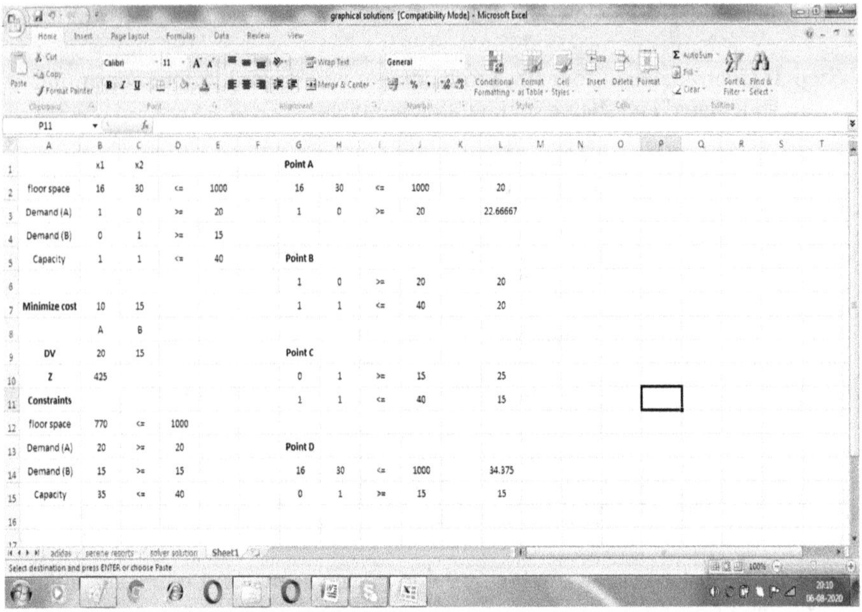

FIGURE 2.10 Solving of simultaneous equations by Excel.

the graph. The corner or extreme points are marked by A, B, C and D. Solutions for decision variables for all these points are detailed by using Excel (Figure 2.10):

Point A is intersection of constraints 1 and 2, i.e.

$$16x_1 + 30x_2 \le 1{,}000$$

$$1x_1 + 0x_2 \ge 20$$

This would give $x_1 = 20$ and $x_2 = 22.67$ by applying the formula =MMULT (MINVERSE(G2:H3), J2:J3) in cell L2 as shown in Figure 2.10. To get the results before applying the formula cells L2 and L3 where values of x_1 and x_2 would be obtained should be selected simultaneously. After inputting the formula instead of pressing "enter", one needs to press ctrl+shift+enter to get the answer.

$$Z(A) = 10x_1 + 15x_2$$

$$= 10*20 + 15*22.67$$

$$= 540.5$$

Point B is intersection of constraints 2 and 4, i.e.

$$1x_1 + 0x_2 \le 20$$

$$1x_1 + 1x_2 \le 40$$

This would give $x_1 = 20$ and $x_2 = 20$ by applying the formula = MMULT (MINVERSE(G6:H7), J6:J7) in cell L6 as shown in Figure 2.10. Cells L6 and L7 where values of x_1 and x_2 would be obtained should be selected simultaneously. Then, press ctrl+shift+enter to get the answer.

$$Z\,(B) = 10*20 + 15*20$$
$$= 500$$

Point C is intersection of constraints 3 and 4, i.e.

$$0x_1 + 1x_2 \geq 15$$

$$1x_1 + 1x_2 \leq 40$$

This would give $x_2 = 15$ and $x_1 = 25$ by applying the formula = MMULT (MINVERSE(G10:H11), J10:J11) in cell L10 as shown in Figure 2.10. Cells L10 and L11 should be selected simultaneously. Then, press ctrl+shift+enter to get the answer.

$$Z\,(C) = 10*25 + 15*15$$
$$= 475$$

Point D is intersection of constraints 1 and 3, i.e.

$$16x_1 + 30x_2 \leq 1{,}000$$

$$0x_1 + 1x_2 \geq 15$$

This would give $x_2 = 15$ and $x_1 = 34.37$ by applying the formula =MMULT (MINVERSE(G14:H15), J14:J15) in cell L14 as shown in Figure 2.10. Cells L14 and L15 should be selected simultaneously. Then, press ctrl+shift+enter to get the answer.

$$Z\,(D) = 10*34.37 + 15*15$$
$$= 568.7$$

Minimum objective function obtained is of \$475 with $x_1 = 25$ and $x_2 = 15$. This implies that manager should arrange 25 of A-type and 15 of B-type seating arrangement, which would result in minimum cost of \$475.

This is the optimal solution. To check for its feasibility, the calculated values of decision variables are put in the constraint equations.

Constraint 1 (floor space):

$$16x_1 + 30x_2 \leq 1{,}000$$

$$16*25 + 30*15 = 850$$

which is less than right-hand side so constraint is not violated.

However, it is less than maximum available space of 1,000 ft² by 150, which remains vacant if 25 seats of A and 15 seats of B are arranged. This indicates resource remaining idle, which is represented by a variable called as *slack variable.* To convert inequality into equality in constraint 1, a slack variable s_1 is added into the equation. This is shown as:

$$16x_1 + 30x_2 + 1s_1 = 1{,}000$$

Constraint 2 (Demand of A):

$$1x_1 + 0x_2 \geq 20$$

$$1*25 + 0*15 = 25$$

which is greater than right-hand side resource, thus satisfying the constraint.

This situation is opposite to above one. Minimum demand that can be fulfilled is of 20, i.e. manager would install at least 20 seating arrangements of A. However, to achieve the objective of minimum cost solution says that at least 25 seats should be installed. More than minimum required is represented by *surplus variable.* To convert inequality into equality in constraint 2, a surplus variable s_2 is subtracted from the equation. This is shown as:

$$1x_1 + 0x_2 - 1s_2 = 20$$

Constraint 3 (Demand of B):

$$0x_1 + 1x_2 \geq 15$$

$$0*25 + 1*15 = 15$$

which is equal to right-hand side resource
Constraint 4 (Capacity):

$$1x_1 + 1x_2 \leq 40$$

$$1*25 + 1*15 = 40$$

which is equal to right-hand side resource
As all the constraints are satisfied, thus optimal solution is also feasible.

Formulation of model in Excel worksheet:

Step 1: Constraints data are entered firstly in the worksheet.
Cell B2:C3 shows the constraint of floor space
Cell B3:C3 shows the constraint of demand of A

Cell B4:C4 shows the constraint of demand of B

Cell B5:C5 shows the constraint of capacity

Cell B7:C7 shows cost per unit for serving each kind of seating arrangement

Cell D2:D5 indicates inequality of constraints

Cell E2:E5 indicates maximum resource available pertaining to each constraint, respectively

Step 2: Cell locations for decision variables are specified.

Cell B9 and C9 are earmarked for decision variables

Step 3: Objective function of minimization is entered in cell I10. Formula for its calculation was

Cell B10: =B7*B9+C7*C9

Step 4: Formulae for computing left side of constraint equation were:

Cell B12: = B2*B9 + C2*C9

Cell B13: = B3*B9 + C3*C9

Cell B14: = B4*B9 + C4*C9

Cell B15: = B5*B9 + B5*B9

Step 5: Formulae for computing right side of constraint equations were:

Cell D12: = E2

Cell D13: = E3

Cell D14: = E4

Cell D15: = E5

Excel Solution

Step 1: Select Data menu in the toolbar.

Step 2: In Data menu select Solver application.

Step 3: Open Solver application. In Solver parameters dialog box:

- Enter B10 in set target cell.
- Select purpose of minimization.
- Enter B9:C9 in By changing cells box.
- To enter constraint equations click on Add button.

Step 4: In Add constraint dialog box

- To enter floor space constraint enter B12 in cell reference box; enter inequality of ≤ and D12 in constraint box. Then click on Add for adding more constraints.
- For Demand (A) constraint, enter B13; inequality ≥ and D13 as constraint.
- For Demand (B) constraint, enter B14; inequality ≥ and D14 as constraint.
- Finally, for capacity, enter B15; ≤ and B15.
- After entering all constraints, click OK.

Step 5: Now choose Options. Select Assumptions of non-negativity and model to be linear. Click OK.

Step 6: When Solver parameters dialog box appear, click on Solve.

Step 7: Finally when final solution appears on the worksheet, select Keep Solver Solution and click OK.

The solution is shown in Figure 2.11.

Figure 2.11 gives final result with values of decision variables in cell I9 and J9. The number of seating arrangements of type A was 25 and that of B was 15.

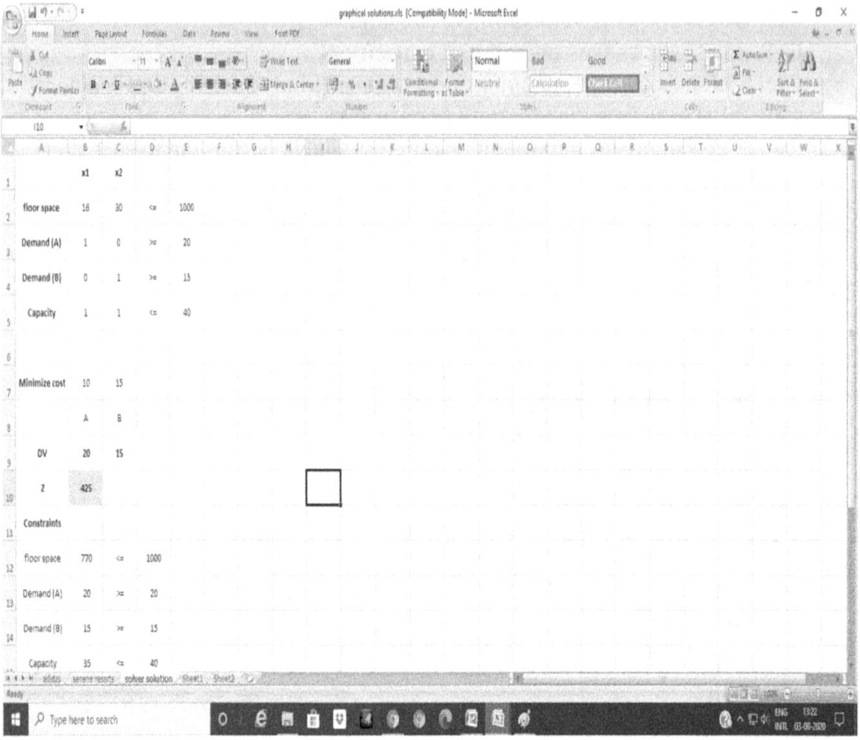

FIGURE 2.11 Excel solution by using solver.

Minimum cost would be

$$Z = 10 * 20 + 15 * 15$$

$$= \$425$$

Feasibility of solution is indicated by fulfilment of constraints. Right side of constraint equations are shown in cells B17, B18 and B19 for three constraints. All of them are fulfilled as right side values are either less or equal to left side.

2.6 SUMMARY

This chapter introduced the concept of LP utilized in solving a variety of business problems. The chapter included the understanding of elements of such models that are required for their formulation. Elements included an objective function that needs to be minimized or maximized; decision variables that would govern the achievement of objective function; set of constraints that would restrict achievement of objective to certain level and coefficients of both objective and constraint equations. The next section discussed assumptions of such models that are utilized in their formation. It was emphasized that LPP models should follow conditions of proportionality, additivity, divisibility and certainty. Finally, numerous applications of LPP models were illustrated, and formation of models was presented.

The next section was devoted to graphical method of solving LPP models. The limitation of such method is that only problems involving two decision variables can be solved as a two-dimensional graph can only show two axis. The procedure involves constructing two-dimensional graph with x_1 and x_2 as axis. This is followed by the identification of values of these decision variables that are permitted by restrictions. The understanding of inequalities in these restrictions is very important as they are used to deduce the common area that is termed as feasible area of solution. Any value of x_1 and x_2 within this area forms a part of feasible value of objective function. However, intersection points in this area determine most optimal solution. For maximization, problem highest intersection point would indicate the value of objective function, whereas the lowest intersection point would provide value of minimization objective. This chapter has illustrated the process by discussing various examples.

2.7 CASE 1: FEDERAL MOGUL CORPORATION

Federal Mogul motor parts, an independent division of Federal Mogul Corporation based in Michigan, USA, manufactures and supplies various automotive parts and components such as pistons, piston rings, shocker tubes, etc. The corporation provides parts to various industries such as aerospace, agriculture vehicles, commercial vehicles, etc. It operates manufacturing plants in various countries including Britain and India. One of the plants producing pistons for commercial vehicles intended to minimize storage cost incurred in storing piston rings. Three types of piston rings need to be packed and stored in the packaging section of a piston manufacturing company. The rings are differentiated based on their size. ‘A’ type of pistons involved all pistons with diameter 30–60 mm and ‘B’ type of pistons categorized pistons with diameter 61–85 mm. Specified racks were earmarked for storage of each type of pistons. ‘A’ type of piston rings was stored in rack R1 occupying 0.27 m^2 of floor space and per rack can store maximum of 3,800 rings. Similarly, ‘B’ type of piston rings was stored in rack R2 occupying 0.54 m^2 of floor space and per rack can store maximum of 6,600 rings. Previous 3 months data estimated that at any given time maximum of 1,00,000 units of all type of pistons are stored in the warehouse. Number of each type of piston rings stored also depended on space occupied by each type of rack. The warehouse had storage space of 360 m^2. The packaging manager wants to minimize the storage cost by estimating the number of racks of each type to be used in the warehouse. The cost of each type of rack was: R1 at \$30 and R2 at \$32 per unit. The variation in costs was because of different number and size of blocks in each rack for storing different sized rings.

- Formulate a LP model specifying objective function, decision variables and constraints.
- Find optimal solution by using graphical procedure and check its feasibility.

2.8 CASE 2: TOYOTA MOTORS

Toyota Motors is considered as pioneers of lean production system, which has proven to be a very effective system in guiding an organization to overcome the unpredictable competition in industrial environment. One of the essential tools in implementation of entire gamut of lean system is providing defect free items as and when required without keeping a stock of products. Kanban and Just-in-time are some of the methods that facilitate low inventory of products. Reliability of supplies plays an essential role in providing products at a varying rate keeping in view fluctuating demand.

A car requires more than 10,000 parts assembled into various components, which are supplied to original equipment manufacturers (OEMs) like Toyota. One of the finished parts that are sourced from various vendors is seats of the car. Competition between suppliers and long-term contract with them provides reliability. Toyota for its Camry range of car sources upholstery from Boshoku and Johnson Controls, Inc., in which the company had a major stake. To create competition at the Tier 1 level, company also selected Lear Corporation to supply seats to various manufacturing plants in the US. Toyota places top priority to maintain long-term stable contract with its supplier.

Toyota has earmarked seats into three categories as two front and one back seats. Thus, seats are shipped as one set consisting of three seats. One production schedule assembles seats of all three types in 20 cars making a demand of total 60 seats per schedule. Due to capacity constraints, no vendor can fulfil the entire demand, so two vendors Boshoku (V1) and Lear (V2) on the condition that each vendor can produce a combination of three seats as one set rather than providing individual seats. The order of a fixed number of seats required per production schedule is placed with each vendor 1 day before assembly. Thus, each vendor has a maximum of 21 hours of production time to fulfil the order. Vendor V1 takes 0.7 hours to manufacture one seat, whereas V2 takes 0.3 hours in this maximum time available. Cost minimization is the main purpose in selection of vendors. Per unit cost of manufacturing from V1 was $30 and from V2 was $35. Formulate a LPP indicating in the case of multiplicity of vendors how many units should be ordered from each vendor so that overall cost is optimized, which would help in deciding which vendor should be selected.

2.9 GLOSSARY

- **Linearity**: implies decision variables with degree of one. It further infers that a unit change in independent, i.e. decision variable would cause a unit change in objective function.
- **Proportionality**: a change in decision variables by a fixed amount would cause proportional change in both objective function and constraint equation.
- **Additivity**: decision variables have a combined effect on objective function.
- **Divisibility**: means decision variables can take both integer and non-integer values. However, they should be divisible and finite.

- **Certainty**: implies that per unit consumption of resources indicated by coefficients in objective function and constraints are constant and known with certainty.
- **Feasible solution**: means those of values of decision variables, which satisfies simultaneously and does not violate constraint equations.
- **Optimal solution**: means values of decision variables, which would provide a maximum or minimum solution. Though these values fulfil objective function, they might not satisfy all constraint equations making solution infeasible.
- **Graphical method**: provides pictorial representation of a LP problem with two constraints. The method is very effective in showing feasible region and optimal solution.
- **Slope**: of linear equation means proportion change in one variable when other variable is changed to some extent. Slope remains the same on a line and changes only at corner. That is why while finding solutions only corner also known as extreme points are considered.

2.10 MODEL QUESTIONS

2.10.1 A survey was conducted by the manager of a telecom company, which changed its logo elements as explained in a previous example. Manager emphasized that success of rebranding could be measured if changed elements have increased the awareness level about reasons of rebranding. To measure change in awareness level for colour aspect of logo, manager asked respondents to compare colour of old and new logo, and it was inferred that on an average, a score of 8.2 out of 10 was obtained regarding the awareness level of a new logo. Similarly, a score of 7.9 and 7.5 was measured for design and font of new logo. In addition to the constraints of maximum score as indicated in the previous example, manager assessed maximum awareness level regarding three combinations of logo elements. Thus, in this case, the purpose was to find out which combination of logo elements would be preferred by how many respondents out of total of 3,000 respondents surveyed. The combinations were decided as:

- Number of respondents preferring colour is more than combined respondents preferring design and font.
- Number of respondents preferring colour and design is more than font only.
- Number of respondents preferring colour and font is more than design only.

Formulate a LPP model that would achieve the purpose of maximizing awareness level.

2.10.2 A promotion manager promotes company's product by using three different types of advertisements. One is colour in print media, second is black and white in vernacular media and third is TV. It was estimated that frequency of ads shown in different media would increase chances of sales, but higher frequency would also entail higher costs. A coloured ad showed in print media once costs $10 (in thousands), whereas the same ad in black and white in vernacular media would result in a cost of \$7 (in thousands). TV media is costliest at \$18 (in thousands). Manager wants to estimate frequency of ads in these three media channels with the objective of minimizing the spend and increase ad appeal and sales. A consumer survey conducted by manager had indicated that coloured ad had an appeal of 4, black and white had 1.25 and TV ad had an appeal of 3. A minimum of 4 score was considered as required for manager to spend more money on increasing frequency of ads in particular channel. Second, level of frequency tends to increase awareness, which affects sales of product. Past experience had shown that a coloured ad shown once in print media generates an average sale of \$11 (in thousands), a black and white ad \$10 (in thousands) and TV ad \$12 (in thousands). A minimum of \$13 (in thousands) in sales should be obtained by spending in each channel. Formulate LPP to determine number of times an ad should be shown in a media channel so that ad spend is minimum.

2.10.3 A retail investor wants to maximize returns by investing a maximum of \$10,000 in two stocks A and B. The returns are characterized by riskiness of stock. Predominantly riskier the stock higher are the returns and vice versa. This riskiness is generally measured by average price of stock and rate of its price fluctuation. It was determined that average price of stock A is \$20 with fluctuation of 1.5 gives an average return of \$300. On the other hand, average price of stock B is \$25 with fluctuation of 3.4 units gives an average return of \$400. Maximum acceptable price fluctuation was estimated to be of 10 units. Formulate LPP to determine number of two stocks to be bought, which would maximize returns.

2.10.4 Toys R Us in their assembly plant is assembling a given ordered quantity of three different toys (T1, T2 and T3). It was estimated that to fulfil the order, a maximum of 1,800 kg of raw material is required. One unit of T1, T2 and T3 requires 3.2, 2.6 and 1.5 kg of raw material. Also, maximum of 4,500 hours of labour hours are earmarked for their production. One quantity of three types of toys requires 6, 4 and 6 hours of production. Maximum cost of production to manufacture the order is \$300. One unit of T1, T2 and T3 takes \$5.5, \$3.6 and \$4.7 as cost of production. Profit earned from each type of toy per unit is \$7, \$5.3 and \$5.4. How many units of each type of toy should be manufactured which would maximize profit? Formulate LPP.

2.10.5 Solve example 2.10.4 if further data regarding maximum demand of each type of toy is given. Maximum demand of T1, T2 and T3 is 950, 900 and 1,000 units. What would be the changes in number of units being manufactured? What would be the change in profits?

2.10.6 Forecast error indicates accuracy of forecasting technique applied for prediction purposes. A restaurant manager wants to check the accuracy of three forecasting techniques (A, B and C) applied to predict demand of restaurants' dining hall for next year. Accuracy of these techniques is direct function of forecast error. The lesser the error, The more the accuracy. It was estimated from previous data that on average 'A', 'B' and 'C' techniques provide 70%, 75% and 82% of accuracy in forecasting, respectively.

Forecasting needed to be done under capacity constraints of both kitchen and dining hall. Gap between existing and estimated capacity of kitchen from three techniques was estimated to be of 20,000, 30,000 and 30,000 units. These gap values indicate error from a particular technique for past 1 year. Annual capacity of kitchen was estimated to be maximum of 60,000 meals. Second, dining hall has a maximum capacity of 12,000 meals per year. The error from three forecasting techniques (A, B and C) with regard to capacity of dining hall for the next year was estimated to be 15,000, 5,000 and 5,000 meals, respectively. Finally, manager decided that no technique would be considered appropriate if on an average it gives an error of more than 20%. Average error of A, B and C techniques estimated from past experience was 4%, 10% and 12%. By applying LP model, manager would like to found the rate of error of each technique, which would contribute to maximizing the accuracy of forecasting.

2.10.7 Braiding and extrusion are two main processes required in production of hoses. A leading manufacturer of hoses and belts manufactures two types of hoses H1 and H2. Hoses are manufactured in terms of length measured in metres. Thus, one unit here implies production of 100m of a particular hose. Company on an average receives order of 20,000 units of H1 and 30,000 units of H2. Braiding of one unit of H1 takes 250 minutes, whereas H2 takes 300 minutes. Total braiding time available within which order has to be fulfilled was estimated to be 34,000 minutes. Extrusion involves two processes: tube extrusion and cover extrusion. Tube extrusion takes 100 and 110 minutes for one unit of H1 and H2, whereas cover extrusion takes 95 and 100 minutes for one unit of H1 and H2. Maximum total time available for tube extrusion and cover extrusion is 20,000 and 15,000 minutes, respectively. The cost of production of one unit of H1 in the plant was $15, and if purchased from a vendor, the cost increased to $16. On the other hand, the cost of production of H2 was $17, but if purchased, then cost was estimated to be $17.5. Formulate LPP to be utilized in estimating the number of units to be manufactured and purchased in order to fulfil demand at the lowest possible cost.

2.10.8 A dietician has recommended an adult to take one serving per day of combination of three diets (D1, D2 and D3). Each serving of a diet contains certain units of nutrients, namely, proteins, fats and carbohydrates. Minimum amount of calories of each nutrient that should be taken to fulfil dieticians' recommendation is also given. In considering the intake amount, the user has to bear the cost of each diet. The purpose is to find optimal units of each diet, which would fulfil minimum daily requirement of each nutrient while keeping cost at minimum possible level. Formulate LPP.

	Diet 1	Diet 2	Diet 3	Minimum Daily Requirement of Each Nutrient
Protein	10	12	22	290
Fats	13	30	17	373
Carbohydrates	28	20	29	455
Cost per diet ($)	60	80	120	

2.10.9 A FMCG company categorizes personal care products into perfumes and cosmetics. An independent watchdog interested in reporting the harmful effects of these products reported that company uses different chemicals in manufacture of these products – some of which are shown and some are not shown in labels of bottles of perfumes and cosmetics. The report categorized these chemicals in three groups namely tested and labelled (A), tested not labelled (B) and neither tested nor labelled (C). On an average, a litre of perfume contains maximum 55%, 30% and 20% of A, B and C, respectively. Similarly, a litre of cosmetic contains maximum of 60%, 50% and 20% of A, B and C, respectively. Production schedule of company can produce maximum of 3,500L of A, 3,700L of B and 2,500L of C. Company procures three types of chemicals at a cost of $12, $15 and $19 per litre respectively. A litre bottle of perfume and cosmetic was sold at an average price of $155 and cosmetic at a price of $127. Formulate a LPP to estimate quantity of each chemical that would be used in manufacture of litre of perfume and cosmetic, which would generate maximum profit.

2.10.10 At the end of every semester, students of three departments are asked to provide feedback regarding the performance of teachers, research scholars and technical staff involved in carrying out various functions of teaching. The scores from feedback form are used to measure the productivity level of three types of staff on a scale of 0–1. These scores indicate the productivity level of one member of each staff. The minimum level of productivity required from a combination of activities of each department is also shown in the table. The management wants to use this data to find out number of each type of staff required for upcoming session. The purpose is to allocate optimal number of staff in a way that minimizes salary bill. The data is:

	Teachers	Research Scholars	Technical Staff	Minimum Level of Productivity Required
Department 1	0.4	0.6	0.5	12
Department 2	0.3	0.4	0.5	14
Department 3	0.5	0.2	0.3	18
Salary bill ($)	60,000	70,000	80,000	

2.10.11 A library of MBA department wants to order text books and reference books in the subject of Business Statistics. The subject is taught in both years of the course, with the first year students learning the basics and the second year students learning the advanced topics of the subject. Text books, which are required basic books, are in much demand as compared to reference books, which deal with advanced topics. Depending on budget and strength of students, it was estimated that at least 200 copies of text and 50 copies of reference books are a must to meet requirements of both years of students. Also, library has limited space as it has to store other subject books also. A maximum of 13,500 cm² of space in form of racks was made available for books of statistics out of which one reference book would approximately occupy 50 cm² and one text book will occupy 60 cm². The librarian wants to find number of both types of books, which would minimize the purchase cost when each reference book would cost $800 and a text book $600. Formulate LPP.

2.10.12 A bottling plant quality manager found significant variation in amount of soft drink filled in 200 ml bottles of three types of soft drinks (A, B and C). The estimated loss in revenue per bottle of each type of soft drink was found to be $0.47, 0.46 and 0.49 per bottle. This could lead to lot of losses as thousands of bottles of each type are filled daily. Taking into consideration, the problem manager asked shop floor supervisor to identify and rectify the problem. For this purpose, supervisor decided to conduct an experiment involving filling of bottles at three pressure levels (P1, P2 and P3). The acceptable limits or maximum amount of drink that could be filled was decided to be 200.01, 200.05 and 200.12 mL at pressure levels P1, P2 and P3. It was also decided that out of total sample of 180 bottles of all three types of soft drinks, at least 178 bottles should fulfil the criteria of acceptable limits of amount of drink filled. Formulate a LP model which would find optimal number of bottles of soft drinks A, B and C which would be within variation limit.

	Types of Soft Drinks		
Pressure Levels	**A**	**B**	**C**
P1	1.01	1.03	1.07
P2	1.09	1.21	1.08
P3	1.17	1.06	1.04

2.10.13 Solve graphically

$$\text{Max} \quad Z = 100x_1 + 80x_2$$

Subject to:

$$150\, x_1 + 120\, x_2 \le 180$$

$$700\, x_1 + 500\, x_2 \le 800$$

$$x_1, x_2 \ge 0$$

2.10.14 Solve graphically

$$\text{Max} \quad Z = 5x_1 + 7x_2$$

Subject to:

$$2x_1 + 4x_2 \le 10$$

$$4x_1 + 3x_2 \le 12$$

$$x_1, x_2 \ge 0$$

2.10.15 Solve graphically

$$\text{Max} \quad Z = 4.2x_1 + 3.8x_2$$

Subject to:

$$5x_1 + 6x_2 \le 5{,}500$$

$$2.6x_1 + 1.5x_2 \le 2{,}100$$

$$x_1, x_2 \ge 0$$

2.10.16 Solve graphically

$$\text{Min} \quad Z = 4x_1 + 5x_2$$

Subject to:

$$5x_1 + 3x_2 \le 6$$

$$7x_1 + 8x_2 \le 10$$

$$x_1, x_2 \ge 0$$

2.10.17 Solve graphically

$$\text{Max} \quad Z = 2x_1 + 3x_2$$

Subject to:

$$2x_1 - 4x_2 \le 5$$

$$-7x_1 + 3x_2 \le 8$$

$$x_1, x_2 \ge 0$$

2.10.18 Solve graphically

$$\text{Min} \quad Z = 150x_1 + 175x_2$$

Subject to:

$$0.05x_1 + 0.44x_2 \geq 100$$

$$0.5x_1 + 0.32x_2 \leq 60$$

$$0.35x_1 + 0.24x_2 \leq 80$$

$$x_1, x_2 \geq 0$$

2.10.19 Solve graphically

$$\text{Max} \quad Z = 40x_1 + 35x_2$$

Subject to:

$$2x_1 + 4x_2 \leq 14$$

$$5x_1 + 1x_2 \geq 11$$

$$4x_1 + 5x_2 = 17$$

$$x_1, x_2 \geq 0$$

3 Linear Programming
Simplex Method

3.1 INTRODUCTION

Chapter 2 illustrated various cases from different fields which can be solved quantitatively by formulating a Linear Programming Model. As each objective function, whether maximization or minimization, is limited or bounded by various constraints, achieving the best optimal solution is the safest bet. In this process, the concept of iteration or finding a better solution by conducting algebraic mathematical operations makes the entire process increasingly complex. Linear algebra involves solving of simultaneous linear equations. The problem becomes very difficult with multiple equations as discussed in the case of OR problems. Solving such simultaneous equations requires development of algorithms and the simplex method is one such algorithm. Developed by George Dantzig in 1947, it has proven over time to be a remarkably effective algebraic method in resolving such problems. These problems, as illustrated by various cases in Chapter 2, involve multi-criteria decision making. Such problems, involving numerous constraints, become very difficult to solve manually. The application of computer software executing the simplex method makes it easier to reach an optimal solution. This chapter has discussed the cases of maximization and minimization of objective function by applying the simplex method. Also, these cases are solved by using Excel.

3.2 ILLUSTRATION OF MAXIMIZATION

3.2.1 Case: Woodland Biomass Power (US)

Woodland Biomass Power, located in Woodland, CA, USA, has been in business since 1989. The plant uses agricultural waste for the generation of clean energy. The main fuel for the power plant is paddy straw, wheat straw, cane trash, mustard, maize stalk, and cotton stalks. The electricity generated is sold directly to the grid and the rate of sale is decided according to government norms. The success of projects in terms of generation of profits is a high priority as such projects are very helpful in reducing polluting greenhouse gases.

Availability of different types of fuels is a limitation in the generation of electricity as fuels are selected according to their moisture content and weight. High moisture content would ask for more pre-preparation in terms of its drying. On the other hand, because some fuels, such as rice husk, are light in weight, they tend to float in water rather than being mixed up when poured in the boiler. Fuels which are heavier, such as dry leaves or sugarcane waste, tend to settle down in water. Thus, a proper mixture of both heavy and light fuels is necessary. Finally, the availability of

DOI: 10.1201/9781003212966-3

certain fuels is seasonal in nature. As a result of such fuel variations, management must maintain a proper mixture of various types of fuels. Presently, power plants use two categories of fuels: A and B. Type A is a mixture of paddy straw, fire wood and cotton stalks, whereas B type is a mixture of mustard husk, wheat nods, maize husk and mentha husk.

Both types of fuels, i.e. A and B, are used to generate electricity. It was estimated that power sold to the government at said rates would give a profit of \$0.2 per unit if it is generated by using fuel A and \$0.3 from fuel B. The average production of electricity varies on an almost daily basis because of particular variation in the availability of fuel. Thus, in this illustration, the average data of a particular month has been used regarding the amount of electricity generated and the amount of fuel of each type required. In the selected month, the maximum amount of power generated was 1,539 mw. Forty-eight percent of this electricity was generated by using fuel A, whereas fuel B contributed 41% and the rest was generated from other agriculture waste. A maximum of 440,000 kg of combined fuel A and B was required to generate this electricity. Out of this total fuel quantity, 160 kg of fuel A and 120 kg of B generated a unit of electricity. To earn profits, company wanted to generate the maximum amount of electricity from limited availability of fuel. To produce the mentioned amount of power (1,539 mw) with available fuel (440,000 kg), an estimated amount of \$7,452 was spent on fuel. Hundred kilogram of fuel considered as one unit was bought at an average price of \$1.58 and similarly, a unit of B was bought at a price of \$2.77. The company manager under such a scenario intends to maximize profits by estimating the number of units of both types of fuel as profits depend on the amount of power produced which is directly proportional to amount of fuel available.

3.2.1.1 Simplex Method

The simplex method requires the following steps for solving a linear programming problem (LPP).

Step 1: Model Construction

The case clearly illustrates that the amount of fuel A and B are variables which would decide the profit of a company. Suppose in a given month, x_1 units of A and x_2 units of B are used in power generation. So if one unit of A gives the profit of \$0.2 and one unit of B \$0.3, then the objective function can be formulated as follows:

Objective function:

$$\textbf{Maximize}\quad Z = 0.2x_1 + 0.3x_2$$

Subject to the following constraints:

Constraint 1: Cost of fuel

Company had resources of \$7,452 to spend on fuel for power production. One unit of A, i.e. 100 kg, was purchased at a price of \$1.58 and B at \$2.77. Thus, when x_1 and x_2 quantities of fuel A and B types are purchased, it would result in:

$$1.58x_1 + 2.77x_2 \leq 7{,}452$$

Constraint 2: Amount of electricity generated

With given resources, a maximum of 1,539 mw of electricity was generated out of which fuel A contributed 48% and B 41%. Thus,

$$0.48x_1 + 0.41x_2 \le 1539$$

Constraint 3: Amount of fuel

A maximum of 440,000 kg of combined fuel A and B was required to generate this electricity. Out of this total fuel quantity, 160 kg of fuel A and 120 kg of B generated a unit of electricity. Thus,

$$1.6x_1 + 1.2x_2 \le 4{,}400$$

Constraint 4: The number of units of fuels A and B

$$x_1, x_2 \ge 0$$

Constraint 4 completes a linear programming formulation indicating non-negativity of decision variables. It shows that even if the plant is not working and no resources are consumed, then the amount of both types of fuel consumed would be zero. When generation starts, then consumption of resources begins, implying a positive value of decision variables. This constraint is very important as it implies that these variables cannot be negative.

It is important to understand that constraints, in addition to being understood as limitations, should also be understood as resources available for fulfilment of an indicated task. For instance, constraint 1 implies that the plant has the maximum of \$7,452 capital to generate the maximum power. Out of this, one unit of A is bought for \$1.58 and one unit of B is purchased for \$2.77. So if only one unit of both fuels has been purchased, then out of \$7,452 available, \$4.35 (\$1.58 + \$2.77) is consumed. So, the right-hand side of every equation shows availability of resource and the left-hand side shows consumption of that resource.

The inequalities in all constraints are *less than or equal to* implying that the left-hand side is less than or equal to the right-hand side. Thus, to convert inequality into equality, another variable called a *slack variable* has to be added to all three equations. Slack variables indicate idle resource of a particular activity, and they are symbolized as s_1, s_2 and s_3 for three constraints. These slack variables are termed as *basis variables.*

This leads to the formulation of the model shown in Table 3.1.

TABLE 3.1
Model

Maximize	$Z = 0.2x_1 + 0.3x_2 + 0s_1 + 0s_2 + 0s_3$
Constraint 1	$1.58x_1 + 2.77x_2 + 1s_1 + 0s_2 + 0s_3 = 7{,}452$
Constraint 2	$0.48x_1 + 0.41x_2 + 0s_1 + 1s_2 + 0s_3 = 1{,}539$
Constraint 3	$1.6x_1 + 1.2x_2 + 0s_1 + 0s_2 + 1s_3 = 4{,}400$

Step 2: Finding initial basic feasible solution (IBFS)

The term IBFS implies two aspects: first, whether the solution for a given case is feasible or not and second, what the initial solution is.

Feasibility of solution: Equations of constraints show that there are five variables: x_1 and x_2 are decision variables (non-basic) and s_1, s_2 and s_3 are basic variables. These five variables are calculated according to the model (Table 3.1) by only three equations. Thus, a solution can only be reached if two out of five variables are assumed to be zero. This condition is fulfilled by non-negativity criteria, i.e. all variables in the objective function should be non-negative. Generally, x_1 and x_2 are assigned as zero, indicating initial situations, implying that production is yet to start and in the beginning, the quantity of two products is zero. So by assigning $x_1=x_2=0$; we get values of basic variables as

Non-basic variables	$x_1=x_2=0$		
Basic variables	$s_1=7{,}452$;	$s_2=1{,}539$;	$s_3=4{,}400$

Thus, all *basis variables* are positive, making the initial solution feasible. This indicates all resources allocated shown in the form of constraints are yet to be used.

Initial solution: By putting feasible values in the objective function of maximization, Z would come out to be zero

$$\begin{aligned} Z &= 0.2x_1+0.3x_2 \\ &= 0.2*0+0.3*0 \\ &= 0 \end{aligned}$$

Thus, the initial basic solution would be zero. So, mathematically, we are starting from the beginning and would try to find a better solution than zero profits. This is the initial solution where 'initial' implies first or beginning of the solution where production or any other function is yet to start.

Step 3: Improvement in solution

In this step, the first, model as shown in Table 3.1 is modified to create a simplex table. It shows decision variables, their coefficients in objective function, basis variables and their coefficients in objective function, constraints and maximum or minimum resources available (Table 3.2).

Iteration 0:

Improvement in the solution process involves the selection of *key column* and *key row*. The detailed process is explained below:

i. **Find initial profit**: Initial profit (denoted by z_j) for both types of fuel, when none is consumed, is calculated in the following way:

TABLE 3.2
Iteration 0

Decision Variables		x_1	x_2	s_1	s_2	s_3		
Coefficients of Decision Variables in Objective Function		**0.2**	**0.3**	**0**	**0**	**0**		
Basis Variables	**Coefficients of Basis Variables c_b in Objective Function**	**Coefficients of Constraint Functions**					**Maximum Available Resource**	**Ratio**
s_1	0	1.58	2.77	1	0	0	7,452	2,690.25
s_2	0	0.48	0.41	0	1	0	1,539	3,753.66
s_3	0	1.6	1.2	0	0	1	4,400	3,666.67
	z_j	0	0	0	0	0		
Net evaluation row	$c_j - z_j$	0.2	0.3	0	0	0		

- Multiply the coefficient of each decision variable in the constraint function with the coefficient of the basic variable in the objective function. For instance, the coefficient of x_1 decision variable for first constraint is 1.58 and is multiplied by 0, which is the coefficient of basic variable s_1. Similarly, it is done for all constraints and corresponding basic variable coefficients.
- Add all the values obtained in the above step. Thus,

$$z_1 = 1.58*0 + 0.48*0 + 1.6*0 = 0$$

$$z_2 = 2.77*0 + 0.41*0 + 1.2*0 = 0$$

$$z_3 = 1*0 + 0*0 + 0*0 = 0$$

$$z_4 = 0*0 + 1*0 + 0*0 = 0$$

$$z_5 = 0*0 + 0*0 +1*0 = 0$$

The value of every objective function is zero as no consumption of resources has started yet.

- Finally, the net evaluation row denoted by $c_j - z_j$ is calculated. It indicates a change in objective function from zero units produced to when one unit is produced. The value of objective function when zero units are produced is calculated for fuel A as 0.2. When one unit is produced, the objective function indicates that one unit of A would consume $0.2. So, a change in objective function would be

$$c_1 - z_1 = 0.2 - 0 = 0;$$

$$c_2 - z_2 = 0.3 - 0 = 0;$$

$$c_3 - z_3 = 0 - 0 = 0;$$

$$c_4 - z_4 = 0 - 0 = 0;$$

$$c_5 - z_5 = 0 - 0 = 0$$

ii. **Column selection**: Net evaluation row ($c_j - z_j$) shows that with per unit consumption of fuel A, profit will increase by $0.2, whereas with consumption of a unit of B, profit will increase by $0.3. The objective is to maximize profits, so a column with maximum positive value would be selected. Thus, in the case of maximization, the key column is selected on the basis that *net evaluation row is maximum positive.* So, in this case, column with $0.3 would be selected. Column selection indicates *incoming variable.* It means that from the initial solution of usage of zero units of A and B now in the first step, fuel B would be consumed and a profit of $0.3 generated.

iii. **Row selection**: Start of operations would cause usage of resources. This depletion of resources indicated as an *outgoing variable* is reflected in the selection of row. In this case, out of three resources which would be used first would be known by dividing the maximum resource available for each constraint with coefficients of incoming variable. For instance, for first basic variable s_1, the maximum resource available was 7,452 and in the selected column known as key column coefficient was 2.77. The ratio would be 7,452/2.77 = 2,690.25. Similarly, other ratios are calculated (Table 3.2). Out of all ratios, *row with a minimum ratio* would be selected as it would indicate minimum or limiting resource.

iv. So, in this case, s_1 is the outgoing variable and x_2 would be the incoming. This would be done by applying mathematical functions in such a manner that coefficients of x_2 column are replaced by coefficients of s_1 column. These are shown in the following table:

x_2 Column Coefficients	Would Change into	s_1 Column Coefficients
2.77		1
0.41		0
1.2		0

This is achieved by applying the following mathematical functions (Table 3.3).

The above calculations will provide another simplex tableau indicating an improved solution and whether the solution can be improved further (Table 3.4).

Iteration 1:

As x_2 has entered the solution with a value of 2,690; thus the objective function has improved from zero to $Z = 0.2*0 + 0.3*2,690 = 807$.

TABLE 3.3
Calculations for New Simplex Tableau

Selected row

$$1.58x_1+2.77x_2+1s_1+0s_2+0s_3=7{,}452$$

To get the coefficient of x_2 as 1 key row would be divided by 2.77. Thus, the key row becomes

Row 1

$$0.57x_1+1x_2+0.361s_1+0s_2+0s_3=2{,}690.25 \quad (3.1)$$

Row 2

$$0.48x_1+0.41x_2+0s_1+1s_2+0s_3=1{,}539 \quad (3.2)$$

To get the coefficient of x_2 as zero multiply equation (3.1) with 0.41 and subtract it from row 2

$$0.234x_1+0.41x_2+0.15s_1+0s_2+0s_3=1{,}103 \quad (3.3)$$

(3.2) – (3.3) will give

$$0.146x_1+0x_2-0.15s_1+1s_2+0s_3=436$$

Row 3

$$1.6x_1+1.2x_2+0s_1+0s_2+1s_3=4{,}400 \quad (3.4)$$

Equation (3.1) is now multiplied with 1.2 and subtracted from row 3

$$0.68x_1+1.2x_2+0.43s_1+0s_2+0s_3=3{,}228 \quad (3.5)$$

(3.4) – (3.5) will give

$$0.92x_1+0x_2-0.43s_1+0s_2+1s_3=1{,}172$$

TABLE 3.4
Iteration 1

Decision Variables		x_1	x_2	s_1	s_2	s_3		
Coefficients of Decision Variables in Objective Function		**0.2**	**0.3**	**0**	**0**	**0**		
Basis Variables	**Coefficients of Basis Variables c_b in Objective Function**	**Coefficients of Constraint Functions**					**Maximum Available Resource**	**Ratio**
x_2	0.3	0.57	1	0.36	0	0	2,690	4,719.2
s_2	0	0.15	0	0.15	1	0	436	2,906.6
s_3	0	0.92	0	−0.43	0	1	1,172	1,273.9
	z_j	1.71	3	1.08	0	0		
Net evaluation row	$c_j - z_j$	0.03	0	−1.08	0	0		

In iteration 1, applying the same method to find key column and key row s_3 would be replaced by x_1. Replacing coefficients of x_1 by s_3 by applying the above-discussed procedure would give us the following new simplex table (Table 3.5).

TABLE 3.5
Iteration 2

Decision Variables		x_1	x_2	s_1	s_2	s_3	
Coefficients of Decision Variables in Objective Function		**0.2**	**0.3**	**0**	**0**	**0**	
Basis Variables	**Coefficients of Basis Variables c_b in Objective Function**	**Coefficients of Constraint Functions**					**Maximum Available Resource**
x_2	0.3	0	1	0.64	0	−0.62	1,963
s_2	0	0	0	0.07	1	−0.16	245
x_1	0.2	1	0	−0.47	0	1.08	1,274
	z_j	0.2	0.3	0.98	0	0.3	
Net evaluation row	$c_j - z_j$	0	0	−0.98	0	−0.3	

Iteration 2:

All values in the net evaluation row are either zero or negative, implying that no further improvement in solution can be achieved. Thus,

$$x_1 = 1{,}274 \text{ kg}$$

$$x_2 = 1{,}964 \text{ kg}$$

$$Z = 0.2x_1 + 0.3x_2$$

$$= 0.2 * 1{,}274 + 0.3 * 1{,}963$$

$$= \$844 /-$$

Thus, the objective function has improved from 807 to 844.

This implies that 1,274 kg of A fuel and 1,864 kg of B fuel are required to generate the mentioned amount of electricity and it would result in earning a profit of $844.

Appropriateness of solution is checked by putting values of decision variables in constraints.

Constraint 1: Cost of fuel

$$1.58x_1 + 2.77x_2 \leq 7{,}452$$

$$1.58 * 1{,}274 + 2.77 * 1{,}963 \leq 7{,}452$$

$$7{,}450 \leq 7{,}452$$

Thus, the constraint is accepted, implying that the company still has $2 with which more fuel can be bought for power production.

Constraint 2: Amount of electricity generated

$$0.48x_1 + 0.41x_2 \leq 1{,}539$$

$$0.48 * 1{,}274 + 0.41 * 1{,}963 \leq 1{,}539$$

$$1{,}416 \leq 1{,}539$$

Non-violation of constraint indicates appropriateness of solution, implying that with available units of fuel, 1,416 units of electricity is generated.

Constraint 3: Amount of fuel

$$1.6x_1 + 1.2x_2 \leq 4{,}400$$

$$1.6 * 1{,}274 + 1.2 * 1{,}963 \leq 4{,}400$$

$$4{,}394 \leq 4{,}400$$

This indicates that 1,416 units of electricity is generated by using 4,394 kg of fuel and this would provide optimal profits by taking into consideration all constraints simultaneously.

Excel Solution:

Any linear programming model when formulated in Excel requires the following steps.

Step 1: Constraints data are entered first in the worksheet.

Cell B5:C5 shows the constraint of the cost of fuel

Cell B6:C6 shows the constraint of the amount of electricity generated

Cell B7:C7 shows the constraint of the amount of fuel

Cell B9:C9 shows the profit per unit

Cell D5:D7 indicates the inequality of constraints

Cell E5:E7 indicates the maximum resource available for each constraint

Step 2: Cell locations for decision variables are specified.

Cells B12 and B13 are earmarked for decision variables.

Step 3: Objective function data of maximization is entered below the constraints data, and a formula for computing the objective function is created in a cell (Figure 3.1).

Cell B14 would give the final value of maximum profit earned, i.e. the objective function. For its calculation, a formula has to be created which is the sum of product of profit per unit from each ward and the number of units sold from that ward. Thus:

Cell B14: =B9*B12+C9*C12

Step 4: The formula for computing the left side of the constraint equation is formulated. In the Excel sheet, select a cell for entering the formula of each constraint. Thus:

	A	B	C	D	E	F
1	Profit Maximization					
2						
3		Per unit data			Max. resource available	
4	Operation	Fuel A	Fuel B			
5	*Cost of fuel*	1.58	2.77	<=	7452	
6	*Amount of electricity generated*	0.48	0.41	<=	1539	
7	*Amount of fuel*	1.6	1.2	<=	4400	
8						
9	Profit per unit	0.2	0.3			

FIGURE 3.1 Model formulation in excel.

Cell B17: = B5*B12 + C5*C12
Cell B18: = B6*B12 + C6*C12
Cell B19: = B7*B12 + C7*C12

Step 5: Similarly, formulas for computing the right side of constraint equations are created in the same row in which the formulas are created in step 5. Thus in the Excel worksheet:
Cell D17: = E5
Cell D18: = E6
Cell D19: = E7

In Microsoft Excel, after entering entire linear programming data in the worksheet, the following steps would lead to a solution.

Step 1: Select Data menu in the toolbar.
Step 2: In Data menu, select Solver application.
Step 3: Open Solver application.
In Solver parameters dialog box
Enter B14 in set target cell.
Select purpose of maximization.
Enter B12:C12 in by changing cell box.
To enter constraint equations, click on Add button.
Step 4: When the Add constraint dialog box opens, it would have three boxes: first, cell reference; second, inequalities of ≤ and lastly, constraint box.
To enter cost of fuel constraint, enter B17 in cell reference box; enter inequality of ≤ and D17 in the constraint box. Then click on Add to add more constraints.
For amount of electricity generated constraint enter B18; inequality ≤ and D18 as constraint.

	A	B	C	D	E
11		**Fuel A**	**Fuel B**		
12	**Number of units**	1274	1963		
13					
14	**Objective function (Z)**	843.7			
15					
16	**Constraints**				
17	***Cost of fuel***	7450.43	<=	7452	
18	***Amount of electricity generated***	1416.35	<=	1539	
19	***Amount of fuel***	4394	<=	4400	

FIGURE 3.2 Excel solution.

Finally, for amount of fuel, enter B19; ≤ and D19.
After entering all constraints, click OK.

Step 5: Now choose Options. Select Assumptions of non-negativity and model to be linear. Click OK.

Step 6: When Solver parameters dialog box appear, click on Solve.

Step 7: Finally, when final solution appears on the worksheet, select Keep Solver Solution and click OK.

Figure 3.2 gives the final result with values of decision variables in cells B12 and C12. The number of units of fuel A shown in cell B12 would be 1,274 units and of fuel B would be 1,963 units (after rounding off to the closest decimal place).

$$\text{Maximum profit would be} = 0.2*1{,}274 + 0.3*1{,}963$$

$$= \$844/-$$

The feasibility of a solution is indicated by fulfilment of constraints. The right side of constraint equations are shown in cells B17, B18 and B19 for three constraints. All of them are fulfilled as right side values are either less or equal to the left side.

3.2.2 Case: Adidas AG Retail Stores

This case has been explained in Chapter 1 and solved graphically in Chapter 2 under Section 2.5.1. Here the model would be solved by using the simplex method. The model has been reproduced below:

3.2.2.1 Simplex Method

$$\textbf{Maximize Profit} \quad Z = 80x_1 + 40x_2$$

Constraint 1: Accessibility:

$$9x_1 + 4x_2 \leq 14$$

Constraint 2: Population density:

$$2x_1 + 7x_2 \leq 10$$

Constraint 3: Median Income:

$$5x_1 + 10x_2 \leq 20$$

$$x_1, x_2, x_3 \geq 0$$

Step 1: Model Construction (Table 3.6)

Step 2: Finding IBFS

In this case, also x_1 and x_2 are decision variables (non-basic) and s_1, s_2, and s_3 are basic variables. By assigning zero to non-basic variables, the initial solution would be:

Non-basic variables		$x_1 = x_2 = 0$	
Basic variables	$s_1 = 14$;	$s_2 = 10$	$s_3 = 20$

Thus, all *basis variables* are positive, making initial solution feasible. This indicates all resources allocated shown in the form of constraints are yet to be used.

By putting feasible values in the objective function of minimization, Z would come out to be zero.

$$\begin{aligned} Z &= 80x_1 + 40x_2 \\ &= 80*0 + 40*0 \\ &= 0 \end{aligned}$$

Thus, the initial basic solution would be zero.

TABLE 3.6
Model

Maximize	$Z = 80x_1 + 40x_2 + 0s_1 + 0s_2 + 0s_3$
Constraint 1	$9x_1 + 4x_2 + 1s_1 + 0s_2 + 0s_3 = 14$
Constraint 2	$2x_1 + 7x_2 + 0s_1 + 1s_2 + 0s_3 = 10$
Constraint 3	$5x_1 + 10x_2 + 0s_1 + 0s_2 + 1s_3 = 20$

TABLE 3.7
Iteration 0

Decision Variables		x_1	x_2	s_1	s_2	s_3		
Coefficients of Decision Variables in Objective Function		80	40	0	0	0		
Basis Variables	Coefficients of Basis Variables c_b in Objective Function	Coefficients of Constraint Functions					Maximum Available Resource	Ratio
s_1	0	9	4	1	0	0	14	1.56
s_2	0	2	7	0	1	0	10	5
s_3	0	5	10	0	0	1	20	4
	z_j	0	0	0	0	0		
Net evaluation row	$c_j - z_j$	80	40	0	0	0		

Step 3: Improvement in solution

Iteration 0:

So in this case, s_1 is the outgoing variable and x_1 would be the incoming (Table 3.7). This would be done by applying mathematical functions in such a manner that coefficients of x_1 column are replaced by coefficients of s_1 column. These are shown in the following table.

x_1 Column Coefficients	Would Change into	s_1 Column Coefficients
9		1
2		0
5		0

This is achieved by applying the following mathematical functions (Table 3.8).

The above calculations will provide another simplex tableau, indicating an improved solution and whether the solution can be improved further (Table 3.9).

Iteration 1:

So, in this case, s_2 is the outgoing variable and x_2 would be the incoming. This would be done by applying mathematical functions in such a manner that coefficients of x_2 column are replaced by coefficients of s_2 column. These are shown in the following table.

x_1 Column Coefficients	Would Change into	s_1 Column Coefficients
0.44		0
6.12		1
7.78		0

TABLE 3.8
Calculations for New Simplex Tableau

Selected row

$$9x_1 + 4x_2 + 1s_1 + 0s_2 + 0s_3 = 14$$

To get the coefficient of x_1 as 1 key row would be divided by 9. Thus, the key row becomes

Row 1

$$1x_1 + 0.44x_2 + 0.11s_1 + 0s_2 + 0s_3 = 1.56 \quad (3.6)$$

Row 2

$$2x_1 + 7x_2 + 0s_1 + 1s_2 + 0s_3 = 10 \quad (3.7)$$

To get the coefficient of x_1 as zero multiply equation (3.6) with 2 and subtract it from row 2

$$2x_1 + 0.88x_2 + 0.22s_1 + 0s_2 + 0s_3 = 3.12 \quad (3.8)$$

(3.7) – (3.8) will give

$$0x_1 + 6.12x_2 - 0.22s_1 + 1s_2 + 0s_3 = 6.88$$

Row 3

$$5x_1 + 10x_2 + 0s_1 + 0s_2 + 1s_3 = 20 \quad (3.9)$$

Equation (3.6) is now multiplied with 5 and subtracted from row 3

$$5x_1 + 2.22x_2 + 0.56s_1 + 0s_2 + 0s_3 = 7.78 \quad (3.10)$$

(3.9) – (3.10) will give

$$0x_1 + 7.78x_2 - 0.56s_1 + 0s_2 + 1s_3 = 12.22$$

TABLE 3.9
Iteration 1

Decision Variables		x_1	x_2	s_1	s_2	s_3		
Coefficients of Decision Variables in Objective Function		**80**	**40**	**0**	**0**	**0**		
Basis Variables	**Coefficients of Basis Variables c_b in Objective Function**	**Coefficients of Constraint Functions**					**Maximum Available Resource**	**Ratio**
x_1	80	1	0.44	0.11	0	0	1.56	3.54
s_2	0	0	6.12	−0.22	1	0	6.88	1.12
s_3	0	0	7.78	−0.56	0	1	12.22	1.57
	z_j	80	35.2	8.8	0	0		
Net evaluation row	$c_j - z_j$	0	4.8	−8.8	0	0		

This is achieved by applying the following mathematical functions (Table 3.10).

The above calculations will provide another simplex tableau indicating an improved solution and whether the solution can be improved further (Table 3.11).

TABLE 3.10
Calculations for New Simplex Tableau

Selected row

$$0x_1 + 6.22x_2 - 0.22s_1 + 1s_2 + 0s_3 = 6.88$$

To get the coefficient of x_2 as 1 key row would be divided by 6.22. Thus, the key row becomes

Row 2

$$0x_1 + 1x_2 - 0.04s_1 + 0.16s_2 + 0s_3 = 1.12 \qquad (3.11)$$

Row 1

$$1x_1 + 0.44x_2 + 0.11s_1 + 0s_2 + 0s_3 = 1.56 \qquad (3.12)$$

To get the coefficient of x_2 as zero multiply equation (3.11) with 0.44 and subtract it from row 2

$$0x_1 + 0.44x_2 - 0.017s_1 + 0.07s_2 + 0s_3 = 0.50 \qquad (3.13)$$

(3.12) – (3.13) will give

$$1x_1 + 0x_2 + 0.127s_1 - 0.07s_2 + 0s_3 = 1.06$$

Row 3

$$0x_1 + 7.78x_2 - 0.56s_1 + 0s_2 + 1s_3 = 12.22 \qquad (3.14)$$

Equation (3.11) is now multiplied with 7.78 and subtracted from row 3

$$5x_1 + 7.78x_2 - 0.31s_1 + 1.24s_2 + 0s_3 = 8.76 \qquad (3.15)$$

(3.14) – (3.15) will give

$$0x_1 + 0x_2 - 0.25s_1 - 1.24s_2 + 1s_3 = 3.46$$

TABLE 3.11
Iteration 2

Decision Variables		x_1	x_2	s_1	s_2	s_3	
Coefficients of Decision Variables in Objective Function		**80**	**40**	**0**	**0**	**0**	
Basis Variables	**Coefficients of Basis Variables c_b in Objective Function**	**Coefficients of Constraint Functions**					**Maximum Available Resource**
x_1	80	1	0	0.127	−0.07	0	1.06
x_2	40	0	1	−0.04	0.16	0	1.12
s_3	0	0	0	−0.25	−1.24	1	3.46
	z_j	80	40	8.56	0.8	0	
Net evaluation row	$c_j - z_j$	0	0	−8.56	−0.8	0	

Iteration 2:

All values in the net evaluation row are either zero or negative, implying that no further improvement in solution can be achieved. Thus,

$$x_1 = 1.06 = 1.06*1,000 = 1,060 \text{ units}$$

$$x_2 = 1.12 = 1.12*1{,}000 = 1{,}120 \text{ units}$$

$$\begin{aligned} Z &= 80x_1 + 40x_2 \\ &= 80*1{,}060 + 40*1{,}120 \\ &= \$129{,}600. \end{aligned}$$

To check the feasibility of solution, all constraints should be satisfied simultaneously. This is calculated as follows:

$$\text{Constraint 1:} \quad 9x_1 + 4x_2 \leq 14$$

$$9*1.06 + 4*1.12 = 14$$

$$\text{Constraint 2:} \quad 2x_1 + 7x_2 \leq 10$$

$$2*1.06 + 7*1.12 = 9.96$$

$$\text{Constraint 3:} \quad 5x_1 + 10x_2 \leq 20$$

$$5*1.06 + 10*1.12 = 16.5$$

$$\text{Constraint 4:} \quad x_1 \text{ and } x_2 \geq 0$$

Thus, the solution is optimal and feasible.

3.3 AN ILLUSTRATION OF MINIMIZATION

3.3.1 Case: Federal-Mogul

Federal-Mogul has been a leading supplier of products and services worldwide to various vehicle manufacturers. In 2018, in major deal, it was acquired by Tenneco of USA. The company's portfolio includes some of the world' most recognized brands in the global vehicle aftermarket. The Goetze brand of products is used by various original equipment manufacturers. The production plant in Friedberg, Germany excels in developing highly efficient and durable pistons. Pistons made of aluminium-based alloys are of different size, shape and strength for different vehicles. The plant produces 1 lakh pistons every month for the North American aftermarket. Two categories of pistons, namely P1 and P2, are being manufactured. The company faced with a problem of high scrap during the casting stage of these pistons. Twenty percent of pistons were found to be defective at the casting stage of the production, resulting in less productivity, excess labour cost and material loss accounting for approx. 30.7 lakhs per annum. One defective item of P1 caused a scrap cost of $1.65, whereas one defective item of P2 resulted in a loss of $1.82. Under this scenario, management decided to launch a project aiming to reduce rejection in pistons P1 and P2 at least by 10%. The first step was to identify the types of defects during the casting stage so as to examine the reasons for scrap. At casting stage, types of defects that could occur were identified as: bottom thickness variation, blow holes, cold lap, depression, hydrogen porosity, shrinkage and defective pin hole. Pareto chart analysis indicated that bottom thickness was predominant (36%) scrap type in the case of pistons P1 whereas shrinkage (49%) was in the case of pistons P2.

Rating of importance of scrap type: A customer survey was also conducted at this stage to confirm the major scrap type to be reduced from all other types. Here a customer is one who consumes both types of pistons. In the survey, a customer was asked to rate the importance of the above-mentioned defects in particular piston type on a scale of 1–10, where a score of 1–4 was considered noise or unworthy of corrective measure, a score of 5–8 was considered having moderate risk, and a score of 9–10 meant items should be scrapped. An average score of 7.3 was obtained regarding bottom thickness (B.T.) variation in pistons P1, which was the highest among all other scrap types. In the case of pistons, P2, highest priority was given to shrinkage scrap type with an average score of 7.9.

The next step was to identify the reasons for the occurrence of such defects. Brainstorming sessions involving all stakeholders such as casting shop manager and workforce indicated a number of reasons for various steps in the casting process (Table 3.12). Quantitative analysis found that die coating thickness (step 1); worker skill set (step 5) and cooling water discharge (step 6) required in judging the proper speed of pouring molten metal were significant reasons, resulting in defects of B.T. variation in P1 and shrinkage in P2. It is also important to mention that both scrap types, i.e. B.T. variation in P1 and shrinkage in P2, were found to be caused for the same mentioned reasons. These reasons for the selected production schedule are discussed further.

Production schedule: Plant is manufacturing a combined quantity of 1 lakh pistons per month. The production depends on demand and other variables vary from month to month. On an average combined demand of both pistons from North American customers was 1 lakh units. This production was achieved by running the plant in three shifts i.e. night, morning and afternoon. Each shift was of 420 minutes duration and the plant worked 6 days a week. In one shift, a quality check is made after every 60 minutes where a sample of 50 units is analysed for mentioned defects. Thus, seven quality checks would be made per shift. The data used in the case is the scrap rate of two types of pistons for one sample.

TABLE 3.12
Steps in Casting Process

Steps	1	2	3	4	5	6	7	8
Process	Die coating	Furnace charge	Molten metal temperature	Density index	Pouring speed	Water cooling system	Inspection	Hardness and chemical analysis

Reasons resulting in defects:

Die coating thickness: Dies are used for casting purpose. Casting involves pouring hot metal through a runner into the piston die, cooling it down with circulating water, and finally removing the metal piston after it has solidified. In this process, it is important to coat dies with proper material through spray guns so as to provide a protective barrier between the die and the casting and to provide control over the solidification process. In the manufacturing process, dies were coated with a thickness of 50 μm. A sample of 50 pistons was taken per hour and their bottom thickness measured. Fourteen percent of P1 pistons, and 12% of P2 pistons were found to vary significantly from tolerable bottom thickness, making them scrap. It was estimated that not more than 15% of pistons produced per shift would be scrapped.

Worker skill set: Skilled operators are required to pour molten metal into the die because the process requires metal to be at the proper temperature and speed when poured into the die. In a sample of 50 units of P1, 11% of units were found to have variation in bottom thickness resulting in scrap. Similarly, in the case of P2 pistons, the scrap rate was 10%. On an average, a maximum of 15% of items to be considered as defective were found at the end of shift.

Cooling water discharge: The solidification process after pouring of molten metal requires circulation of cooling water. It is a critical step in the casting process because more or less water would significantly affect the quality of piston. A quantity of 7–9 l per minute poured into the die of P1 resulted in 12% scrap items, whereas in the case of P2, the same method produced 13% of defective pistons. Per shift, a maximum of 16% of pistons (total of both P1 and P2) were found to be defective.

Under given production constraints, management intended to find the minimum cost that would be incurred due to defective items.

This illustration is solved by the simplex method.

3.3.1.1 Simplex Method

The simplex method requires the following steps for solving a LPP.

Step 1: Model Construction

In this case, scrap cost is a direct function of the number of defective items of P1 and P2. Suppose a sample of 50 units is inspected after every 60 minutes of production,

x_1 units of P1 and

x_2 units of P2 were found to be defective due to B.T. variation and shrinkage, respectively.

So, if one scrap unit of P1 causes a loss of \$1.65, then x_1 units would cause a loss of $1.65x_1$. Similarly, one scrap unit of P2 would cause a loss of $1.82x_2$. This led to the formulation of the following objective function denoted by Z. In this case, Z would be the minimum cost incurred by company due to the production of x_1 and x_2 defective items per sample of 50 units each.

Objective function:

$$\textbf{Minimize } Z = 1.65x_1 + 1.82x_2$$

No manufacturing can work under ideal conditions or no constraints. The production of pistons was carried under certain limitations, resulting in a high level of scrap. Significant reasons of causing defective items were identified and highlighted in the case. The reasons were die coating thickness, worker skill set and cooling water discharge. Also a rating among customers was carried out to understand which scrap type is most important and should ask for corrective measures.

Constraint 1: Die coating thickness

$$14x_1 + 12x_2 \leq 15$$

Constraint 2: Worker skill set

$$11x_1 + 10x_2 \leq 15$$

Constraint 3: Cooling water

$$12x_1 + 13x_2 \leq 16$$

Constraint 4: Rating of importance

$$7.3x_1 + 7.9x_2 \geq 9$$

Constraint 5: Number of scrap items

$$x_1, x_2 \geq 0$$

Constraint 4 indicates that a minimum rating of 9 is required to result in corrective action. Also, response regarding the importance of bottom thickness variation in P1 was found to be on an average 7.3, whereas in the case of P2, it was 7.9. It can be argued that these values are less than the required critical value of 9, so said sample should not be put under corrective action. But this was an average score and some of the items in the sample might have fallen into the category of high risk. Thus, this constraint was included to understand whether scrap or defective units require corrective action or not.

It is important to understand here that constraints, in addition to being understood as limitations, should also be understood as resources available for fulfilment of the indicated task. For instance, constraint 1 relating to die coating thickness implies that the maximum scrap that would be acceptable under given or available resources of coating a die is 15%. It could be possible that with better dies, i.e. better technology resulting in higher cost, a lesser number of units is getting defective. But this would require more resources. So right-hand side of every constraint means maximum or minimum resource available under which objective has to be fulfilled.

TABLE 3.13
Model

Minimize	$Z = 1.65x_1 + 1.82x_2 + 0s_1 + 0s_2 + 0s_3 + 0s_4$
Constraint 1	$14x_1 + 12x_2 + 1s_1 + 0s_2 + 0s_3 + 0s_4 = 15$
Constraint 2	$11x_1 + 10x_2 + 0s_1 + 1s_2 + 0s_3 + 0s_4 = 15$
Constraint 3	$12x_1 + 13x_2 + 0s_1 + 0s_2 + 1s_3 + 0s_4 = 16$
Constraint 4	$7.3x_1 + 7.9x_2 + 0s_1 + 0s_2 + 1s_3 - 1s_4 = 9$

The inequalities in the first three constraints are *less than or equal to* implying that the left-hand side is less than or equal to the right-hand side. Thus, to convert inequality into equality, another variable called a *slack variable* has to be added to all three equations. Slack variables indicate idle resource of particular activity and they are symbolized as s_1, s_2 and s_3 for three constraints. These slack variables are termed as *basis variables.* Constraint 4, on the other hand, has an inequality which is of *greater than.* This implies that the left-hand side could be greater than or equal than the right-hand side. So, to make equations equal, a variable has to be subtracted, which is termed as *surplus variable.* In this case, constraint 4 implies that ratings can be greater than 9, but it has to be minimum of 9 to be considered defects as critical. The surplus variable for constraint 4 is denoted as s_4.

This results in the following linear programming model (Table 3.13).

Step 2: IBFS:

The term initial basic feasible solution implies two aspects: first, whether the solution for a given case is feasible or not and second, what the initial solution is.

Feasibility of solution: Equations of constraints show that there are seven variables: x_1 and x_2 are decision variables (non-basic) and s_1, s_2, s_3, s_4 and a_4 are basic variables. These seven variables are calculated according to the model (Table 3.13) by only four equations. Thus, the solution can only be reached if three out of seven variables are assumed to be zero. This condition is fulfilled by non-negativity criteria, i.e. all variables in the objective function should be non-negative. In general, x_1 and x_2 are assigned as zero, indicating initial situations, implying that production is yet to start and in the beginning, the quantity of two products is zero. So by assigning

$$x_1 = x_2 = 0;$$

we get values of basic variables as

$$s_1 = 15;\ s_2 = 15;\ s_3 = 16 \text{ and } s_4 = -9.$$

Thus, the non-negativity criterion is *violated by surplus variable.* This is rectified by adding an artificial variable a_4 in constraint 4 which would always be greater than s_4 so as to make it positive. Thus,

$$a_4 \geq s_4$$

The addition of an artificial variable to constraint 4 and putting s_4 as zero would result

$$s_4 = 0 \text{ and } a_4 = 9$$

Thus, non-basic variables are those variables which are assumed to be zero. In this case, x_1, x_2 and s_4 are put as zero making

$$s_1 = 15;\ s_2 = 15;\ s_3 = 16 \text{ and } a_4 = 9$$

Thus, all *basis variables* are positive making initial solution as feasible. This indicates all resources allocated shown in the form of constraints for manufacturing of P1 and P2 are yet to be used.

Initial solution: By putting feasible values in the objective function of minimization, Z would come out to be zero.

$$\begin{aligned} Z &= 1.65x_1 + 1.82x_2 \\ &= 1.65*0 + 1.82*0 \\ &= 0 \end{aligned}$$

Thus, the initial basic solution would be zero. So mathematically, we are starting from the beginning and would try to find a better solution than zero profits. This is the initial solution where 'initial' implies first or beginning of the solution where production or any other function is yet to start.

Taking three slack, one surplus and one artificial variable into consideration, a complete linear programming model is formulated as shown in Table 3.14. An *important point to note is the coefficient of artificial variable a_4 in objective function.* The coefficient is denoted by M. The usage of M as a coefficient of artificial variable is called as Big M method.

Big M method: M is assumed to carry very big values so that artificial variable associated with it goes out of the solution as early as possible. This is done because an artificial variable is an imaginary variable used for making a solution feasible. It does not contribute in any way to the objective function. *That is why this method is also called as Big M method.* Another point to be noted is that an

TABLE 3.14
Model with M Coefficient

Minimize	$Z=1.65x_1+1.82x_2+0s_1+0s_2+0s_3+0s_4+Ma_4$
Constraint 1	$14x_1+12x_2+1s_1+0s_2+0s_3+0s_4+0a_4=15$
Constraint 2	$11x_1+10x_2+0s_1+1s_2+0s_3+0s_4+0a_4=15$
Constraint 3	$12x_1+13x_2+0s_1+0s_2+1s_3+0s_4+0a_4=16$
Constraint 4	$7.3x_1+7.9x_2+0s_1+0s_2+1s_3-1s_4+1a_4=9$

TABLE 3.15
Iteration 0

Decision Variables		x_1	x_2	s_1	s_2	s_3	s_4	a_4	
Coefficients of Decision Variables in Objective Function		1.65	1.82	0	0	0	0	M	
Basis Variables	**Coefficients of Basis Variables c_b in Objective Function**	**Coefficients of Constraint Functions**							**Maximum Available Resource**
s_1	0	14	12	1	0	0	0	0	15
s_2	0	11	10	0	1	0	0	0	15
s_3	0	12	13	0	0	1	0	0	16
a_4	M	7.3	7.9	0	0	0	−1	1	9
	z_j	7.3M	7.9M	0	0	0	−M	M	
Net evaluation row	$c_j - z_j$	1.65–7.3M	1.82–7.9M	0	0	0	M	0	

artificial variable is added to the objective function if the function is of minimization, but it is subtracted if the objective function is of maximization. The addition of very large values is done for the following reasons: the objective function indicates minimization of cost, etc. If a very huge value gets added to it, then it would increase the cost to a very large extent, signifying the importance that should be assigned to it. So, predominantly, in most cases involving artificial variables, it gets the utmost attention and is removed first. Similarly, in the case of maximization objective negative coefficient of artificial variable is added to the objective function of profit or sales. If a very large value is subtracted from the objective function of profit, then profit will fall substantially asking for attention. Thus, in the first step of finding a better solution, mostly artificial variables are focused upon and are replaced by a relevant basis variable.

Step 3: Improvement in solution

In this step, method of finding an improved solution from the initial solution would be discussed. Initially, we found that loss in revenue when none of the units are produced is zero. But production has to start and when it gets started, some of the items will be defective. The purpose is to find a number of defective items of both P1 and P2 types of pistons which would result in a minimum loss in revenue. To accomplish this, Table 3.2 is first modified to create a simplex table. It shows decision variables, their coefficients in objective function, basis variables and their coefficients in objective function, constraints and maximum or minimum resources available. Table 3.15 represents a simplex table.

Iteration 0:

Find initial losses*:* Initial losses (denoted by z_j) for both types of pistons when none are produced are calculated in the following way:

- Multiply the coefficient of each decision variable in the constraint function with the coefficient of basic variable in the objective function. For instance, the coefficient of x_1 decision variable for first constraint is 14 and is multiplied by 0, which is the coefficient of basic variable s_1. Similarly, it is done for all constraints and corresponding basic variable coefficients.
- Add all the values obtained in the above step. Thus,

$$z_1 = 14*0 + 11*0 + 12*0 + 7.3*M = 7.3M$$

$$z_2 = 12*0 + 10*0 + 13*0 + 7.9M = 7.9M$$

$$z_3 = 1*0 + 0*0 + 0*0 + 0*0 = 0$$

$$z_4 = 0*1 + 1*0 + 0*0 + 0*0 = 0$$

$$z_5 = 0*1 + 0*0 + 1*0 + 0*0 = 0$$

$$z_6 = 0*0 + 0*0 + 0*0 - 1*M = -M$$

$$z_7 = 0*M + 0*M + 0*M + 1*M = M$$

z_1, z_2, z_6 and z_7 have some values, whereas as the initial solution implies no production, so the value of every objective function should be zero. These values represent values of artificial variables which are imaginary variables having no contribution to objective function. Thus, logically, the initial value of an objective function is zero.

- Finally, the net evaluation row denoted by $c_j - z_j$ is calculated. It indicates a change in objective function from zero units produced to when one unit is produced. The value of objective function when zero units are produced is calculated for P1 as 7.3M. When one unit is produced, the objective function indicates that loss per unit of P1 is \$1.65. So, a change in objective function would be

$$c_1 - z_1 = 1.65 - 7.3M;$$

$$c_2 - z_2 = 1.82 - 7.9M$$

$$c_3 - z_3 = 0;$$

$$c_4 - z_4 = 0$$

$$c_5 - z_5 = 0;$$

$$c_6 - z_6 = M$$

$$c_7 - z_7 = 0$$

Column selection: Net evaluation row shows that with each defective unit of P1 revenue decreased by 1.65–7.3M and with P2 by 1.82–7.9M. The purpose is to reduce loss of revenue by maximum. This can be calculated by assuming a very high value for M as indicated in the Big M method. In this case, suppose the value of M is 100, then the value of:

$$c_1 - z_1 = 1.65 - 7.3M = 1.65 - 7.3 * 100 = -728.35 \text{ and}$$

$$c_2 - z_2 = 182 - 7.9M = 182 - 7.9 * 100 = -608$$

Thus, the second column regarding P2 would be selected as by introducing decision variable x_2 in the solution, loss in revenue would decrease more than if x_1 was introduced. *So, in the case of minimization, the column with maximum negative value would be selected for the new solution (Table 3.16).* Now if one variable has entered into the solution, then another has to leave. Identification of it is done through row selection.

Row selection: The variable that would move out of the equation and be replaced by x_2 is identified by dividing the maximum resource available for each constraint with coefficients of incoming variable. For instance, for the first basic variable, the maximum resource available was 15 and the coefficient of the selected column was 12. The ratio would be 15/12 = 1.25. Similarly, other ratios are calculated. Logically, this ratio implies per unit of resource. *Thus, out of all ratios, the row with minimum ratio would be selected as it would indicate a minimum or limiting resource (Table 3.16).* In this case, row 4 would be selected as its ratio is minimum (1.13).

At this stage, the value of objective function Z is zero as no decision variable has entered the solution yet.

TABLE 3.16
Row and Column Selection

Decision Variables		x_1	x_2	s_1	s_2	s_3	s_4	a_4		
Coefficients of Decision Variables in Objective Function		1.65	1.82	0	0	0	0	M		
Basis Variables	Coefficients of Basis Variables c_b in Objective Function	Coefficients of Constraint Functions							Maximum Available Resource	Ratio
s_1	0	14	12	1	0	0	0	0	15	1.25
s_2	0	11	10	0	1	0	0	0	15	1.5
s_3	0	12	13	0	0	1	0	0	16	1.23
a_4	M	7.3	7.9	0	0	0	−1	1	9	1.13
	z_j	7.3M	7.9M	0	0	0	−M	M		
Net evaluation row	$c_j - z_j$	1.65–7.3M	1.82–7.9M	0	0	0	M	0		

Also, it is worthy to note that in the case of a model having an artificial variable the first outgoing variable would be the artificial variable because of the very high penalty associated with it. If it remains in the final solution, then most likely solution would not be optimal.

It is important to understand that the selected column represents the incoming variable, whereas selected row indicates the outgoing variable. So, in this case, a_4 would be replaced by x_2. This would be done by applying mathematical functions in such a manner that coefficients of x_2 column are replaced by coefficients of a_4 column. These are shown in the following table.

x_2 Column Coefficients	Would Change into	a_4 Column Coefficients
12		0
10		0
13		0
7.9		1

This is achieved by applying the following mathematical functions (Table 3.17).

TABLE 3.17
Calculations for New Simplex Table

Selected row

$$7.3x_1 + 7.9x_2 + 0s_1 + 0s_2 + 0s_3 - 1s_4 + 1a4 = 9$$

To get the coefficient of x_2 as 1 key row would be divided by 7.9. Thus, the key row becomes

$$0.92x_1 + 1x_2 + 0s_1 + 0s_2 + 0s_3 - 0.12s_4 + 0.12a_4 = 1.14 \qquad (3.17)$$

Row 1

$$14x_1 + 12x_2 + 1s_1 + 0s_2 + 0s_3 + 0s_4 + 0a_4 = 15 \qquad (3.18)$$

To get the coefficient of x_2 as zero multiply equation 3.17 with 12 and subtract it from row 1

$$11.04x_1 + 0x_2 + 0s_1 + 0s_2 + 0s_3 - 1.44s_4 + 1.44a_4 = 13.68 \qquad (3.19)$$

(3.18) – (3.19) will give

$$2.96x_1 + 0x_2 + 1s_1 + 0s_2 + 0s_3 + 1.44s_4 + 1.44a_4 = 1.32$$

Row 2

$$11x_1 + 10x_2 + 0s_1 + 1s_2 + 0s_3 + 0s_4 + 0a_4 = 15 \qquad (3.20)$$

Equation (3.17) is now multiplied with 10 and subtracted from row 2

$$9.2x_1 + 10x_2 + 0s_1 + 0s_2 + 0s_3 - 1.2s_4 + 1.2a_4 = 11.4 \qquad (3.21)$$

(3.20) – (3.21) will give

$$1.8x_1 + 0x_2 + 0s_1 + 1s_2 + 0s_3 - 1.24s_4 - 1.2a_4 = 3.6$$

Row 3

$$12x_1 + 13x_2 + 0s_1 + 0s_2 + 1s_3 + 0s_4 + 0a_4 = 16 \qquad (3.22)$$

Equation (3.17) is now multiplied with 13 and subtracted from row 2

(Continued)

TABLE 3.17 (*Continued*)
Calculations for New Simplex Table

$$11.96x_1 + 13x_2 + 0s_1 + 0s_2 + 1s_3 - 1.56s_4 + 1.56a_4 = 14.82 \quad (3.23)$$

$$(3.22) - (3.23)$$

$$0.04x_1 + 0x_2 + 0s_1 + 0s_2 + 1s_3 - 0.12s_4 + 0.12a_4 = 1.14$$

Row 4

$$0.92x_1 + 1x_2 + 0s_1 + 0s_2 + 0s_3 - 0.12s_4 + 0.12a_4 = 1.14$$

TABLE 3.18
Iteration 1

Decision Variables		x_1	x_2	s_1	s_2	s_3	s_4		
Coefficients of Decision Variables in Objective Function		**1.65**	**1.82**	**0**	**0**	**0**	**0**		
Basis Variables	**Coefficients of Basis Variables c_b in Objective Function**	**Coefficients of Constraint Functions**						**Maximum Available Resource**	**Ratio**
s_1	0	2.96	0	1	0	0	1.44	1.32	0.45
s_2	0	1.8	0	0	1	0	1.2	3.6	2
s_3	0	0.04	0	0	0	1	1.56	1.18	29.5
x_2	1.82	0.92	1	0	0	0	−0.12	1.14	1.23
	z_j	1.67	1.82	0	0	0	−0.22		
Net evaluation row	$c_j - z_j$	−0.02	0	0	0	0	0.22		

Iteration 1:

The above calculations will provide another simplex tableau indicating an improved solution and whether the solution can be improved further (Table 3.18). It is important to note that in this table artificial variable is not shown as it has been replaced and is an artificial or imaginary variable which has no contribution.

Table 3.18 shows that decision variable x_2 has entered the solution with a value of 1.14, whereas x_1 is still zero. Thus, the value of the objective function would be $Z = 1.65*0 + 1.82*1.14 = 2.07$, which is an improvement from zero.

Applying the rule of column and row selection column of x_1 decision variable and row of s_1 basic variable are selected, implying that x_1 is the incoming and s_1 is the outgoing variable in a new solution. Thus, x_1 column coefficients would be replaced by s_1 column coefficients.

x_1 Column Coefficients	Would Change into	s_1 Column Coefficients
2.96		1
1.8		0
0.04		0
0.92		0

TABLE 3.19
Calculations for New Simplex Table

Selected row (Row 1)

$$2.96x_1 + 0x_2 + 1s_1 + 0s_2 + 0s_3 + 1.44s_4 = 1.32$$

To get the coefficient of x_1 as 1 key row would be divided by 2.96. Thus, the key row becomes

$$1x_1 + 0x_2 + 0.34s_1 + 0s_2 + 0s_3 + 0.48s_4 = 0.45 \qquad (3.24)$$

Row 2

$$1.8x_1 + 0x_2 + 0s_1 + 1s_2 + 0s_3 + 1.2s_4 = 3.6 \qquad (3.25)$$

To get the coefficient of x_1 as zero multiply equation 3.24 with 1.8 and subtract it from row 1

$$1.8x_1 + 0x_2 + 0.61s_1 + 0s_2 + 0s_3 + 0.86s_4 = 0.81 \qquad (3.26)$$

(3.25) – (3.26) will give

$$0x_1 + 0x_2 - 0.61s_1 + 1s_2 + 0s_3 + 0.34s_4 = 1.32$$

Row 3

$$0.04x_1 + 0x_2 + 0s_1 + 0s_2 + 1s_3 + 1.56s_4 = 1.18 \qquad (3.27)$$

Equation (3.24) is now multiplied with 0.04 and subtracted from row 3

$$0.04x_1 + 0x_2 + 0.01s_1 + 0s_2 + 0s_3 + 0.22s_4 = 0.02 \qquad (3.28)$$

(3.27) – (3.28) will give

$$0x_1 + 0x_2 - 0.01s_1 + 0s_2 + 1s_3 + 1.34s_4 = 1.16$$

Row 4

$$0.92x_1 + 1x_2 + 0s_1 + 0s_2 + 0s_3 - 0.22s_4 = 1.14 \qquad (3.29)$$

Equation (3.24) is now multiplied with 0.92 and subtracted from row 4

$$0.92x_1 + 0x_2 + 0.32s_1 + 0s_2 + 0s_3 + 0.44s_4 = 0.41 \qquad (3.30)$$

(3.29) – (3.30) will give

$$0x_1 + 1x_2 - 0.32s_1 + 0s_2 + 0s_3 - 0.66s_4 = 0.73$$

This is achieved by applying mathematical functions as explained. This would give us the following simplex table (Table 3.19).

Iteration 2:

Table 3.20 shows that now decision variable x_1 has entered the solution with a value of 0.45, whereas x_2 has changed to 0.73.

As *all values in the net evaluation row are either zero or positive*; thus, this is the final and most optimal solution. It implies that further introduction of any other variable would have no influence or would increase rather than minimize the objective

TABLE 3.20
Iteration 2

Decision Variables		x_1	x_2	s_1	s_2	s_3	s_4	
Coefficients of Decision Variables in Objective Function		1.65	1.82	0	0	0	0	
Basis Variables	Coefficients of Basis Variables c_b in Objective Function	Coefficients of Constraint Functions						Maximum Available Resource
x_1	1.65	1	0	0.34	0	0	0.48	0.45
s_2	0	0	0	−0.61	1	0	0.34	1.32
s_3	0	0	0	−0.01	0	1	1.34	1.16
x_2	1.82	0	1	−0.32	0	0	−0.66	0.73
	z_j	1.65	1.82	−0.02	0	0	−0.41	
Net evaluation row	$c_j - z_j$	0	0	0.02	0	0	0.41	

function. Importantly, the solution is indicated as optimal also by removal of all artificial variables in the final solution. By definition, artificial variables are added only to fulfil non-negativity criteria without being represented by consumption of any resources.

Thus, from the optimal solution:

$$x_1 = 0.45 \text{ and } x_2 = 0.73$$

Putting these values in the objective function

$$Z = 1.65x_1 + 1.82x_2$$

$$= 1.65 * 0.45 + 1.82 * 0.73$$

$$= \$2.07$$

Thus, the minimum loss in revenue from a sample of 50 units due to defective item on each of P1 and P2 pistons is $2.07. This loss in revenue can be calculated for per shift, per day or for the entire production schedule of a month in which demand of 1 lakh units needs to be met. For instance, quality check is made every hour, so in a shift of 7 hours, the number of scrap items of P1 and P2 would be 0.45*7 = 3.15 and 0.73*7 = 5.11 units, respectively. Similarly, loss in revenue would be:

$$Z = 1.65 * 3.15 + 1.82 * 5.11$$

$$= \$14.49$$

Appropriateness of solution is checked by putting values of decision variables in constraints.

Constraint 1: Die coating thickness

$$14x_1 + 12x_2 \leq 15$$

$$14 * 0.45 + 12 * 0.73 = 15$$

which is approximately equal to the maximum resource acceptable, which is 15. Thus, this condition is validated. This also indicates that no slack is available.

Constraint 2: Worker skill set

$$11x_1 + 10x_2 \leq 15$$

$$11 * 0.45 + 10 * 0.73 = 12.25$$

This condition is again validated. In this case, a slack of 15 – 12.25 = 2.75 is available, implying though interestingly but applicable in this case that worker skill set can be further relaxed by 2.75 units, and if that happens, the number of defective items would still be within unacceptable limits.

Constraint 3: Cooling water

$$12x_1 + 13x_2 \leq 16$$

$$12 * 0.45 + 13 * 0.73 = 14.89$$

A positive slack of 16–14.89 = 1.11 imply feasibility of solution. This also means that cooling water variation is within acceptable limits as the number of scrap items due to such variation is less than the acceptable limit of 16%.

Constraint 4: Rating of importance

$$7.3x_1 + 7.9x_2 \geq 9$$

$$7.3 * 0.45 + 7.9 * 0.73 = 9.05$$

The rating is approximately equal to 9, which is the minimum value required for initiation of corrective action. Thus, the solution indicates that management must apply measures to reduce loss in revenue due to scrap items as customers consider scrap type of B.T. variation in P1 and shrinkage in P2 to be critical reasons for casing scrap.

Constraint 5: Number of scrap items

$$x_1, x_2 \geq 0$$

As $x_1 = 0.45$ and $x_2 = 0.73$ so this condition is also met.

Thus, the final solution obtained is feasible.

	A	B	C	D	E
1		Minimize: Scrap reduction			
2	Constraints	P1	P2	Resources	
3	Die thickness	14	12	15	
4	Skill set	11	10	15	
5	Cooling water	12	13	16	
6	Rating of importance to customer(P1)	7.3	7.9	9	
7	Scrap cost	1.65	1.82		
8					
9	Decision variables	0.46	0.72		
10	Minimization	2.06			
11	Constraints				
12	Die thickness	15	<=	15	
13	Skill set	12.20	<=	15	
14	Cooling water	14.80	<=	16	
15	Rating of importance to customer(P1)	9	>=	9	

FIGURE 3.3 Excel solution.

Excel Solution

As discussed in maximization illustration, the minimization problem is also solved by using the Solver application in Excel. Construction of model and formula creation is similar, so only the final solution is shown in Figure 3.3.

3.3.2 Case: Rose's Luxury Restaurant

The case discussed in Section 2.5.2 involves the following model:

Objective function

$$\textbf{Minimize} \quad Z = 10x_1 + 15x_2$$

Constraint 1 (Floor space):

$$16x_1 + 30x_2 \leq 1{,}000$$

Constraint 2 (Demand of A):

$$1x_1 + 0x_2 \geq 20$$

Constraint 3 (Demand of B)

$$0x_1 + 1x_2 \geq 15$$

Constraint 4 (Capacity):

$$1x_1 + 1x_2 \leq 40$$

3.3.2.1 Simplex Method

Step 1: Model Construction (Table 3.21)

Step 2: IBFS:

Non-basic variables are those variables which are assumed to be zero. In this case, x_1, x_2, s_2 and s_3 are put as zero making

$$s_1 = 1{,}000;\ a_2 = 20;\ a_3 = 15 \text{ and } s_4 = 40$$

Thus, all *basis variables* are positive, making initial solution feasible.

Initial solution: By putting feasible values in the objective function of minimization, Z would come out to be zero.

Z	$= 10x_1 + 15x_2$
	$= 10*0 + 15*0$
	$= 0$

Step 3: Improvement in solution

Table 3.22 represents the simplex table.

Iteration 0:

In the case of minimization, the column with maximum negative value is selected, leading to selection of the second column pertaining to decision variable x_2. So x_2 would be an incoming variable. Outgoing variable ratios indicating consumption of resources per unit were calculated as shown in Table 3.23.

The rule of minimum ratio implies that the artificial variable a_3 would be the outgoing variable. So a_3 would be replaced by x_2. To do so, the following calculations (Table 3.24) are carried out.

TABLE 3.21
Model

Minimize	$Z = 10x_1 + 15x_2 + 0s_1 + 0s_2 + Ma_2 + 0s_3 + Ma_3 + 0s_4$
Constraint 1	$16x_1 + 30x_2 + 1s_1 + 0s_2 + 0a_2 + 0s_3 + 0a_3 + 0s_4 = 1{,}000$
Constraint 2	$1x_1 + 0x_2 + 0s_1 + -1s_2 + 1a_2 + 0s_3 + 0a_3 + 0s_4 = 20$
Constraint 3	$0x_1 + 1x_2 + 0s_1 + 0s_2 + 0a_2 + -1s_3 + Ma_3 + 0s_4 = 15$
Constraint 4	$1x_1 + 1x_2 + 0s_1 + 0s_2 + 0a_2 + 0s_3 + 0a_3 + 1s_4 = 40$

TABLE 3.22
Iteration 0

Decision Variables		x_1	x_2	s_1	s_2	a_2	s_3	a_3	s_4	
Coefficients of Decision Variables in Objective Function		10	15	0	0	M	0	M	0	
Basis Variables	Coefficients of Basis Variables c_b in Objective Function	Coefficients of Constraint Functions								Maximum Available Resource
s_1	0	16	30	1	0	0	0	0	0	1,000
a_2	M	1	0	0	−1	1	0	0	0	20
a_3	M	0	1	0	0	0	−1	1	0	15
s_4	0	1	1	0	0	0	0	0	1	40
	z_j	M	M	0	−M	M	−M	M	0	
Net evaluation row	$c_j - z_j$	10−M	15−M	0	M	0	M	0	0	

TABLE 3.23
Row and Column Selection

Decision Variables		x_1	x_2	s_1	s_2	a_2	s_3	a_3	s_4		
Coefficients of Decision Variables in Objective Function		10	15	0	0	M	0	M	0		
Basis Variables	Coefficients of Basis Variables c_b in Objective Function	Coefficients of Constraint Functions								Maximum Available Resource	Ratios
s_1	0	16	30	1	0	0	0	0	0	1,000	33.3
a_2	M	1	0	0	−1	1	0	0	0	20	–
a_3	M	0	1	0	0	0	−1	1	0	15	15
s_4	0	1	1	0	0	0	0	0	1	40	40
	z_j	M	M	0	−M	M	−M	M	0		
Net evaluation row	$c_j - z_j$	10−M	15−M	0	M	0	M	0	0		

TABLE 3.24
Calculations for New Simplex Table

Selected row (Row 3)

$$0x_1 + 1x_2 + 0s_1 + 0s_2 + 0a_2 - 1s_3 + 1a_3 + 0s_4 = 15 \quad (3.31)$$

Coefficient of x_2 is already 1 so no division is required and it becomes key row

(Continued)

TABLE 3.24 (*Continued*)
Calculations for New Simplex Table

To get coefficients of a_3 in other rows following operations would be carried out

Row 1

$$16x_1 + 30x_2 + 1s_1 + 0s_2 + 0a_2 + 0s_3 + 0a_3 + 0s_4 = 1{,}000 \qquad (3.32)$$

To get the coefficient of x_2 as zero multiply equation 3.31 with 30 and subtract it from row 1

$$0x_1 + 30x_2 + 0s_1 + 0s_2 + 0a_2 - 30s_3 + 30a_3 + 0s_4 = 450 \qquad (3.33)$$

(3.32) – (3.33) will give

$$16x_1 + 0x_2 + 1s_1 + 0s_2 + 0a_2 + 30s_3 - 30a_3 + 0s_4 = 550$$

Row 4

$$1x_1 + 1x_2 + 0s_1 + 0s_2 + 0a_2 + 0s_3 + 0a_3 + 1s_4 = 40 \qquad (3.34)$$

Equation (3.31) is now multiplied with 1 and subtracted from row 4

$$0x_1 + 1x_2 + 0s_1 + 0s_2 + 0a_2 - 1s_3 + 1a_3 + 0s_4 = 15 \qquad (3.35)$$

(3.34) – (3.35) will give

$$1x_1 + 0x_2 + 0s_1 + 0s_2 + 0a_2 + 1s_3 - 1a_3 + 1s_4 = 25$$

Row 2 remain same as its coefficient already is zero

Iteration 1:

Table 3.25 shows x_1 to be incoming and a_2 to be an outgoing variable. The following calculations (Table 3.26) were made to result in such a replacement.

Iteration 2:

As all values in the net evaluation row are zero or positive, so an optimal solution has been reached with $x_1 = 20$ and $x_2 = 15$ (Table 3.27). Also, all artificial variables were replaced, indicating the solutions' optimality.

TABLE 3.25
Iteration 1

Decision Variables		x_1	x_2	s_1	s_2	a_2	s_3	s_3		
Coefficients of Decision Variables in Objective Function		**10**	**15**	**0**	**0**	**M**	**0**	**0**		
Basis Variables	**Coefficients of Basis Variables c_b in Objective Function**	**Coefficients of Constraint Functions**							**Maximum Available resource**	**Ratios**
s_1	0	16	0	1	0	0	30	0	550	34.37
a_2	M	1	0	0	−1	1	0	0	20	20
x_2	15	0	1	0	0	0	−1	0	15	–
s_4	0	1	0	0	0	0	1	1	25	25
	z_j	M	15	0	−M	M	−15	0		
Net evaluation row	$c_j - z_j$	10−M	0	0	M	0	15	0		

TABLE 3.26
Calculations for New Simplex Table

Selected row (Row 2)

$$1x_1 + 0x_2 + 0s_1 - 1s_2 + 1a_2 + 0s_3 + 0s_4 = 20 \quad (3.36)$$

Coefficient of x_2 is already 1 so no division is required and it becomes key row

To get coefficients of a_2 in other rows following operations would be carried out

Row 1

$$16x_1 + 0x_2 + 1s_1 + 0s_2 + 0a_2 + 30s_3 + 0s_4 = 550 \quad (3.37)$$

To get the coefficient of x_1 as zero multiply equation 3.36 with 16 and subtract it from row 1

$$16x_1 + 0x_2 + 0s_1 - 16s_2 + 16a_2 + 0s_3 + 0s_4 = 320 \quad (3.38)$$

(3.37) – (3.38) will give

$$0x_1 + 0x_2 + 1s_1 + 16s_2 - 16a_2 + 30s_3 + 0s_4 = 230$$

Row 4

$$1x_1 + 0x_2 + 0s_1 + 0s_2 + 0a_2 + 1s_3 + 1s_4 = 25 \quad (3.39)$$

Equation (3.36) is now multiplied with 1 and subtracted from row 4

$$1x_1 + 0x_2 + 0s_1 - 1s_2 + 1a_2 + 0s_3 + 0s_4 = 20 \quad (3.40)$$

(3.39) – (3.40) will give

$$0x_1 + 0x2 + 0s1 + 1s2 - 1a2 + 1s3 + 1s4 = 5$$

Row 3 remain same as its coefficient already is zero

TABLE 3.27
Iteration 2

Decision Variables		x_1	x_2	s_1	s_2	s_3	s_3	
Coefficients of Decision Variables in Objective Function		**10**	**15**	**0**	**0**	**0**	**0**	
Basis Variables	**Coefficients of Basis Variables c_b in Objective Function**	**Coefficients of Constraint Functions**						**Maximum Available Resource**
s_1	0	0	0	1	16	30	0	230
x_1	10	1	0	0	−1	0	0	20
x_2	15	0	1	0	0	−1	0	15
s_4	0	0	0	0	1	1	1	5
	z_j	10	15	0	−10	−15	0	
Net evaluation row	$c_j - z_j$	0	0	0	10	15	0	

So, the objective function would be:

$$\textbf{Minimize } Z = 10x_1 + 15x_2$$

$$Z = 10*20 + 15*15$$

$$= \$425$$

Putting values of decision variables in constraint equations would imply feasibility of solution:

Constraint 1 (Floor space):

$$16x_1 + 30x_2 \leq 1{,}000$$

$$16 * 20 + 30 * 15 = 770$$

So s_1 would be 1,000 – 770 = 230, implying the amount of floor space lying unused. The space can be utilized only if demand increases. A total of 770 units of space, given the demand, incur a minimum cost.

Constraint 2 (Demand of A):

$$1x_1 + 0x_2 \geq 20$$

$$1 * 20 + 0 * 15 = 20$$

Constraint 3 (Demand of B):

$$0x_1 + 1x_2 \geq 15$$

$$0 * 20 + 1 * 15 = 15$$

Demand for both types of seating capacity is completely fulfilled.

Constraint 4 (Capacity):

$$1x_1 + 1x_2 \leq 40$$

$$1 * 20 + 1 * 15 = 35$$

So s_4 would be 5 units, indicating unutilized capacity.

As all constraints are satisfied simultaneously, so it can be said that achieving an optimal solution is also feasible.

3.4 ILLUSTRATION OF MAXIMIZATION PROBLEM WITH GREATER THAN EQUAL TO CONSTRAINTS

To illustrate the procedure of greater than equal to constraint by introducing surplus variables, the case in 3.1 is modified by including the following constraints.

Constraint 1 regarding the cost of fuel has marginally increased to $7,500. Thus,

$$1.58x_1 + 2.77x_2 \leq 7{,}500$$

Constraint 2 restricted the maximum amount of power to be generated to 1,539 units whereas now if there is limitation in terms of minimum amount of

TABLE 3.28
Model

Maximize	$Z = 0.2x_1 + 0.3x_2 + 0s_1 + 0s_2 + 0s_3 + 0s_4 - Ma_2 - Ma_3 - Ma_4$
Constraint 1	$1.58x_1 + 2.77x_2 + 1s_1 + 0s_2 + 0s_3 + 0s_4 + 0a_2 + 0a_3 + 0a_4 = 7{,}500$
Constraint 2	$0.48x_1 + 0.41x_2 + 0s_1 - 1s_2 + 0s_3 + 0s_4 + 1a_2 + 0a_3 + 0a_4 = 2{,}000$
Constraint 3	$1x_1 + 0x_2 + 0s_1 + 0s_2 - 1s_3 + 0s_4 + 0a_2 + 0a_3 + 0a_4 = 150$
Constraint 4	$0x_1 + 1x_2 + 0s_1 + 0s_2 + 0s_3 - 1s_4 + 0a_2 + 0a_3 + 1a_4 = 430$

power, i.e. due to the increased demand, the plant has to produce a minimum of 2,000 units of electricity with one kg of fuel A producing 0.48 units and 1 kg of B producing 0.41 units. Thus, constraint 2 would be

$$0.48x_1 + 0.41x_2 \geq 2{,}000$$

Finally, constraint 3 in previous case restricted availability of fuel to 4,400 units but with increased demand for power, the amount of fuel required also increases. So, plant tends to keep stock of both types of fuel. If we add two more constraints regarding the opening stock of fuel A and fuel B, i.e. opening stock of fuel A is 150 kg and that of fuel B is 430 kg. Thus,

$$\text{Constraint 3} \quad 1.x_1 + 0.x_2 \geq 150$$

$$\text{Constraint 4} \quad 0.x_1 + 1.x_2 \geq 430$$

Objective function

$$\textbf{Maximize} \quad Z = 0.2x_1 + 0.3x_2$$

3.4.1 Simplex Method

Step 1: Model Construction

In this case, constraint 1 includes one slack variable, and constraints 2, 3 and 4 include three surplus and artificial variables. It is important to notice here that in the case of maximization, coefficients of artificial variables, i.e. M are subtracted from the objective function. Thus, model becomes (Table 3.28).

Step 2: Finding IBFS

The model shown in Table 3.28 indicates four equations are formed to find values of nine variables; two decision, one slack, three surplus and three artificial variables. To find the initial solution, decision variables x_1 and x_2, s_2, s_3 and s_4 are put to zero. This would give us the following:

Non-basic variables (which are assumed to be zero)	$x_1 = x_2 = s_2 = s_3 = s_4 = 0$			
Basic variables	$s_1 = 7{,}500$	$a_2 = 2{,}000$	$a_3 = 150$	$a_3 = 430$

TABLE 3.29
Iteration 0

Decision Variables		x_1	x_2	s_1	s_2	s_3	s_4	a_2	a_3	a_4		
Coefficients of Decision Variables in Objective Function		0.2	0.3	0	0	0	0	−M	−M	−M		
Basis Variables	Coefficients of Basis Variables c_b in Objective Function	Coefficients of Constraint Functions									Maximum Available Resource	Ratio
s_1	0	1.58	2.77	1	0	0	0	0	0	0	7,500	4,746.8
a_2	−M	0.48	0.41	0	−1	0	0	1	0	0	2,000	4,166.6
a_3	−M	1	0	0	0	−1	0	0	1	0	150	150
a_4	−M	0	1	0	0	0	−1	0	0	1	430	430
	z_j	−1.48M	−1.41M	0	M	M	M	−M	−M	−M		
Net evaluation row	$c_j - z_j$	0.2+1.48M	0.3+1.41M	0	−M	−M	−M	0	0	0		

Step 3: Improvement in solution

Iteration 0:

At this stage, the objective function is zero.

For improvement of solution column that contributes maximum to objective function of maximization would be selected (Table 3.29). Supposing M to be 100, i.e. an assumed high value:

$$c_1 - z_1 = 0.2 + 1.48M = 0.2 + 1.48*100 = 148.2 \text{ and;}$$

$$c_2 - z_2 = 0.3 + 1.41M = 0.3 + 1.41*100 = 141.3$$

So, for the first iteration, column 1 is selected. Thus, x_1 would be an incoming variable. Row 3 with a minimum ratio is selected, representing a_3 as the outgoing variable. This leads to the improved solution for which the new simplex table is shown below (Table 3.30).

Iteration 1:

With $x_1 = 150$ and $x_2 = 0$ objective function

$$Z = 2x_1 + 3x_2$$

$$Z = 2*150 + 3*0 = 300$$

TABLE 3.30
Iteration 1

Decision Variables		x_1	x_2	s_1	s_2	s_3	s_4	a_2	a_4		
Coefficients of Decision Variables in Objective Function		**0.2**	**0.3**	**0**	**0**	**0**	**0**	**−M**	**−M**		
Basis Variables	**Coefficients of Basis Variables c_b in Objective Function**	**Coefficients of Constraint Functions**								**Maximum Available Resource**	**Ratio**
s_1	0	0	2.77	1	0	1.58	0	0	0	7,263	2,622
a_2	−M	0	0.41	0	−1	0.48	0	1	0	1,928	4,702.4
x_1	0.2	1	0	0	0	−1	0	0	0	150	–
a_4	−M	0	1	0	0	0	−1	0	1	430	430
	z_j	2	−1.41M	0	M	−2−0.48M	M	−M	−M		
Net evaluation row	$c_j - z_j$	0	0.3+1.41M	0	−M	2+0.48M	−M	0	0		

TABLE 3.31
Iteration 2

Decision Variables		x_1	x_2	s_1	s_2	s_3	s_4	a_2		
Coefficients of Decision Variables in Objective Function		**0.2**	**0.3**	**0**	**0**	**0**	**0**	**−M**		
Basis Variables	**Coefficients of Basis Variables c_b in Objective Function**	**Coefficients of Constraint Functions**							**Maximum Available Resource**	**Ratio**
s_1	0	0	0	1	0	1.58	2.77	0	6,071.9	3,842.9
a_2	−M	0	0	0	−1	0.48	0.41	1	1,751.7	3,649.4
x_1	0.2	1	0	0	0	−1	0	0	150	−150
x_2	0.3	0	1	0	0	0	−1	0	430	–
	z_j	2	3	0	M	−2−0.48M	−3−0.41M	−M		
Net evaluation row	$c_j - z_j$	0	0	0	−M	2+0.48M	3+0.41M	0		

Iteration 2:

With $x_1 = 150$ and $x_2 = 430$ objective function (Table 3.31)

$$Z = 2x_1 + 3x_2$$

$$Z = 2*150 + 3*430 = 1{,}590$$

TABLE 3.32
Iteration 3

Decision Variables		x_1	x_2	s_1	s_2	s_3	s_4		
Coefficients of Decision Variables in Objective Function		**0.2**	**0.3**	**0**	**0**	**0**	**0**		
Basis Variables	**Coefficients of Basis Variables c_b in Objective Function**	**Coefficients of Constraint Functions**						**Maximum Available Resource**	**Ratio**
s_1	0	0	0	1	3.36	0	1.43	305.9	91.1
s_3	0	0	0	0	−2.13	1	0.85	3,649.3	−1,713.3
x_1	0.2	1	0	0	−2.13	0	0.85	3,799.3	−1,783.7
x_2	0.3	0	1	0	0	0	−1	430	–
	z_j	0.2	0.3	0	−4.26	0	−1.3		
Net evaluation row	$c_j - z_j$	0	0	0	4.26	0	1.3		

Iteration 3:

With $x_1 = 3{,}799.3$ and $x_2 = 430$ objective function (Table 3.32)

$$Z = 2x_1 + 3x_2$$

$$Z = 2*3{,}799.3 + 3*430 = 8{,}888.6$$

Iteration 4:

Iteration 4 indicates the achievement of an optimal solution as all values in net evaluation row are either zero or negative, implying that any introduction of variable will either result in no change or will reduce the objective function of maximization (Table 3.33).

According to the solution decision variable, $x_1 = 3{,}993.4$ and $x_2 = 430$. Thus, the objective function would be:

$$Z = 2x_1 + 3x_2$$

$$Z = 2*3{,}993.4 + 3*430 = 9{,}276.8$$

Feasibility of solution is checked by putting values of decision variables in constraint equations.

Constraint 1 $\quad 1.58x_1 + 2.77x_2 \leq 7{,}500$

$$1.58*3{,}993.4 + 2.77*430 \leq 7{,}500$$

$$7{,}500 = 7{,}500$$

TABLE 3.33
Iteration 4

Decision Variables		x_1	x_2	s_1	s_2	s_3	s_4	
Coefficients of Decision Variables in Objective Function		0.2	0.3	0	0	0	0	
Basis Variables	Coefficients of Basis Variables c_b in Objective Function	Coefficients of Constraint Functions						Maximum Available Resource
s_2	0	0	0	0.29	1	0	0.42	91.1
s_3	0	0	0	0.62	0	1	1.74	3,843.4
x_1	0.2	1	0	0.62	0	0	1.74	3,993.4
x_2	0.3	0	1	0	0	0	−1	430
	z_j	2	3	1.24	0	0	3.48	
Net evaluation row	$c_j - z_j$	0	0	−1.24	0	0	−3.48	

Thus, constraint is satisfied.

Constraint 2 $0.48x_1 + 0.41x_2 \geq 2{,}000$

$$0.48*3{,}993.4 + 0.41*430 \geq 2{,}000$$

$$2{,}093.13 \geq 2{,}000$$

Thus, constraint is satisfied.

Constraint 3 $1.x_1 + 0.x_2 \geq 150$

$$1*3{,}993.4 + 0*430 \geq 150$$

$$3{,}993.4 \geq 150$$

Thus, constraint is satisfied.

Constraint 4 $0.x_1 + 1.x_2 \geq 430$

$$0*3{,}993.4 + 1*430 \geq 430$$

$$430 = 430$$

Thus, constraint is satisfied.

Hence, solution is feasible.

3.5 SPECIAL VERSIONS OF LPP SOLVED BY SIMPLEX METHOD

3.5.1 DEGENERACY

A solution is said to have degeneracy if there is a tie in two or more basic variables. In such cases, it becomes important to understand which row would be selected as the outgoing basic variable. This is for two reasons. First, the other row (unselected row) would change in similar fashion as the ratio of selected row. In all possibilities, the ratio value would be either zero or undefined, making unselected row *degenerate*. Second, if optimal solution is reached at this iteration, then the final solution is reported. However, if there are subsequent iterations and a degenerate row is selected as outgoing variable in any of those iterations, then there is always the risk that the solution will enter a loop of coming in and out of degeneracy without affecting the objective function. Degeneracy does not indicate faulty model formulation. But care should be taken during the solving process to avoid repeating the same sequence of iterations without practically affecting the objective function.

The following illustration explains the concept clearly:

$$\textbf{Minimize} \quad 2x_1 + 8x_2 + 9x_3 + 4x_4 + 7x_5 + 3x_6$$

Subject to:

$$\text{Constraint 1:} \quad x_1 + x_2 + x_3 \geq 1$$

$$\text{Constraint 2:} \quad -x_1 + x_4 + x_5 = 0$$

$$\text{Constraint 3:} \quad -x_2 - x_4 + x_6 = 0$$

$$\text{Constraint 4:} \quad x_3 + x_5 + x_6 \leq 1$$

$$x_1, x_2, x_3, x_4, x_5, x_6 \geq 0$$

Solution: As constraints include equality constraint so in such cases only artificial variables would be added. Thus model would be:

$$x_1 + x_2 + x_3 + 0x_4 + 0x_5 + 0x_6 - 1s_1 + 0s_4 + 1a_1 + 0a_2 + 0a_3 = 1$$

$$-x_1 + 0x_2 + 0x_3 + x_4 + x_5 + 0x_6 + 0s_1 + 0s_4 + 0a_1 + 1a_2 + 0a_3 = 0$$

$$0x_1 - x_2 + 0x_3 - x_4 + 0x_5 + x_6 + 0s_1 + 0s_4 + 0a_1 + 0a_2 + 1a_3 = 0$$

$$0x_1 + 0x_2 + x_3 + 0x_4 + x_5 + x_6 + 0s_1 + 1s_4 + 0a_1 + 0a_2 + 0a_3 = 1$$

and objective function would be:

$$\textbf{Minimize} \quad Z = 2x_1 + 8x_2 + 9x_3 + 4x_4 + 7x_5 + 3x_6 + 0s_1 + 0s_4 + Ma_1 + Ma_2 + Ma_3$$

TABLE 3.34
Iteration 1

		2	8	9	4	7	3	0	0	M	M	M		
Basis	c_b	x_1	x_2	x_3	x_4	x_5	x_6	s_1	s_4	a_1	a_2	a_3		**Ratio**
a_1	M	1	1	1	0	0	0	−1	0	1	0	0	1	1
a_2	M	−1	0	0	1	1	0	0	0	0	1	0	0	0
a_3	M	0	−1	0	−1	0	1	0	0	0	0	1	0	0
s_4	0	0	0	1	0	1	1	0	1	0	0	0	1	1
$c_j - z_j$		2	8	9−M	4	7−M	3−M	M	0	0	0	0		

Iteration 1:

In this case, there is a tie in row selection (Table 3.34). The row selection rule states that the row should be chosen if it indicates an outgoing variable with a minimum positive value. In this case, rows have either zeroes or ones. The rule further states that in the case of tie, select uppermost row which is row 1. Only rows with ones have been considered at this stage for selection instead of rows with zeroes. This is done for two reasons. First, as discussed in the case of minimization column with maximum negative value would be selected as it indicates cost reduction by maximum. Thus, column three represented by x_3 is selected. The coefficients in this column further indicate that by including one unit of the incoming variable (x_3), the basic variable will be decreased by the corresponding value in the selected column. Thus, if one unit of x_3 is added to the new solution, then basic variable a_1 will decrease by one unit (value of coefficient in that column). Second, right-hand side values in constraints indicate maximum or minimum resource available. For instance, suppose right-hand side value is 10 hours, implying that the maximum number of production hours available is 10 hours, and the coefficient in the selected column represents the time required to produce one unit, which is assumed to be 2 hours. Then, the ratio would imply that five units can be produced in 10 hours. Thus, in this case, value one in the ratio column would have some meaning that by introducing x_3 in the new solution, the objective function would change by one unit, whereas if a row with ratio zero would have been selected, then there would be no change in Z, which would be exercise in futility.

Iteration 2:

In iteration 2, incoming variable is x_5 as it would decrease the objective function by maximum (Table 3.35). But ratio values are only zero and the row selection rule states that row with a minimum positive greater than zero should be selected. So the solution would end here with $x_3 = 1$. With this value objective function Z would be 9.

For solution purposes, even if iteration is solved further by selecting the uppermost row with zero ratio value, then the solution will remain the same. This is illustrated as follows.

Iteration 3:

See Table 3.36.

TABLE 3.35
Iteration 2

		2	**8**	**9**	**4**	**7**	**3**	**0**	**0**	**M**	**M**		
Basis	c_b	x_1	x_2	x_3	x_4	x_5	x_6	s_1	s_4	a_2	a_3		**Ratio**
x_3	9	1	1	1	0	0	0	−1	0	0	0	1	–
a_2	M	−1	0	0	1	1	0	0	0	1	0	0	0
a_3	M	0	−1	0	−1	0	1	0	0	0	1	0	0
s_4	0	−1	−1	0	0	1	1	−1	1	0	0	0	–
$c_j - z_j$		M−7	8+M	0	4	7−M	3−M	9	0	0	0		

TABLE 3.36
Iteration 3

		2	**8**	**9**	**4**	**7**	**3**	**0**	**0**	**M**		
Basis	c_b	x_1	x_2	x_3	x_4	x_5	x_6	s_1	s_4	a_3		**Ratio**
x_3	9	1	1	1	0	0	0	−1	0	0	1	–
x_5	7	−1	0	0	1	1	0	0	0	0	0	0
a_3	M	0	−1	0	−1	0	1	0	0	1	0	0
s_4	0	0	−1	0	−1	0	1	−1	1	0	0	0
$c_j - z_j$		0	M−1	0	M−3	0	3−M	9	0	0		

TABLE 3.37
Iteration 4

		2	**8**	**9**	**4**	**7**	**3**	**0**	**0**	
Basis	c_b	x_1	x_2	x_3	x_4	x_5	x_6	s_1	s_4	
x_3	9	1	1	1	0	0	0	−1	0	1
x_5	7	−1	0	0	1	1	0	0	0	0
x_6	3	0	−1	0	−1	0	1	0	0	0
s_4	0	0	0	0	0	0	0	−1	1	0
$c_j - z_j$		0	3	0	0	0	0	9	0	

Iteration 4:

See Table 3.37.

Thus $x_3 = 1$, $x_5 = 0$, $x_6 = 0$ and $s_4 = 0$.

Putting these values in objective function and constraints would give $Z = 9$, which is the same as in iteration 2.

$$\text{Constraint 1: } x_1 + x_2 + x_3 \geq 1$$

$$1 = 1; \text{ this is correct}$$

$$\text{Constraint 2:}\quad -x_1 + x_4 + x_5 = 0$$

$$0 = 0;\ \text{this is correct}$$

$$\text{Constraint 3:}\quad -x_2 - x_4 + x_6 = 0$$

$$0 = 0;\ \text{this is correct}$$

$$\text{Constraint 4:}\quad x_3 + x_5 + x_6 \le 1$$

$$1 = 1;\ \text{this is correct}$$

Simultaneous fulfilment of constraints makes a solution feasible.

3.5.2 Unbounded

A solution is termed as unbounded when objective function Z can be changed, i.e. increased in the case of maximization to positive infinity level and decreased in the case of minimization to negative infinity level. This implies that considered constraints in solution are unable to bind the change in Z. Ideally, every firm would want to have infinite profits, but practically it is not possible, so an unbounded solution would infer erroneous formulation of constraints. In other words, an unbounded solution can be explained as a solution when an outgoing variable cannot be selected because of negative or undefined ratio values. This happens when the incoming variable with a maximum positive value is selected but the coefficients in pivot column are either zero or negative.

This is illustrated by the following example:

$$\textbf{Maximize}\quad Z = 120x_1 + 180x_2$$

Subject to:

$$10x_1 + 20x_2 \le 1{,}000$$

$$1x_1 + 1x_2 \le 350$$

Iteration 1:

See Table 3.38.

Iteration 2:

See Table 3.39.

Iteration 3:

See Table 3.40.

Thus, as ratios are undefined or negative, the outgoing basic variable cannot be selected even if the value of incoming variable is increased to any value, making objective function unbounded.

TABLE 3.38
Iteration 1

		120	180	0	0		
Basis	c_b	x_1	x_2	s_1	s_2	b	Ratio
s_1	0	10	20	1	0	1,000	50
s_2	0	1	1	0	1	350	350
$c_j - z_j$		120	180	0	0		

TABLE 3.39
Iteration 2

		120	180	0	0		
Basis	c_b	x_1	x_2	s_1	s_2	b	Ratio
x_2	180	0.5	1	1/20	0	50	100
s_2	0	0.5	0	−1/20	1	300	600
$c_j - z_j$		30	0	−9	0		

TABLE 3.40
Iteration 3

		120	180	0	0		
Basis	c_b	x_1	x_2	s_1	s_2	b	Ratio
x_1	120	1	2	0	0	100	–
s_2	0	0	−1	−1/10	0	250	−2,500
$c_j - z_j$		0	Negative	1/10	0		

3.5.3 Infeasibility

When no solution can fulfil all the constraints simultaneously, then solution is considered to be infeasible. It is important to note that infeasibility does not imply wrong formulation of linear programming model. On the other hand, it helps managers to improve solutions by concentrating on constraints which are binding for fulfilment of conditions of other constraints. There are two indications which help to judge the infeasibility of a solution. First, as suggested, when conditions of all constraints whether greater or less than are not met simultaneously and second, when an artificial variable remains in the final solution and is positive. Artificial variable by nature does not contribute in any fashion to objective function, so they should be removed from solution at the earliest possible. If that does not happen, it results in infeasibility. Also, artificial variables are required only to resolve greater than or equal to constraints. Thus, as a problem with less than or equal to inequality does not include artificial variable, so it does not suffer from the condition of infeasibility pertaining to artificial variables.

Illustration of infeasibility is provided below.

$$\textbf{Minimize} \quad Z = -4x_1 + 5x_2 + 5x_3$$

Subject to:

$$0x_1 + 1x_2 + 1x_3 \geq 2$$

$$-1x_1 + 1x_2 + 1x_3 \geq 1$$

$$0x_1 + 0x_2 - 1x_3 \geq 1$$

$$x_1, x_2, x_3 \geq 0$$

Model:

See Table 3.41.

Iteration 1:

See Table 3.42.

TABLE 3.41
Model

		−4	5	5	0	0	0	M	M	M		
Basis	c_b	x_1	x_2	x_3	s_1	s_2	s_3	a_1	a_2	a_3	**b**	**Ratio**
a_1	M	0	1	1	−1	0	0	1	0	0	2	2
a_2	M	−1	1	1	0	−1	0	0	1	0	1	*1*
a_3	M	0	0	−1	0	0	−1	0	0	1	1	–
$c_j - z_j$		−4+M	5−M	5	M	M	M	0	0	0		

TABLE 3.42
Iteration 1

		−4	5	5	0	0	0	M	M		
Basis	c_b	x_1	x_2	x_3	s_1	s_2	s_3	a_1	a_3	**b**	**Ratio**
a_1	M	1	0	0	−1	1	0	1	0	1	*1*
x_2	5	−1	1	1	0	−1	0	0	0	1	−1
a_3	M	0	0	−1	0	0	1	0	1	1	–
$c_j - z_j$		1−M	0	M	M	5	M	0	0		

TABLE 3.43
Iteration 2

		−4	5	5	0	0	0	M	
Basis	c_b	x_1	x_2	x_3	s_1	s_2	s_3	a_3	**b**
x_1	−4	1	0	0	−1	1	0	0	1
x_2	5	0	1	1	−1	0	0	0	2
a_3	M	0	0	−1	0	0	−1	1	1
$c_j - z_j$		0	0	M	1	4	M	0	

Iteration 2:

See Table 3.43.

As all values in net evaluation row are either zero or positive so optimal solution has been reached with $x_1 = 1$ and $x_2 = 2$. Putting these values in objective function and constraints would give us:

$$\textbf{Minimize} \quad Z = -4x_1 + 5x_2 + 5x_3$$

$$Z = -4*1 + 5*2 + 5*0$$

$$= 6$$

Subject to:

$$0x_1 + 1x_2 + 1x_3 \geq 2$$

$$0*1 + 1*2 + 1*0 = 2 \text{ which is} \geq 2$$

$$-1x_1 + 1x_2 + 1x_3 \geq 1$$

$$-1*1 + 1*2 + 1*0 = 1 \text{ which is} \geq 1$$

$$0x_1 + 0x_2 - 1x_3 \geq 1$$

$$0*1 + 0*2 - 1*0 = 0 \text{ which is not} \geq 1$$

$$x_1, x_2, x_3 \geq 0$$

Constraint 3 is violated and also positive artificial variable in the form of a_3 remains in final solution implying that all constraints are not accepted simultaneously, making solution infeasible.

3.5.4 Multiple Optimal Solutions

The simplex method stops after providing one basic feasible solution. But sometimes, there could be other factors which have not been considered in the original model influencing the final solution. These other factors would not change the value of objective function as was in the original solution but do significantly influence the practical aspect of problem. For instance, a company manufacturing three types of pens (A, B and C) gets maximum profit from production and selling of only two types of pens (A and B). The production capacity can also only manufacture any two types of pens. But because of change in competitive condition management found that third type of pen (C) has shown increased demand in place of B. Due to limited production capacity, resources required to manufacture B are now used to manufacture C. The profits have remained the same. Thus, another solution has been found to resource allocation by maintaining no change in objective function.

The simplex method indicates possibility of alternative solutions if at least one non-basic variable in net evaluation row ($c_j - z_j$) is zero. Conceptually, introducing a non-basic variable in the solution with zero value would imply that inclusion of its one unit would have no influence on objective function though solution might change. This has been illustrated in following example:

$$\textbf{Minimize} \quad Z = 3x_1 + 3x_2$$

Subject to:

$$2x_1 + 0.5x_2 \geq 10$$

$$2x_1 + 0x_2 \geq 4$$

$$4x_1 + 4x_2 \geq 32$$

$$x_1, x_2 \geq 0$$

Model:
See Table 3.44.

Iteration 1:
See Table 3.45.

TABLE 3.44
Model

		3	3	0	0	0	M	M	M		
Basis	c_b	x_1	x_2	s_1	s_2	s_3	a_1	a_2	a_3	b	Ratio
a_1	M	2	0.5	−1	0	0	1	0	0	10	5
a_2	M	2	0	0	−1	0	0	1	0	4	2
a_3	M	4	4	0	0	−1	0	0	1	32	8
$c_j - z_j$		3–8M	3–4.5M	M	M	M	0	0	0		

TABLE 3.45
Iteration 1

		3	3	0	0	0	M	M		
Basis	c_b	x_1	x_2	s_1	s_2	s_3	a_1	a_3	**b**	**Ratio**
a_1	M	0	0.5	−1	1	0	1	0	6	12
x_1	3	1	0	0	−0.5	0	0	0	2	–
a_3	M	0	4	0	2	−1	0	1	24	6
$c_j - z_j$		0	3–4.5M	M	1.5−3M	M	0	0		

Iteration 2:
See Table 3.46.

Iteration 3:
See Table 3.47.

Iteration 4:
See Table 3.48.

TABLE 3.46
Iteration 2

		3	3	0	0	0	M		
Basis	c_b	x_1	x_2	s_1	s_2	s_3	a_1	**b**	**Ratio**
a_1	M	0	0	−1	0.75	1/8	1	3	24
x_1	3	1	0	0	−0.5	0	0	2	–
x_2	3	0	1	0	0.5	−1/4	0	6	−24
$c_j - z_j$		0	0	M	−0.75M	¾ −M/8	0		

TABLE 3.47
Iteration 3

		3	3	0	0	0		
Basis	c_b	x_1	x_2	s_1	s_2	s_3	**b**	**Ratio**
s_3	0	0	0	−8	6	1	24	4
x_1	3	1	0	0	−0.5	0	2	−4
x_2	3	0	1	−2	2	0	12	6
$c_j - z_j$		0	0	6	−4.5	0		

TABLE 3.48
Iteration 4

		3	3	0	0	0		
Basis	c_b	x_1	x_2	s_1	s_2	s_3	**b**	**Ratio**
s_2	0	0	0	−4/3	1	1/6	4	−3
x_1	3	1	0	−2/3	0	1/12	4	−6
x_2	3	0	1	2/3	0	−1/3	4	6
$c_j - z_j$		0	0	0	0	3/4		

All the values in net evaluation row ($c_j - z_j$) are either zero or positive making solution optimal with $x_1 = 4$ and $x_2 = 4$.

$$\textbf{Minimize} \quad Z = 3x_1 + 3x_2$$

$$= 3 * 4 + 3 * 4 = 24$$

Subject to:

i. $2x_1 + 0.5x_2 \geq 10$
$2{*}4 + 0.5{*}4 = 10$ which is ≥ 10

ii. $2x_1 + 0x_2 \geq 4$
$2{*}4 + 0{*}4 = 8$ which is ≥ 4

iii. $4x_1 + 4x_2 \geq 32$
$4{*}4 + 4{*}4 = 32$ which is ≥ 32

$$x_1, x_2 \geq 0$$

Now, in the final iteration, which gives the final optimal solution, we have x_1, x_2 and s_2 as basic variables. One of the non-basic variables, i.e. s_1 in net evaluation row has a value of zero, making solution having another optimal solution. By introducing this variable into the solution, we get the following solution.

All the values in net evaluation row ($c_j - z_j$) are either zero or positive, making solution optimal with $x_1 = 8$ and $x_2 = 0$ (Table 3.49).

TABLE 3.49
Final Solution

		3	3	0	0	0		
Basis	c_b	x_1	x_2	s_1	s_2	s_3	**b**	**Ratio**
s_2	0	0	2	0	1	−1/2	10	5
x_1	3	1	1	0	0	−1/4	8	8
s_1	0	0	3/2	1	0	−1/2	6	4
$c_j - z_j$		0	0	0	0	3/4		

$$\textbf{Minimize} \quad Z = 3x_1 + 3x_2$$
$$= 3 * 8 + 3 * 0 = 24$$

Subject to:

i. $2x_1 + 0.5x_2 \geq 10$
   $2*8 + 0.5*0 = 16$ which is ≥ 10
ii. $2x_1 + 0x_2 \geq 4$
   $2*8 + 0*0 = 16$ which is ≥ 4
iii. $4x_1 + 4x_2 \geq 32$
   $4*8 + 4*0 = 32$ which is ≥ 32

$$x_1, x_2 \geq 0$$

It can be seen that the value of the objective function remains same, but in previous solution, conceptually, both x_1 and x_2 were manufactured if the question is regarding two types of products being manufactured. But, in the second solution, only one type of product is being manufactured. As a result, the company would achieve the same amount of objective but through different operations. Thus, it depends on the manager which solution is more suitable for the plant.

3.6 SUMMARY

This chapter introduced a simplex method for solving a LPP. Formulation of the LPP model was understood in Chapter 2. However, those models were in non-standard form as constraints consisted of inequalities. Thus, the first step in solving LPP was to convert it into standard form by incorporating slack or surplus variables. Slack variables are added if inequality is less than or equal to, whereas surplus variables are subtracted if inequalities are greater than or equal to or are equality constraints. It was also understood in the creation of the standard form that in order to fulfil non-negativity criteria of the initial solution, artificial variables are added in constraints that involve surplus variables where the artificial variable is assumed to have higher value than surplus. To remove artificial variable from the system as early as possible, an assumed very high value denoted by M is associated with them and introduced in the objective function. Such a method is called as Big M method. Feasibility of the final solution is assessed by removal of all artificial variables from the final solution.

It was also emphasized in Chapter 2 that LPP involves identifying an optimal solution to the problem of either maximization or minimization. This chapter began with an illustration of maximization, which was solved by the simplex method. To further the understanding of methodology of simplex another example of maximization was explained and solved. A similar process was adopted for minimization problem.

Finally, special cases of degeneracy, unboundedness, infeasibility and multiple solutions were discussed in detail, and it was explicitly illustrated how such situations can occur when solving a LPP using the simplex method.

The summarized form of solving LPP through the simplex method is presented below:

Step 1: Formulate a LPP.
Step 2: Construct its standard form by introducing slack, surplus and artificial variables depending on the inequalities.
Step 3: Find IBFS by fulfilling non-negativity criteria and identifying basic and non-basic variables.
Step 4: Construct simplex tableau clearly showing basic and non-basic variables as well as coefficients of objective and constraint functions and right-hand side value of constraint equations. Proper construction is necessary as same would be used for every iteration.
Step 5: Find values of net evaluation row to check the optimality of solution. Depending on the objective of minimization or maximization, incoming non-basic variable is selected.
Step 6: Select the key row by finding the ratio of right-hand side value and coefficients of the selected column. The row with a minimum positive ratio is selected. This row would indicate the outgoing basic variable.
Step 7: Perform necessary mathematical operations in such a way to convert the selected column into unit column.
Step 8: Check optimality of solution by finding values in the net evaluation row. For minimization if all these values are greater than or equal to zero and for maximization if these are less than or equal to zero then optimality is said to be reached. If not, return to step 5.

3.7 CASE STUDY: JOHNSON CONTROLS

The growing market of electric vehicles has increased demand for lithium-ion batteries. Companies, in addition to manufacturing new batteries, are banking on recycling spent batteries in order to fulfil demand. More so with rising concerns of batteries polluting environment and depleting rare earth metals recycling market has grown to be a favourable business proposition. According to a report, battery recycling market is estimated to grow from USD 17.2 billion in 2020 to USD 23.2 billion by 2025. Various leading companies like Exide and Johnson Controls have used this opportunity to augment their businesses.

Johnson Controls supplies lithium ion batteries to various major car manufactures around the world, including BMW, Toyota and GM. With an increasing focus on recycling of used batteries, the company has dedicated a plant in Florence, South Carolina. The operations in this facility reclaim lead from spent batteries and refine them for proper usage in new batteries. This manufacturing process is completed in three stages. In stage 1, after receiving and processing of raw material, the metal is melted in a furnace. In stage 2, the molten metal is cast into a particular type of lead family. The final product can be in the shape of pigs, billets and slugs. As the demand for pigs is maximum, so major portion of production is dedicated to the manufacturing of pigs. In last stage, after casting, the cooling is performed, and this would result in the final product. Due to increased demand for recycled lead to be used in automotive batteries, the production manager of the company wants to increase production by minimizing idle time at three stages of production. To understand the minimization of idle time, production schedule of 1 day was broken down into an hourly basis. The plant runs three shifts each for 8 hours. Idle time was found to be a function of production hours. The manager was interested in understanding the number of hours that production should be allocated at every stage so as to minimize idle time. Past data showed that out of total hourly idle time, 15% was generated at stage 1, 24% and 37% at stages 2 and 3, respectively.

Reduction of idle time was considered to be influenced by the amount of units produced per hour at every stage, inventory produced, number of operators required and utilization of machines. The higher the idle time, the lower would be the production, so it was important to examine the rate of production on an hourly basis. It was estimated that metal to the tune of manufacturing 10,000 units of lead blocks was produced per hour. The second process of casting had the capacity of producing 8,000 casts of lead blocks per hour and, lastly, cooling of 6,000 units could be done on an hourly basis. The production capacity can produce a maximum of 7,000 units of final product per hour. The second constraint that was examined was work-in-process (WIP) inventory produced. Due to the varying capacity of production at every stage, WIP inventory was created that needed to be stored on the shop floor. The material has to be adjusted or moved, resulting in idle time for the next process. Three thousand units of lead block were stored per hour as WIP inventory after stage 1, 5,000 units after stage 2 whereas 4,000 units per hour after stage 3. Due to the differentiating rate of availability of raw material and production capacity of three stages, the manager has decided to keep a minimum of 4,000 units as inventory. Third, the manager would like to minimize the number of operators required at every stage because a higher number would result in jumbled flow and obstruction among operators resulting in higher idle time. Thus, the manager estimated a minimum of six operators to be applied on the shop floor at any given time. Presently, four, three and six operators are performing at stage 1, 2 and 3 respectively on an hourly basis. Finally, machine utilization was studied due to the application of the process of mass production in batches. In this process, the machine keeps on producing till raw material is available irrespective of demand. If a high quantity of raw material is available, then more units would be produced than required, whereas lower availability would result in non-fulfilment of demand. This aspect has an effect on inventory, quantity produced, employees required and also the productivity of machines, which would in turn influence the creation of idle time. For effective production, the manufacturing line should have utilization of minimum 80%. Breaking down machine-wise, it was found that the furnace utilized in stage 1 and casting in stage 2 had utilization rate of 90%, whereas stage 3 machine had utilization rate 100% on an hourly basis. Taking these four constraints into consideration, formulate a linear model which would minimize idle time.

3.8 GLOSSARY

- **Simplex method**: It is an algorithm designed to solve multiple simultaneous linear equations. The method is very effective in finding an optimal solution by iterating from one solution to other.
- **Non-basic variables**: The variables which are assumed to be zero to fulfil non-negativity criteria of linear programming.
- **Basic variables**: The variables which assume certain positive values obtained after non-basic variables are put to zero. A linear programming model in structured form if has *n* variables and *m* constraints then *n-m* would be non-basic variables and remaining would be basic variables.
- **Slack variable:** In order to construct a linear programming model, constraints with less than or equal to inequalities must be converted into equalities by adding slack variables to the left side of equation. Logically, they indicate idle resources available which are yet to be utilized.
- **Surplus variable**: In constraints with greater than or equal to inequalities, the left side is made equal to the right side by subtracting a variable. These variables are termed as surplus variables as they indicate excess resources available than required.
- **Artificial variable**: The variable that is added only in the case of constraints with greater than or equal to constraints to fulfil non-negativity criterion. It does not carry any tangible meaning as it does not possess any resource.
- **Simplex tableau**: A table representing non-basic variables in columns and basic variables in rows with maximum resource available in the last column.
- **Net evaluation row**: Row indicated by $c_j - z_j$ implying per unit increase in profit or per unit decrease in loss.
- **Key column**: The column indicating incoming non-basic variable. In the case of maximization case key column is with maximum positive value and in the case of minimization it is with maximum negative value.
- **Key row**: The row representing outgoing basic variable. The criterion of selecting key row is with minimum positive value indicating minimum possible resource required to achieve objective function.
- **Degeneracy**: A solution is said to have degeneracy if there is a tie in two or more basic variables.
- **Unbounded**: A solution when an outgoing variable cannot be selected because of negative or undefined ratio values.
- **Infeasibility**: When no solution can fulfil all the constraints simultaneously, then the solution is considered to be infeasible.
- **Multiple optimal solutions**: Solutions which present the possibility of alternative solutions if at least one non-basic variable in net evaluation row $(c_j - z_j)$ is zero.

3.9 MODEL QUESTIONS

3.9.1 Solve by Using Simplex method:

$$\text{Max} \quad Z = 250x_1 + 175x_2 + 90x_3$$

Subject to:

$$0.5x_1 + 0.3x_2 + 0.1x_3 \le 240$$

$$2x_1 + 1x_2 + 0.7x_3 \le 1{,}000$$

$$x_1, x_2, x_3 \ge 0$$

3.9.2 Solve by using the simplex method:

$$\text{Min} \quad Z = 8x_1 + 3.5x_2$$

Subject to:

$$5x_1 + 1.5x_2 \ge 30$$

$$4x_1 + 2.5x_2 \ge 40$$

$$x_1, x_2 \ge 0$$

3.9.3 Solve by using the simplex method:

$$\text{Max} \quad Z = 8x_1 + 7x_2 + 3x_3$$

Subject to:

$$2.5x_1 + 1.3x_2 + 0.7x_3 \le 7$$

$$2x_1 + 3x_2 + 0.6x_3 \le 11$$

$$3x_1 + 2x_2 + 4x_3 = 14$$

$$x_1, x_2 x_3 \ge 0$$

3.9.4 Solve by using the simplex method:

$$\text{Min} \quad Z = 1.7x_1 + 1.5x_2 + 0.3x_3$$

Subject to:

$$3x_1 + 3x_2 + 2.6x_3 \le 60$$

$$7x_1 + 4x_2 + 2x_3 = 50$$

$$2x_1 + 9x_2 + 3x_3 \ge 40$$

$$x_1, x_2, x_3 \le 0$$

3.9.5 4,3,3 are weights given to three criteria namely scores in an aptitude test, post graduate grade point average (GPA) and undergraduate GPA. These criteria are used to select candidates for jobs in an MBA institute. The scores were calculated on a scale of 1–5 and indicated contribution of each criterion in maximizing chances of selection. But from present year institute and companies decided to calculate these scores on a scale of ten. To understand the effect of changed system on job selection chances scores regarding three criterions were calculated from three disciplines of Finance, Marketing and Operations. Companies also decided minimum required score for each discipline. The data is shown in the following table:

Discipline	Scores in Aptitude Test	Scores in Post Graduate GPA	Scores in under Graduate GPA	Minimum Score Required
Finance	7	7.3	6.8	8
Marketing	6.8	7.4	6.6	8
Operations	6.9	6.7	7.8	8

Using the given solve by using the simplex method percentage change in each criteria which would maximize chances of selection.

3.9.6 In a training program organized by Business Solutions Ltd. on management education out of maximum 100 sessions at least 7 and 8 sessions be devoted to research methodology (RM) and business statistics (BS), respectively. To conduct these sessions, two specialists, A for RM and B for BS, are available. A would like to conduct a minimum of ten sessions so as to deliver the course effectively. Similarly, B would conduct a minimum of four sessions. The cost incurred per session to deliver a course on RM is $120 and for BS is $150. Formulate a linear programming model to plan the optimal number of sessions of RM and BS, which would minimize total cost.

3.9.7 An illustration of storing three types of piston rings (A, B and C) in three different types of racks (R1, R2 and R3) was given in the introduction section of Chapter 1. Try to resolve the given modified problem using the simplex method. Storage space should be developed in such a manner that it can store at least 150,000 of total rings. One rack of R1 type can store 2,590 rings, whereas those of R2 and R3 types can store 4,600 and 3,100. Due to high demand and storage required for A type of rings, the racks should be ordered at a ratio of 3:1:2. Taking into consideration the availability of space and funds, it was estimated that a maximum of 72 racks could be purchased. Also, these racks have to be placed in the maximum available space of 100 m^2. R1, R2 and R3 occupy 2.7, 2.4 and 4.4 m^2 of space. One unit of each of these types of racks would cost $30, $13.8 and $60, respectively. The manager would like to estimate the minimum number of racks of each type that would result in the minimum total purchase cost. Check the appropriateness of solution and infer the results.

3.9.8 A media house prints two weekly magazines, namely political news and a business magazine. The major source of revenue from these magazines is thorough advertisements. Predominantly, the magazines contain three categories of advertisements, i.e. cosmetic product, electronic product and food product advertisements though in varying quantities. As cosmetic advertisements are in higher demand and generate greater revenue, so publisher tends to give more space to these advertisements. The percentage allocation is shown in the following table. But the entire space earmarked for ads cannot be allocated to only one type of product, so there is capacity constraint indicated in terms of the number of ads. A maximum of 20 cosmetic product, 12 electronic products and 8 food product ads could be placed in both types of magazines per issue combined. Profit earned by publisher by allocating space to different types of ads is \$2.5 and \$2.0 per issue of each magazine. Formulate LPP to find the optimal number of ads of each type so as to maximize profit.

	Cosmetic Products Ads	Electronic Products Ads	Food Products Ads
Political news magazine	At least 50%	Maximum of 25%	Maximum of 10%
Business news magazine	At least 40%	Maximum of 30%	Maximum of 10%

3.9.9 Assembly of a product requires production of two parts A and B. One unit of part A and B requires 3 hours and 2 hours for production. The maximum production time available to fulfil demand of 10,000 units of A and 9,000 units of B was 3,000 hours. The manager intends to find whether such demand can be fulfilled in available time or whether production of certain units is outsourced to a vendor. It was estimated that the cost of production at the facility of one unit of A was \$10, which increases to \$10.5 if bought from an outside source. Similarly, one unit of B costs \$9.8 in-house, whereas \$10.2 if purchased. The following questions needed to be answered:

i. Is existing production capacity sufficient to fulfil demand for two parts?
ii. How many units of both parts will be manufactured and how many will be outsourced?
iii. What would be proper combination of manufacturing and purchasing which would minimize cost?

3.9.10 Three types of dolls (D1, D2 and D3) manufactured by a toy maker generate a profit of \$1.0, \$1.3 and \$1.1, respectively. Because of the constraints of labour and machine time available, only a limited quantity of dolls can be manufactured, which was estimated to be 16,000 of D1, 25,000 of D2 and 12,000 of D3. To manufacture such quantity, labour time per unit required was calculated to be 0.3, 0.5 and 0.34 minutes, whereas machine time was 0.03, 0.045 and 0.03 minutes for dolls D1, D2 and D3, respectively. The maximum labour time and machine time available to fulfil the mentioned demand were 9,000 minutes and 11,000 minutes. All the three types of dolls undergo the same process for manufacturing. The manager would like to find combination of the number of dolls of each type that could be manufactured under given constraints which would maximize profits.

3.9.11 A retail investor wants to plan his investments for next 5 years in such a manner that at the end of every year a minimum amount of cash is available. At beginning of each year, he receives a certain fixed amount of salary from which a fixed amount is invested in three financial instruments (A, B and C). The returns from these three financial instruments are obtained at the end of each year. So, for next year's investments, the corpus he has includes salary plus returns from three investments. The increase in salary per annum is fixed at 10%. But from this corpus, a good amount goes into meeting various sundry expenses. The remaining amount should be at least the sum which has been estimated for every year. The investor would like to assess the number of units for each investment so as to minimize the cost of meeting yearly obligations. The cost of purchase of one unit of each instrument was $10, $12 and $13.5.

	Return from Instrument A (%)	Return from Instrument B (%)	Return from Instrument C (%)	Minimum Amount Required at End of Every Year
At end of year 1	9.1	7.6	10.01	60,000
At end of year 2	8.8	7.7	10.02	720,00
At end of year 3	8.9	7.4	9.98	80,000
At end of year 4	9.04	7.05	9.84	70,000
At end of year 5	9.01	7.23	10.05	94,000

3.9.12 A fast food restaurant catering to the tastes of residents of two wards in a city wants to understand customers' reaction towards their new food offerings. For this, the manager of the restaurant conducted a survey. It was decided that a total of 1,200 respondents would be interviewed. For fair and representative survey, the total number of males should be at least 60% of total females. Due to cost constraints, a maximum of 500 households from both wards 1 and 2 were to be taken. From census data, it was hypothesized that each household would have more than one male or female member. Finally, the surveys have to be conducted within a distance of 5,000 m. The cost of conducting an interview from ward 1 was estimated to be $0.5 per respondent and from ward 2 to be $0.6. The manager wants to estimate the number of male and female respondents from each geographic area who would be part of survey so that cost is minimized.

4 Sensitivity Analysis and Duality Theory

4.1 INTRODUCTION

As discussed in previous chapters, Linear Programming is an Operations Research technique applied in solving multi-criteria decision-making problems. The main emphasis is on finding an optimal solution that could be either of maximization or of minimization. Any technique of OR adopts a process of formulating given problem in the form of model, subjecting it to certain mathematical operations which at every iteration tends to provide a better solution than the previous one and reaching to the most optimal and feasible solution. The issue with this approach is that real-time problems are modified into some small-scale models, and the solution of these models is generalized to real-time problems. Often a mismatch is found between them, resulting in an incorrect optimal solution and starting again from scratch. There is no guarantee that the devised model always represents the real problem due to too many uncertainties in implementation of a deduced solution that must be done in the near future. Therefore, it is necessary to properly formulate the model, taking into consideration as much as possible all variables that effect the objective function. Second, decision maker should not focus on demanding a perfect single solution rather than emphasize on finding a 'good enough' solution, which satisfies numerous criteria and is able to achieve various long-term goals.

The above discussion thus strongly underlines the importance of finding a solution that is as close as possible to the real problem. This requires post-optimality analysis, which has become a very important part of an OR technique. 'What-if analysis' is a very constructive post-optimality method, which addresses certain questions about what would happen to the most optimal solution if the model is subjected to changes in future conditions or addition of other constraints. These questions consider relevant constraints that might emerge in the near future, thus giving managers various alternative solutions.

Another important post-optimality analysis is conducting 'sensitivity analysis', which involves identifying sensitive parameters out of numerous parameters effecting objective function. Sensitive parameters are considered to be those parameters whose value cannot be changed without causing a change in the optimal solution. Identification of such parameters is very important as then a decision maker would put extra focus and effort in estimating them. As OR techniques use models for problem solving, so these parameters are only *estimates* and not true values. The model is considered to be good enough if the estimated value of parameters provide a solution, which is close enough to represent true values. Thus, in such a process, a model can generate a number of parameter values. Sensitivity analysis helps in finding out a range of these values of each sensitive parameter. It is important to rephrase here that sensitive parameters are those which would change the optimal solution. So, utmost

DOI: 10.1201/9781003212966-4

care should be taken in their identification and application. Sensitivity analysis helps in selecting a combination of parameters that makes a solution a good one.

For instance, in location analysis, a store manager needs to select a suitable location from multiple sites for setting up a new store. The selection decision is governed by the sales performance of the store at a particular selection. Sales would be higher if customers considered the store as attractive to visit and purchase. The manager intends to devise which variables would make the store attractive. The LPP technique of OR was found to provide a model to achieve the purpose of maximization of sales in the presence of a number of variables. These variables could be accessibility, availability of parking space, availability of merchandize and many more. The initial OR model would be able to provide a single optimal solution. However, post-optimality analysis such as sensitivity analysis would, in addition to finding which parameters are sensitive and most important in effecting objective function, would also tell the range of values of these parameters over which identified solution would be optimal and feasible. This range of values of selected sensitive parameters would take into consideration all relevant goals in endeavour of providing a valid solution.

Post-optimality analysis also provides a sequence of solutions where each solution is better than the previous one. In this process, shortcomings in various parameters of objective and constraint function are identified, which helps manager to decide what to improve. This results in another improved solution. The process is repeated until improvement is so minimal that it does not warrant any tangible action in the form of allocation of precious resources. Sensitivity analysis helps in testing the validity of each solution.

This chapter explains the concept of duality and its important implications after discussing sensitivity analysis as a post-optimality technique. It was deduced that every linear programming problem has its dual. The optimal solution of this dual reveals values over which prime solution is feasible. The finding of these values of selected parameters is discussed as function of sensitivity analysis. Thus, the important issue of duality theory is interpretation and implementation of sensitivity analysis. It is important to understand the concept of optimality and the feasibility of solution. As said, the parameter values in any linear programming model are only estimates, so there is every possibility of finding a variety of parameter values which would fulfil the objective of either maximization or minimization. These solutions are considered as feasible if all constraints within which an objective has to be fulfilled are satisfied simultaneously. Out of these solutions, the values that accomplish the objective, i.e. either minimize the cost to maximum or provide maximum profit, are considered to be optimal. Thus, there is strong relationship between optimal and feasible solution, which has been further discussed under the concept of duality theory as this theory provides an optimal solution by conducting sensitivity analysis on prime solution.

4.2 FUNDAMENTAL NATURE OF SENSITIVITY ANALYSIS

To illustrate the nature of sensitivity analysis, the following two cases have been discussed.

Case 1: Section 2.5.2 explained an illustration regarding minimization of loss in revenue of a restaurant, which utilized two types of seating arrangements (A and

B). The decision variables indicated the number of seats required of each type of arrangement notated by x_1 and x_2. Loss in revenue if one A type of seat goes vacant was $10, whereas that of B was $15 leading to formation of the objective function:

$$\textbf{Minimize } Z = 10x_1 + 15x_2$$

The final solution implied that value of decision variables x_1 and x_2 were 20 and 15, respectively.

Now, according to the additivity assumption of LPP, as discussed in Section 2.3, every function in a LPP, whether it is an objective function or the right-hand side (RHS) of constrain function, is the sum of contribution of each individual activity. Thus, in this case, loss in revenue is sum of losses due to vacant seats of types A and B. To reduce such losses, the manager decided to use the same space for organizing small parties, mostly on weekdays when demand for restaurant services was less as compared to weekends. The promotion campaign led to more footfalls and down the line losses in revenue showed a downward trend due to increased occupancy. But the manager found that the decrease in loss in revenue was more than just because of higher occupancy of A and B types of seats. This was attributed to the utilization of the same restaurant space as a banquet hall for conduct of small-scale parties. Thus, both types of seating arrangements are complementary to each other in decreasing loss of revenue due to the capacity of the restaurant as well as that of the banquet hall.

Thus, in the case of *complementary* products, the objective function would be $10x_1 + 15x_2 - x_1x_2$. In this case, parameter x_1x_2 has been subtracted as complimentary nature of items, resulting in *decrease* of loss in revenue. So, the value of Z when $x_1 = 20$ and $x_2 = 15$ would be $Z = 10*20 + 15*15 - 20*15 = 125$ instead of $Z = 10*20 + 15*15 = 425$.

Similarly, if two items are *competitive*, then also value of objective function would undergo a change. If demand for any or both of services increases, then they would be competing with each other for the same space at the same time. For instance, demand for parties increases on weekends. In addition to demand for restaurant, the manager has to decide which service to forego. Competition for the same facility between two services could result in an *increase* in loss of revenue. Thus, the objective function would be $Z = 10x_1 + 15x_2 + x_1x_2$. So, the value of Z when $x_1 = 20$ and $x_2 = 15$ would be $Z = 10*20 + 15*15 + 20*15 = 725$ instead of $Z = 10*20 + 15*15 = 425$.

The above case indicates a change in value of an objective function with change in values of parameters indicating their sensitivity. This change can also be observed in the case of *constraint functions* where the additivity assumption of sum of individual contribution is characterized. For instance, the fourth constraint in the same case was that the maximum capacity of seats of both types was 40. To illustrate, just for supposition purposes, if due to rush of customers, 50% more seats were added to the restaurant to make it more congested, then it would lead to increase in capacity. This would make the constraint as $1x_1 + 1x_2 + 0.5x_1x_2 \leq 40$. Putting the values of $x_1 = 20$ and $x_2 = 15$ in the equation $1*20 + 1*15 + 0.5*20*15 = 185$ instead of 35 which was obtained from $1x_1 + 1x_2 \leq 40$.

Therefore, the value of optimal solution is subject to change depending on the selection of parameter value. Thus, conduct of post-optimality analysis by using sensitivity analysis is of prime importance in finding feasible and optimal solution.

Case 2: Another illustration to emphasize the application of sensitivity analysis in determining sensitive parameters has been discussed here which involves a scenario wherein the proportionality assumption of LPP is violated. As discussed in Section 2.3, this assumption implies that the value of an objective function changes in proportion to the change in level of activity represented by decision variables. The same happens for constraint function where value of the left side of the function is proportional to the level of activity. For instance, a toy manufacturer involved in manufacturing of three types of dolls (A, B and C) earns a profit of \$1, \$1.3 and \$1.1 per unit, respectively. Then, if x_1, x_2 and x_3 are the number of dolls of A, B and C, then total profit would be $1x_1 + 1.3x_2 + 1.1x_3$. Three decision variables represent the level of activity, implying that one unit of x_1 would generate profit of \$1, whereas its two units would earn profit of \$2. Thus, change is in proportion to change in level of activity.

But in certain scenarios, this proportionality assumption is violated. In most manufacturing processes, production involves certain set up costs which are incurred due to setting of machinery, raw material, tools, etc. Once the facilities are set up, then they can be utilized for production of the entire lot, and such costs are not incurred after production of each unit. Thus, it is one time cost. If set up cost for each type of doll i.e. from A to B is \$0.1, then profit from manufacturing of A is $1x_1 - 0.1$ rather than $1x_1$.

Another case could be that the profit function is enhanced not only by contribution from decision variable but due to economies of scale obtained from longer production runs, mass production system and offering of quantity discounts. Thus if A type of doll is produced in high volumes making the production process efficient, then profit instead of $1x_1$ could be $1x_1 + y_1$ where y_1 is profit obtained due to economies of scale. Similarly, assumed extra profit due to mass production of B and C types of dolls would be y_2 and $y_{3.}$ Hence, instead of profit being $1x_1 + 1.3x_2 + 1.1x_3$ in the modelled problem, it would be $(1x_1 + y_2) + (1.3x_2 + y_3) + (1.1x_3 + y_3)$ in real future scenario. This clearly violates the principle of proportionality.

Finally, the proportionality assumption might also be violated followed in constraint function. For instance, in above example of doll manufacturing, out of total production time of 440 minutes available, one unit of doll A and C takes 1.8 minutes to manufacture, whereas one unit of doll B takes 2.7 minutes. Thus, the constraint function of production time becomes $1.8x_1 + 2.7x_2 + 1.8x_3$. Now, as both A and C takes same amount of time, so during the changeover from A to B, the spare time for which machine is idle can be used to manufacture C. Thus, the production time of C would be less than 1.8 minutes. If utilized spare time was 0.8 minutes for supposition purposes, then one unit of doll C is manufactured in 1.0 minutes, making constraint function in reality as $1.8x_1 + 2.7x_2 + 1.0x_3$.

These two cases involving different hypothesized examples make clear that parameters both in objective and constraint functions are estimates that are prone to change with changing circumstances. Another reason for their sensitivity could be that members of the OR team have not taken into account all the variables that go into estimating parameters. Hence it becomes increasingly important to conduct sensitivity analysis for the proper evaluation of optimal solution as only then the solution could be considered as valid. The following section helps in interpreting the implications of sensitivity analysis when there is change in:

i. Right-hand side (RHS) value of constraint function;
ii. coefficients of basic variable in objective function; and
iii. coefficients of non-basic variable in objective function.

The following sections discuss both cases by first giving an example where a changed value in parameter value is shown to have an impact on the initial optimal solution. Second, the range of values of changed parameters over which solution would remain optimal is estimated. It is important to mention here that conditions of both feasibility and optimality need to be met at each step of the procedure. The feasibility of a solution is implied by non-negative RHS values of all basic variables in the simplex table. Whereas a solution is considered optimal if values in the net evaluation row are either zero or positive in case of minimization or zero or negative in case of maximization. More importantly, the values of all non-basic variables in z_j row of the simplex table need to be non-negative to make a solution optimal.

4.3 APPLYING SENSITIVITY ANALYSIS

4.3.1 Change in Right-Hand Side (RHS) Values of Constraint Functions (b_i)

RHS values in constraint functions represents resource availability of certain activity. In most cases, the management team provides tentative values for these activities as either they are not sure of the resources' availability as activity has to happen in the near future or it is either underestimated or overestimated due to uncertainties involved. Sensitivity analysis is utilized to figure out change in optimal solution with change in amount of resource availability. For instance, number of production hours available in 8 hour shift after making allowances for food and rest time could be 420 minutes. But if overtime of 1 hour is applied, then the total production time available would be 480 minutes. This increased resource availability would allow an increase in production, but this would be justifiable only if it offsets the cost of overtime. This increased value of resource indicated by 'b' can be used to calculate the new value of objective function 'Z'. The difference between value of 'Z' from original model and new value obtained from marginal increase (i.e. increase of one unit) of 'b' is called as '*shadow price*'. This can be done by considering the marginal increase in RHS values of other activities.

This phenomenon of change in optimal solution with change in RHS value of constraint function is explained by taking the example of maximization of sales of Adidas AG retail stores illustrated in Section 2.5. The case was regarding estimating the number of units to be sold to customers of geographic areas A and B. Sales estimation was dependent on constraints of accessibility, population density and median income. Profit achieved from area A and B by selling per thousand units were \$80 and \$40, respectively, making the objective function to be $Z = 80x_1 + 40x_2$. A constraint of population density discussed in the case implied that the maximum average population that two stores can cater to was 10,000. The final solution table has been reproduced here (Table 4.1) for easy comparison.

The change in optimal solution with increase in population density by one unit, i.e. to 11,000 is discussed here. The modified model would be (Table 4.2):

The final simplex table that would give the optimal solution was (Table 4.3):

TABLE 4.1
Final Solution

Decision Variables		x_1	x_2	s_1	s_2	s_3	
Coefficients of Decision Variables in Objective Function		80	40	0	0	0	Maximum Available Resource
Basis Variables	Coefficients of Basis Variables c_b in Objective Function	Coefficients of Constraint Functions					
x_1	80	1	0	0.127	−0.07	0	1.06
x_2	40	0	1	−0.04	0.16	0	1.12
s_3	0	0	0	−0.25	−1.24	1	3.46
	z_j	80	40	8.56	0.8	0	
Net evaluation row	$c_j - z_j$	0	0	−8.56	−0.8	0	

TABLE 4.2
Model

Maximize	$Z = 80x_1 + 40x_2 + 0s_1 + 0s_2 + 0s_3$
Constraint 1	$9x_1 + 4x_2 + 1s_1 + 0s_2 + 0s_3 = 14$
Constraint 2	$2x_1 + 7x_2 + 0s_1 + 1s_2 + 0s_3 = 11$
Constraint 3	$5x_1 + 10x_2 + 0s_1 + 0s_2 + 1s_3 = 20$

TABLE 4.3
Final Solution of Revised Model

Decision Variables		x_1	x_2	s_1	s_2	s_3	
Coefficients of Decision Variables in Objective Function		80	40	0	0	0	Maximum Available Resource
Basis Variables	Coefficients of Basis Variables c_b in Objective Function	Coefficients of Constraint Functions					
x_1	80	1	0	0.127	−0.07	0	0.92
x_2	40	0	1	−0.04	0.16	0	1.3
s_3	0	0	0	−0.25	−1.24	1	2.18
	z_j	80	40	8.56	0.8	0	
Net evaluation row	$c_j - z_j$	0	0	−8.56	−0.8	0	

Comparing two models would provide the following inferences:

i. *Value of objective function* in original model was: $Z = 80x_1 + 40x_2 = 80 *1{,}060 + 40*1{,}120 = \$129{,}600$, whereas for the revised model, the value of decision variables has changed to $x_1 = 0.92*1{,}000 = 920$ units and $x_2 = 1.3*1{,}000 = 1{,}300$ units. With these values of x_1 and x_2, total profit was $Z = 80x_1 + 40x_2 = 80*920 + 40*1{,}300 = \$125{,}600$. This change in value of the objective function represented by Z* indicates the shadow price of the respective resource representing economic activity. Thus, an increase in one unit of population density (in this case, one unit represents 1,000 households) results in a decrease in economic activity, i.e. sales, by $Z^* = \$129{,}600 - \$125{,}600 = \$4{,}000$.
ii. The solution was still *feasible* as indicated by non-negative RHS values (b_i) in the revised model. Also it is important to put new values of decision variables in constraint functions to suggest the appropriateness of the revised model. Such calculations also help in inferring which resource is used to maximum and which is lying idle. For instance, with $x_1 = 1.06$ and $x_2 = 1.12$ in the original model, all constraint functions were satisfied, resulting in the derived solution as feasible. In the revised model with $x_1 = 0.92$ and $x_2 = 1.3$, all constraints are met as shown below:

$$\text{Constraint 1}: 9x_1 + 4x_2 \le 14$$

$$9*0.92 + 4*1.3 = 13.48$$

$$\text{Constraint 2}: 2x_1 + 7x_2 \le 11$$

$$2*0.92 + 7*1.3 = 10.94$$

$$\text{Constraint 3}: 5x_1 + 10x_2 \le 20$$

$$5*0.92 + 10*1.3 = 17.6$$

$$\text{Constraint 4: } x_1 \text{ and } x_2 \ge 0$$

iii. The final solution from the revised model is still *optimal* as suggested by non-negative coefficients of all non-basic variables in z_j row. In the final simplex table of the revised model, non-basic variables were s_1 and s_2, and their values in z_j row were 8.56 and 0.8, implying optimality of solution. It is important to understand its interpretation. s_1, s_2 and s_3 are slack variables associated with constraints of accessibility, population density and median income, respectively. They represent the amount of resource consumed from the maximum resource available. For instance, if one unit of a product consumes 3 hours out of total maximum available production time of 100 hours, then the value of s_1 would represent the remaining available resource of production time, i.e. 97 hours. So in this case similarly $s_1 = 8.56$ associated with accessibility constraint represents that out of the maximum scope of sales area of 14 miles, a retail store is able to cover only an area of 14–8.56 = 5.44 miles. Hence, sales of store can be increased if a manager employs resources to cater to customers within the remaining area. Thus, importantly, values of slack variables represent contribution to objective function with change in resources. In similar vein, the value of non-basic variables s_2 as 0.8 implies that out of total of retailers' capacity to cater to the demand of 11,000 households (marginal increase in resource from 10 to 11), capacity of 11,000 – 800 = 10,200 was utilized. The capacity of 800 units was lying idle. This concept is important as it is applied in the calculation of the allowable range of b_i in the subsequent section.

iv. Also, values of non-basic variables in net evaluation row ($c_j - z_j$) for both models are negative and non-zero, implying that the optimal solution is reached and no further iterations are required. If, in any further iteration, a negative non-basic variable is selected as an incoming variable, then it would result in a decrease of profits, which is contrary to the objective function of maximization.

The above discussion clearly indicates achieving another optimal solution with an incremental change in value of one of the resources available. The same discussion can be applied to change in values of b_i of other resources. There can be a slew of such cases now. Thus, the question is which b_i to be changed and how much these values can be changed to keep the solution optimal. Duality theory helps in reaching the final solution that is discussed in subsequent sections. For the time being, the next section focusses on identifying the range within which any change in b_i values would result in optimal solution. This range is termed as *allowable range.*

4.3.2 Allowable Range of RHS Values (b_i)

The following illustration explains the calculation of change in b_3 (Δb_3) as contribution to profit function with marginal increase in capacity of retailer to cater to demand (constraint 2). Non-basic variable associated with this constraint is s_2 so in following illustration coefficients of s_2 have been used.

To understand the range over which b_2 can be allowed, the following method should be used.

'b_i' Values in Initial Model		Change in b_2 (Δb_2)		s_2 Column Values in Initial Model		'b_i' Values in Revised Model
1.06	+	1	*	−0.07	=	0.92
1.12				0.16		1.3
3.46				−1.24		2.18

Thus,

$$1.06 - \Delta b_2 * 0.07 = 0.92$$

$$1.12 + \Delta b_2 * 0.16 = 1.3$$

$$3.46 - \Delta b_2 * 1.24 = 2.18$$

For feasibility of solution, the RHS values need to be greater than or equal to zero. Thus,

- $1.06 - \Delta b_2 * 0.07 \geq 0 \; \Delta b_2 \leq 15.14$
- $1.12 + \Delta b_2 * 0.16 \geq 0 \; \Delta b_2 \geq -7$
- $3.46 - \Delta b_2 * 1.24 \geq 0 \; \Delta b_2 \leq 2.79$

Calculation of above three equations helps to decide lower and upper bound of 'b_i' values. It was found that 'b_i' can take minimum of –7 and maximum of 15.14. So, the range of 'b_i' values was found to be:

$$-7 \leq \Delta b_2 \leq 15.14$$

The initial RHS value of constraint 2 was maximum of 10 units. By adding this 10 into minimum and maximum values of Δb_3 range of feasibility can be found. Thus,

$$10 - 7 \leq b_2 \leq 15.14 + 10$$

$$3 \leq b_2 \leq 25.14$$

So, a solution would remain feasible if two retail stores of two geographic areas have the capacity to cater to the demand to the population within the range of minimum value 3*1,000 = 3,000 and maximum value 25.42*1,000 = 25,420 households.

To validate the range of b_2, let's solve the question by increasing b_2 to 30, i.e. capacity has been increase to cater to a population of 30,000 households. The following is the final simplex table (Table 4.4) for such scenario.

TABLE 4.4
Final Solution of Modified Constraint 2

Decision Variables		x_1	x_2	s_1	s_2	s_3	
Coefficients of Decision Variables in Objective Function		**80**	**40**	**0**	**0**	**0**	
Basis Variables	**Coefficients of Basis Variables c_b in Objective Function**	**Coefficients of Constraint Functions**					**Maximum Available Resource**
x_2	40	2.27	1	0.25	0	0	3.52
s_2	0	−12.8	0	−1.75	1	0	5.36
s_3	0	5	0	0	0	1	8.96
	z_j	90.8	40	10	0	0	
Net evaluation row	$c_j - z_j$	10	0	−10	0	0	

The above solution is incomplete as incoming variables of x_1 and x_2 are repeating themselves in the solution. In first iteration, x_1 becomes the incoming variable and then x_2. The solution shown in Table 4.4, if solved further, would show x_1 as an incoming variable again replacing x_2. This shows the solution to be an unbounded.

Hence, it could be stated that values outside range values makes a solution infeasible.

4.3.3 Change in Objective Function Coefficient (Non-basic Variable)

In this section, sensitivity analysis technique is applied to understand the impact on optimality of initial solution with change in coefficient of non-basic variable in the objective function. Every objective function consists of certain decision variables denoted by x_j (where j = 1,2,3...). These are multiplied by some certain constants indicating per unit profit or cost termed as coefficients denoted by c_j. We have seen in solutions of LPP by simplex method that either all or some of decision variables are basic variables. The remaining variables are considered in the non-basic variables category. What happens to initial optimal solution when the coefficient of one of these non-basic variables is changed? This section deals with estimating the range of values of such coefficient within which the solution remains optimal.

Expanding the above illustration of maximization of profit from two areas to three areas, namely area C, the store manager estimated that average profit earned per unit from C was $30. The estimated profit was considered to be less than the other two areas because of it being farther away from the store. The average distance of between potential customers of C and the store was calculated to be 10 miles. Another change in this illustration was RHS value of accessibility constraint. The store manager found that customers located at a distance more than 10 miles from the store were not buying frequently. Thus, the manager decided to focus on customers within an area of 10 miles rather than 14 miles. A median household income of $11,000 and population density of 9,000 households make the store a potential market for store manager. Taking these aspects into consideration, the model would be (Table 4.5):

Final simplex table of above model is (Table 4.6):

Thus, the above modification of maximization problem provides a feasible solution as all values in RHS column are positive. Also, all values in net evaluation row ($c_j - z_j$) being either zero or negative implies that any further iteration would have no influence on the increase in sales. Importantly, the solution indicates x_3 as a non-basic variable. The coefficient of x_3 was 30, meaning that per unit sale from area C

TABLE 4.5
Initial Model

Maximize	$Z = 80x_1 + 40x_2 + 30x_3 + 0s_1 + 0s_2 + 0s_3$
Constraint 1 (Accessibility)	$9x_1 + 4x_2 + 10x_3 + 1s_1 + 0s_2 + 0s_3 = 14$
Constraint 2 (Population density)	$2x_1 + 7x_2 + 9x_3 + 0s_1 + 1s_2 + 0s_3 = 10$
Constraint 3 (Median income)	$5x_1 + 10x_2 + 11x_3 + 0s_1 + 0s_2 + 1s_3 = 20$

TABLE 4.6
Final Solution

Decision Variables		x_1	x_2	x_3	s_1	s_2	s_3	
Coefficients of Decision Variables in Objective Function		80	40	30	0	0	0	
Basis Variables	Coefficients of Basis Variables c_b in Objective Function	Coefficients of Constraint Functions						Maximum Available Resource (b_i)
x_1	80	1	0	0.17	1.14	−0.07	0	1.05
x_2	40	0	1	1.23	−0.33	0.16	0	1.12
s_3	0	0	0	−2.09	−2.5	−1.24	1	3.5
	z_j	80	40	62.8	78	0.8	0	
Net evaluation row	$c_j - z_j$	0	0	−32.8	−78	−0.8	0	

entails a profit of $30. The intention here is to find whether any reasonable change in this coefficient still maintains feasibility and optimality of the original solution. The change has to be realistic. For instance, area C being farthest away from store cannot be assumed to surpass the profits of other areas. Therefore, for assumption purposes, let profit generated from area C be changed to $40 instead of $30. The objective function would change from $Z = 40x_1 + 60x_2 + 30x_3$ to $Z = 40x_1 + 60x_2 + 40x_3$. The final solution is calculated through the Simplex method provided in the following table (Table 4.7):

Hence, as indicated by the above, the original solution is still optimal as no noteworthy changes were observed in either RHS values or values in z_j row. Thus,

TABLE 4.7
Revised Solution

Decision Variables		x_1	x_2	x_3	s_1	s_2	s_3	
Coefficients of Decision Variables in Objective Function		80	40	40	0	0	0	
Basis Variables	Coefficients of Basis Variables c_b in Objective Function	Coefficients of Constraint Functions						Maximum Available Resource (b_i)
x_1	80	1	0	0.17	1.14	−0.07	0	1.05
x_2	40	0	1	1.23	−0.33	0.16	0	1.12
s_3	0	0	0	−2.09	−2.5	−1.24	1	3.5
	z_j	80	40	62.8	78	0.8	0	
Net evaluation row	$c_j - z_j$	0	0	−22.8	−78	−0.8	0	

sensitivity analysis helps in deducing that parameters pertaining to non-basic variables are *insensitive parameters*, and any realistic changes in them does not alter optimality, feasibility and profit function of the initial solution.

However, it is important to find a range of values of this parameter over which the solution remains optimal. To do so, the following method is applied where instead of putting any numeric value to the coefficient of x_3, it is assumed to be c_3. So, the only change in the initial solution would be c_3 instead of 30 in a row pertaining to coefficients of decision variables in objective function. This is shown in Table 4.8.

Now feasibility condition in case of maximization objective function implies that all values in net evaluation row $(c_j - z_j)$ be either zero or negative. Thus, when c_3 is changed, solution would remain optimal as long as:

$$c_3 - 62.8 \leq 0$$

$$c_3 \leq 62.8$$

So, if the coefficient of x_3 remains between 0 and 62.8, then there will be no change in the current solution, though it will result in multiple solutions with no significant effect on the objective function, as indicated by its value when increased from 30 to 40.

But what happens if value of c_3 goes out of estimated range? Suppose value of c_3 is 100, then Table 4.9 gives the final solution.

The revised solution shows that now x_2 instead of x_3 has become non-basic variable. Now a solution is considered feasible if: (i) all RHS values are non-negative and (ii) all constraints are met simultaneously. According to the results in the Table 4.9, one of the conditions of RHS values is met, but when values of decision variables which are $x_1 = 0.45$, $x_2 = 0$ and $x_3 = 1.01$ are put in constraints as given in the model, the following values are obtained:

TABLE 4.8
Allowable Range of Non-basic Variable

Decision Variables		x_1	x_2	x_3	s_1	s_2	s_3	
Coefficients of Decision Variables in Objective Function		**80**	**40**	c_3	**0**	**0**	**0**	**Maximum Available Resource**
Basis Variables	**Coefficients of Basis Variables c_b in Objective Function**	**Coefficients of Constraint Functions**						**(b_i)**
x_1	80	1	0	0.17	1.14	−0.07	0	1.05
x_2	40	0	1	1.23	−0.33	0.16	0	1.12
s_3	0	0	0	−2.09	−2.5	−1.24	1	3.5
	z_j	80	40	62.8	78	0.8	0	
Net evaluation row	$c_j - z_j$	0	0	c_3−62.8	−78	−0.8	0	

TABLE 4.9
Revised Solution Out of Range

Decision Variables		x_1	x_2	x_3	s_1	s_2	s_3	
Coefficients of Decision Variables in Objective Function		**80**	**40**	**100**	**0**	**0**	**0**	
Basis Variables	**Coefficients of Basis Variables c_b in Objective Function**	**Coefficients of Constraint Functions**						**Maximum Available Resource (b_i)**
x_1	80	1	−0.54	0	0.15	−0.16	0	0.44
x_3	100	0	0.88	1	−0.03	0.14	0	1.01
s_3	0	0	2.9	0	−0.39	−0.79	1	6.8
	z_j	80	44.8	100	9	1.2	0	
Net evaluation row	$c_j - z_j$	0	−4.8	0	−9	−1.2	0	

Constraint 1 (Accessibility):

$$9x_1 + 4x_2 + 10x_3 \leq 14$$

$$9*0.45 + 4*0 + 10*1.01 = 14.16$$

This is more than RHS violating the constraint

Constraint 2 (Population density):

$$2x_1 + 7x_2 + 9x_3 \leq 10$$

$$2*0.44 + 7*0 + 9*1.01 = 9.97$$

which is less than RHS satisfying the constraint

Constraint 3 (Median income):

$$5x_1 + 10x_2 + 11x_3 \leq 20$$

$$5*0.44 + 10*0 + 11*1.01 = 13.31$$

which is less than RHS satisfying the constraint

Thus, as the c_3 value keeps going away from maximum range value of 62.8, the solution would increasingly become infeasible.

4.3.4 Change in Objective Function Coefficient (Basic Variable)

The final simplex table indicating the solution of the expanded version of profit maximization case showed that the solution was optimal as all values in net evaluation row were either zero or negative. Thus, any further iteration would not contribute anything to objective function. The solution also identifies x_1 and x_2 as basic variables, whereas x_3 as a non-basic variable. The effect of change in coefficient of non-basic variable was illustrated in the previous section. This section illustrates the change in coefficient of any one of basic variables. Just to recall the model was:

$$\textbf{Maximize } Z = 80x_1 + 40x_2 + 30x_3$$

Subject to:

$$9x_1 + 4x_2 + 10x_3 \leq 14$$

$$2x_1 + 7x_2 + 9x_3 \leq 10$$

$$5x_1 + 10x_2 + 11x_3 \leq 20$$

The range of values that the objective function coefficient of the basic variable can take is calculated by assuming that the coefficient to be c_1 instead of 80 in the case of x_1. Following is the final simplex tableau (Table 4.10) with such change.

According to optimality condition, net evaluation row values needs to be equal to or less than zero. Thus:

i. $-0.17c_1 - 19.2 \leq 0$

$$c_1 \geq -112.94$$

TABLE 4.10
Allowable Range of Basic Variable

Decision Variables		x_1	x_2	x_3	s_1	s_2	s_3	
Coefficients of Decision Variables in Objective Function		c_1	40	30	0	0	0	
Basis Variables	**Coefficients of Basis Variables c_b in Objective Function**	**Coefficients of Constraint Functions**						**Maximum Available Resource (b_i)**
x_1	c_1	1	0	0.17	1.14	−0.07	0	1.05
x_2	40	0	1	1.23	−0.33	0.16	0	1.12
s_3	0	0	0	−2.09	−2.5	−1.24	1	3.5
	z_j	c_1	40	$0.17c_1 + 49.2$	$1.14c_1 - 13.2$	$-0.07c_1 + 6.4$	0	
Net evaluation row	$c_j - z_j$	0	0	$-0.17c_1 - 19.2$	$-1.14c_1 + 13.2$	$0.07c_1 - 6.4$	0	

ii. $-1.14c_1 + 13.2 \leq 0$

$$c_1 \geq 11.57$$

iii. $0.07c_1 - 6.4 \leq 0$

$$c_1 \leq 91.42$$

As c_1 indicates the profit from selling a unit of merchandise to geographic area A, it cannot be negative. Thus, from the above inferences, the value of c_1 would lie between

$$11.57 \leq c_1 \leq 91.42$$

Hence, this illustration entails that the range of objective function coefficient of a basic variable over which current solution would remain optimal is a minimum of 11.57 and a maximum of 91.42. This implies that store would earn profits from area A in the range of \$11.57–\$91.42 by selling one unit while the solution is optimal.

To illustrate optimality of solution, let's *change the profit from area A to 90 instead of 80* and see its effect on objective function as well as optimality of the new solution (Table 4.11).

In the initial solution, the values of decision variables x_1 and x_2 were 1.05 and 1.12 making

$$Z = 80x_1 + 40x_2 + 30x_3$$

$$= 80 * 1{,}050 + 40 * 1{,}120 + 30 * 0 = 128{,}800$$

TABLE 4.11
Revised Solution

Decision Variables		x_1	x_2	x_3	s_1	s_2	s_3	
Coefficients of Decision Variables in Objective Function		**90**	**40**	**30**	**0**	**0**	**0**	
Basis Variables	**Coefficients of Basis Variables c_b in Objective Function**	**Coefficients of Constraint Functions**						**Maximum Available Resource (b_i)**
x_1	90	1	0	0.17	1.14	−0.07	0	1.05
x_2	40	0	1	1.23	−0.33	0.16	0	1.12
s_3	0	0	0	−2.09	−2.5	−1.24	1	3.5
	z_j	90	40	64.5	89.4	0.1	0	
Net evaluation row	$c_j - z_j$	0	0	−34.5	−89.4	−0.1	0	

whereas in the new solution, when c_1 has changed to 90 making

$$Z = 90*1{,}050 + 40*1{,}120 + 30*0 = 139{,}300$$

Thus, total profits have shown a change that, in this case, was an increase. The point to be emphasized is that with the change in coefficient there has been a change in objective function, making it a *sensitive parameter.* However, the initial solution has changed, but the new solution was found to be both optimal and feasible as all RHS values are positive, making the new solution feasible. Also, all values in the net evaluation row are either zero or negative, making the solution optimal. Finally, by putting new values of basic variables in constraint equations, it was found that all constraints were satisfied, substantiating the feasibility of solution.

Now let's again illustrate the importance of finding a *range of optimality by taking c_1 as 100* which is out of allowable range. The final solution would be (Table 4.12):

In this case, the value of Z would be Z = 100*1,560 + 40*0 + 30*0 = 156,000. The solution has changed significantly as the number of units sold in area B has reduced to nil and the optimality of new solution has achieved in a lesser number of iterations, making the initial solution non-optimal. Also, checking the feasibility of solution would reveal that one of the constraints is not met. It has been shown below.

$$\text{Constraint } 1: 9x_1 + 4x_2 + 10x_3 \leq 14$$

$$9*1.56 + 4*0 + 10*0 = 14.10$$

This is more than RHS violating the constraint

TABLE 4.12
Outside Range

Decision Variables		x_1	x_2	x_3	s_1	s_2	s_3	
Coefficients of Decision Variables in Objective Function		**100**	**40**	**30**	**0**	**0**	**0**	**Maximum Available Resource**
Basis Variables	**Coefficients of Basis Variables c_b in Objective Function**	**Coefficients of Constraint Functions**						**(b_i)**
x_1	100	1	0.44	0.71	1	0	0	1.56
s_2	0	0	6.12	7.58	−2	1	0	6.88
s_3	0	0	7.8	7.45	−5	0	1	12.2
	z_j	100	44	71	100	0	0	
Net evaluation row	$c_j - z_j$	0	−4	−41	−100	0	0	

$$\text{Constraint } 2: 2x_1 + 7x_2 + 9x_3 \leq 10$$

$$2*1.56 + 7*0 + 9*0 = 3.12$$

which is less than RHS satisfying the constraint

$$\text{Constraint } 3: 5x_1 + 10x_2 + 11x_3 \leq 20$$

$$5*1.56 + 10*0 + 11*0 = 7.8$$

which is less than RHS satisfying the constraint

Thus, as the c_1 value keeps going away from the maximum range value of 91.42, solution would increasingly become infeasible.

Thus, the above cases do emphasize the importance of conducting post-optimality analysis in the form of sensitivity analysis for primarily identifying sensitive parameters. This was done by taking three cases that were as follows; first, changes in RHS values of a linear programming problem. This case is most important as such values indicate maximum resources available, which are only estimates at initial stages. Knowledge about changes in any of these variables causing significant change in objective function proves immensely helpful in decision-making. The second and third cases were similar as coefficients of objective function were modified to see their effect on optimality and feasibility of initial solution. The only difference was that in the second case, only the coefficient of non-basic variable in the objective function was modified, whereas in the third case, it was the coefficient of basic variable. In all such cases, the procedure involved finding an initial solution, estimating the allowable range within which the initial solution remains feasible and optimal and then finally illustrating the importance of allowable range by finding a solution with a value outside the range. Importantly, sensitivity analysis is performed only by modifying one of the coefficients by keeping other coefficients assumed to be constant.

Excel Solution

Sensitivity analysis through Excel application is explained by using the same case of sales maximization as illustrated in the previous section. The model and its solver solution have been reproduced here (Figure 4.1).

$$\textbf{Maximize Profit} \quad Z = 80x_1 + 40x_2$$

$$\text{Constraint } 1: 9x_1 + 4x_2 \leq 14$$

$$\text{Constraint } 2: 2x_1 + 7x_2 \leq 10$$

$$\text{Constraint } 3: 5x_1 + 10x_2 \leq 20$$

$$x_1, x_2, x_3 \geq 0$$

To get a sensitivity report before achieving the final solution in 'solver results' dialog box, select sensitivity and then click OK as shown in Figure 4.2.

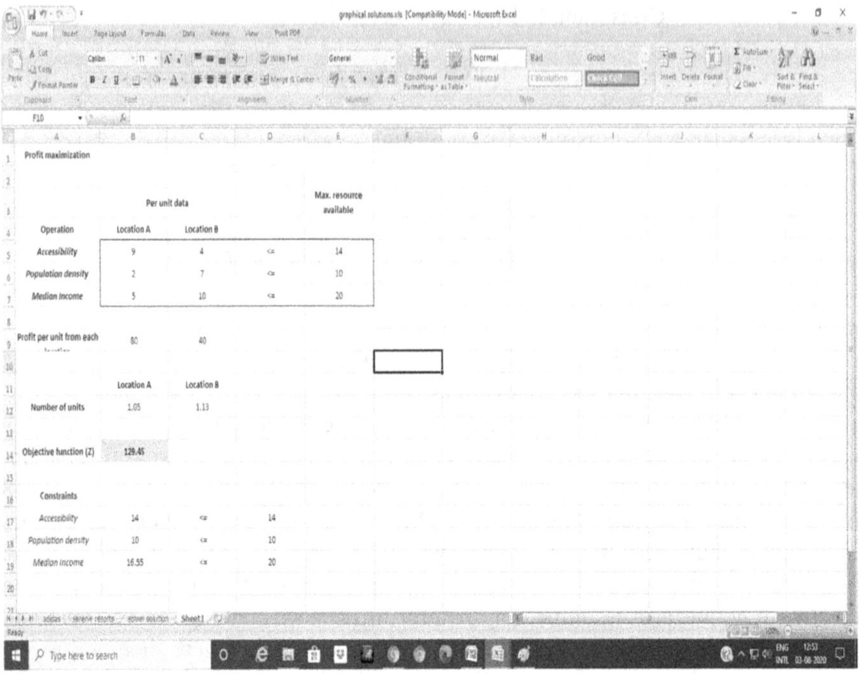

FIGURE 4.1 Excel solution.

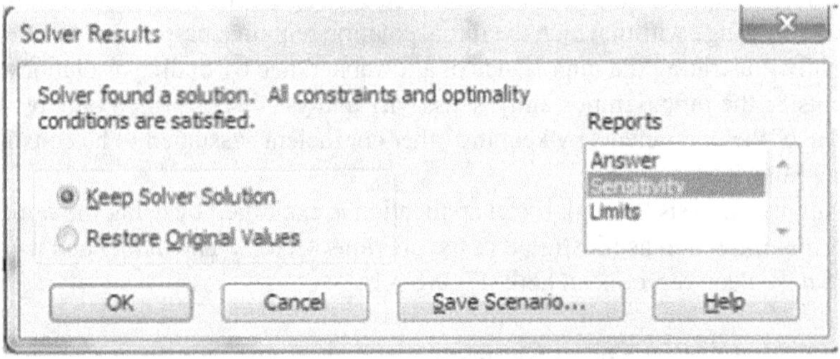

FIGURE 4.2 Solver results.

The sensitivity report generated is shown in Figure 4.3.

The 'Final value' column shows the number of units sold in thousand from two locations, which would result in maximum profit. The reduced cost column implies change in objective function value if corresponding value of decision variable changes. As these values are zero, there would be no change in objective function value even if the number of defective items (decision variables) changes. So, decision variable values are optimal values. Other three column values provide an optimal range of decision variables, i.e. the range of values of variables within which the

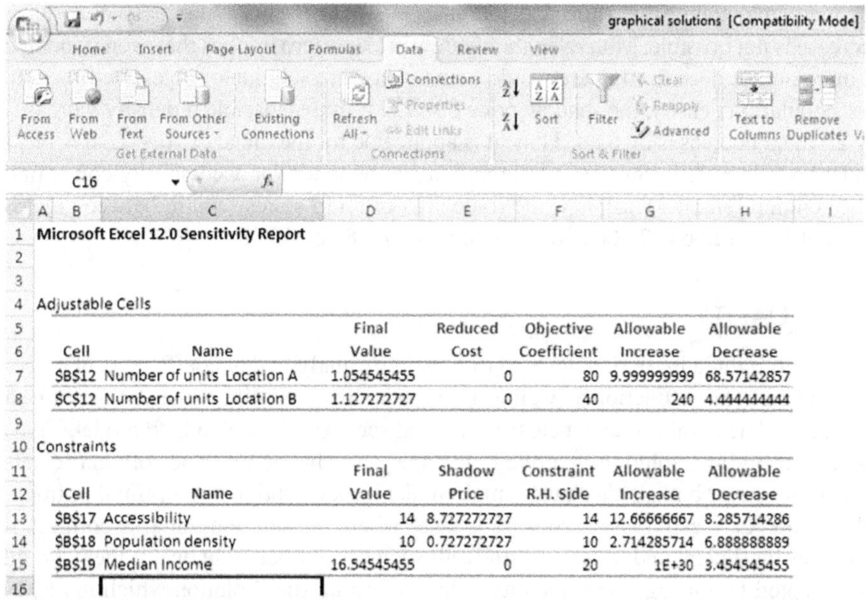

Microsoft Excel 12.0 Sensitivity Report

Adjustable Cells

Cell	Name	Final Value	Reduced Cost	Objective Coefficient	Allowable Increase	Allowable Decrease
$B$12	Number of units Location A	1.054545455	0	80	9.999999999	68.57142857
$C$12	Number of units Location B	1.127272727	0	40	240	4.444444444

Constraints

Cell	Name	Final Value	Shadow Price	Constraint R.H. Side	Allowable Increase	Allowable Decrease
$B$17	Accessibility	14	8.727272727	14	12.66666667	8.285714286
$B$18	Population density	10	0.727272727	10	2.714285714	6.888888889
$B$19	Median Income	16.54545455	0	20	1E+30	3.454545455

FIGURE 4.3 Sensitivity report.

objective function remains optimal. For ward A, allowable increase of 10 (approximated from 9.99) and decrease of 68.57 gives optimal range as:

$$80 + 10 \leq \text{Sales (Ward A)} \leq 80 - 68.57$$

$$90 \leq \text{Sales (Ward A)} \leq 11.43$$

Similarly, for ward B, the optimal range would be:

$$40 + 240 \leq \text{Sales (Ward B)} \leq 40 - 4.44$$

$$280 \leq \text{Sales (Ward B)} \leq 35.56$$

The study of first column of constraint report indicates the maximum usage of resource available to achieve the optimal solution. Accessibility within 14 miles, population density of 10,000 households, and median income of 20,000 per household in a given ward would result in the achievement of maximum possible profits.

The shadow price indicates a change in the value of the solution as the per unit increases on the right side of constraint. For instance, if accessibility increases from 14 to 15, then the effect on profit would be positive 8.72. Similarly, if the population to which stores are catering increases from maximum of 10 to 11, then the effect on objective function would be positive 0.727. However, any change in resources available pertaining to median income would have zero effect on objective function.

The last two columns indicate the range of the right side of the constraint within which the shadow price of the corresponding constraint remains feasible. For instance, for median income, the allowable increase is somewhat frivolous as such a huge increase is not possible. Moreover, the shadow price of zero implies that even if income is increased or decreased, there would be no change in solution or objective function. For population density, the shadow price of 0.727 remains feasible if density increases by 2.71 and decreases by 6.88. Thus, it is applicable for the range of 10 + 2.71 = 12.71 and 10 – 6.88 = 3.12. Similar interpretation can be made for accessibility constraint. The solution remains feasible if the trade area of retail stores remain between maximum of 14 + 12.66 = 26.66 and minimum of 14 – 8.28 = 5.72.

4.4 DUALITY

It was highlighted in section devoted to sensitivity analysis that coefficients of objective and constraint function as well as parameters in constraint equations are at best *estimates*. These values depict close to the real scenario through a LPP model. Thus, the initial optimal solution is prone to change with change in values of coefficients or parameters or both. This change might make initial solution non-optimal or infeasible. Now there can be numerous such cases where the solution remains optimal. It becomes inefficient and tedious to study all such cases when only one of them would be accepted by management. Duality helps in finding such solution which indicates changed values of coefficients and parameters if any, which would provide an optimal and feasible solution. Such a change would have a significant effect on the objective function of either maximization or minimization. The change in these values, as was discussed earlier, represents economic activity in terms of shadow prices. Thus, duality theory helps in analysing and interpreting the optimality of the initial solution by identifying any relevant modification in economic activity. For instance, in the previous section, it was discussed that with increase in marginal value of right side of constraints, there is a significant impact on optimality of solution. So, by application of duality, it is possible to determine which solution is optimal and what changes, if any, should be made to the original solution.

This purpose is attained by expressing the original LPP in another form, which is considered to be the mirror and opposite image of original problem. The original linear programming problem is called *primal* and its mirror image is called as *dual*. To understand duality theory's application, the following sections discuss the structure of dual problem, its relation with primal problem and its economic interpretation.

4.4.1 Construction of Dual Problem

As a dual problem is a mirror image of primal, its construction involves the same elements though used in different formats. The elements of a primal LPP were:

Z: is an objective function of either maximization or minimization

x_j: called as decision variables indicating level of activity where $j = 1, 2, 3...$

c_j: coefficients of decision variables in an objective function implying unit profit or cost from each activity

b_i: are RHS values of constraint equation indicating maximum resource available.

a_{ij}: are coefficients of level of activity in constraint equation indicating the amount of resource i consumed by each unit of activity j.

By using these elements of a primal linear programming model of profit maximization as illustrated in the beginning of the previous section was:

Primal Problem:

$$\textbf{Maximize } Z = 80x_1 + 40x_2$$

Subject to:

$$\text{Accessibility}: 9x_1 + 4x_2 \leq 14$$

$$\text{Population density}: 2x_1 + 7x_2 \leq 10$$

$$\text{Median Income}: 5x_1 + 10x_2 \leq 20$$

and $x_1, x_2 \geq 0$

The construction of a dual problem from a primal problem should follow the following steps:

1. **RHS values** of constraint functions of primal problem are the objective function coefficients of dual problem, i.e. b_i of primal becomes c_j of dual. *Similarly, c_j of primal would become b_i of dual.* As there are only two such coefficients in above primal problem, so in the dual problem, instead of three RHS values of three constraints, only two constraints would be created with coefficients of objective functions (c_j) as right and side values (b_i) of the dual problem.
2. **Maximization primal problem would become minimization dual problem** and vice versa. The reason for this is very simple. The new solution as indicated by solving dual problem would be feasible if the primal problem is optimal. Thus, the optimality of primal corresponds to the feasibility of dual. As a result, the optimal value of Z in primal is *minimum* feasible in dual, so it should be minimized in the case of maximization of primal.
3. **Number of decision variables** in the primal problem would be the same as the number of constraints in the dual problem. In this case, there are two decision variables in the primal problem, i.e. x_1 and x_2. Thus, the dual problem will have two constraints instead of three as were in primal.
4. **Number of constraints** in the primal problem is the same as the number of decision variables in the dual problem. In the primal problem, this case has three constraint functions. Thus, a dual problem will have three variables called *dual variables* in the objective function instead of two. Dual variables are represented as y_1, y_2 and y_3 pertaining to each of the constraints. In the above illustration, y_1 represents accessibility constraint, y_2 population density constraint and y_3 represents median income constraint.
5. The inequalities would be reversed, i.e. as dual is opposite and mirror of primal, less than inequality in the primal would convert to greater than equal to in the dual.

6. The coefficients of decision variables in the constraint functions of the dual problem remains as the coefficients of variables in the constraint functions of primal problem.

The formation of dual from primal as discussed above can be summarized as:

Primal Problem	Dual Problem
Objective function coefficients	Objective function coefficients
Right side values of constraint functions	Right side values of constraint functions
Coefficients of variables in constraint functions	Coefficients of variables in constraint functions
Maximization objective function	Maximization objective function
Minimization objective function	Minimization objective function
Greater than inequalities	Greater than inequalities
Lesser than inequalities	Lesser than inequalities
Number of decision variables	Number of decision variables
Number of constraints	Number of constraints

Understanding of these concepts would lead to formulation of the dual problem which is shown as under.

Dual Problem:

$$\textbf{Minimize}\quad Z = 14y_1 + 10y_2 + 20y_3$$

Subject to:

$$C1: 9y_1 + 2y_2 + 5y_3 \geq 80$$

$$C2: 4y_1 + 7y_2 + 10y_3 \geq 40$$

and $y_1, y_2, y_3 \geq 0$

4.4.2 Relationship between Primal and Dual Problem

The simplex table of the dual problem from primal according to above-mentioned changes can be shown in the following way (Table 4.13):

It is important to note that in simplex table of primal problem:

1. There are three constraint equations, whereas the dual problem has only two.
2. s_1, s_2 and s_3 are slack variables as inequalities were of less than or equal to, whereas in dual, s_4 and s_5 are surplus variables as inequalities were greater than or equal to.
3. **Relationship between basic and non-basic variables of primal and dual***:* As per definition, variables with zero value are termed as non-basic, whereas those with some value in optimal solution are termed as basic

TABLE 4.13
Formation of Dual Problem

	Primal Problem of Maximization							Dual Problem of Minimization					
	x_1	x_2	s_1	s_2	s_3			y_1	y_2	y_3	s_4	s_5	
Coefficients	c_1	c_2	0	0	0	b_i		b_1	b_2	b_3	0	0	b_i
Basic Variables	**Coefficients of Constraint Functions**							**Coefficients of Constraint Functions**					
s_1	a_{11}	a_{12}	1	0	0	b_1	s_4	a_{11}	a_{21}	a_{31}	−1	0	c_1
s_2	a_{21}	a_{22}	0	1	0	b_2	s_5	a_{12}	a_{22}	a_{32}	0	−1	c_2
s_3	a_{31}	a_{32}	0	0	1	b_3							
z_j	z_1	z_2	z_3	z_4	z_5								

variables. Also, according to duality theory, which is mirror image of primal, variables of dual are complementary to primal, i.e. basic variables of primal would be considered as non-basic of dual, whereas non-basic of primal would be basic of dual. The initial solution, as shown in the previous chapter, while calculating LPP by using the Simplex method showed that x_1 and x_2 are non-basic variables. In the dual problem, s_4 and s_5 i.e. surplus variables are basic so coefficients of decision variables (x_1 and x_2) in z_j row of primal (z_1 and z_2) gives corresponding values of surplus variables of dual i.e. s_4 and s_5. This is because, as indicated in iterations 0 of simplex tableau, z_j row contains coefficients of decision variables x_1 and x_2 in the form of z_1 and z_2, respectively, and coefficients of slack variables s_1, s_2 and s_3 in the form z_3, z_4 and z_5, respectively. Similarly, slack variables s_1, s_2 and s_3 in initial primal solution are basic which would transform into non-basic in initial dual solution, i.e. y_1, y_2 and y_3. So coefficients of slack variables in z_j row of primal (z_3, z_4 and z_5) would correspond to objective function values of the dual problem. This is because a slack variable represents the amount of resource consumed or remaining resource. For instance, if out of a maximum of 100 hours of production time, manufacturing of one unit takes 3 hours, so the remaining 97 hours is represented by a slack variable. It can also be seen as amount of production time consumed indicating the level of activity, i.e. a decision variable. Hence, it is safe to assume that slack variables of primal correspond to decision variables of dual. This transformation is shown in Table 4.14.

4. **Feasibility of dual**: Furthermore, as indicated in the above discussion, values of surplus variables s_4 and s_5 are represented by coefficients of decision variables in z_j row of primal. For dual to be feasible, these surplus variables should fulfil criteria of non-negativity, i.e. they must be either zero or greater than zero. So if these coefficients in primal are positive, then it can be inferred that dual is feasible. Also then, values of these coefficients can be put in constraint equations to check whether constraints are satisfied simultaneously.

TABLE 4.14
Transformation of primal into dual

Primal Problem			Corresponding Dual Problem
Decision variables (non-basic)	x_1	s_4	Surplus variables (basic)
	x_2	s_5	
Slack variables (basic)	s_1	y_1	Decision variables (non-basic)
	s_2	y_2	
	s_3	y_3	

As both primal and dual problems are forms of the same LPP, the values of decision variables, objective function and constraint functions should be the same. But if they vary from each other, then the difference between these values, especially objective function, would indicate shadow price as was discussed in sensitivity analysis. Identifying these variations and finding the most suitable one is the function of duality theory. This is done by solving both primal and dual through the simplex method and checking feasibility and optimality of both solutions at each iteration.

This procedure would be understood more by explaining an illustration in the next section.

4.4.3 Dual Problem of Standard LPP

The above-shown model of sales maximization is a standard form of a linear programming problem. A maximization problem involving functional constraints with only less than or equal to ($\leq$) inequalities is considered to be a standard form. As discussed in Chapter 3, on the simplex method, only slack variables are added to constraint equations to convert inequalities to equalities. Furthermore, these slack variables representing the amount of remaining resources should fulfil criteria of non-negativity. On the other hand, a minimization problem cannot be constructed by definition without greater than equal to ($\geq$) or equal to ($=$) inequalities. These inequalities imply a minimum value that an activity can achieve optimally. In addition, surplus variables are subtracted from the constraint equations to make left-hand side equal to RHS and apply the simplex method. But when decision variables are assumed as non-basic variables, then these surplus variables violates criteria of non-negativity. To rectify, this artificial variable is added to the equation, the value of which would be greater than surplus variables. Due to the involvement of artificial variable because of the mentioned reason, the LPP form is considered as non-standard.

Primal problem:

$$\textbf{Maximize } Z = 80x_1 + 40x_2$$

$$\text{Constraint 1}: 9x_1 + 4x_2 \leq 14$$

$$\text{Constraint 2}: 2x_1 + 7x_2 \leq 10$$

$$\text{Constraint } 3: 5x_1 + 10x_2 \leq 20$$

$$x_1, x_2, x_3 \geq 0$$

Dual problem:

$$\textbf{Minimize } Z = 14y_1 + 10y_2 + 20y_3$$

$$\text{Constraint } 1: 9y_1 + 2y_2 + 5y_3 \geq 80$$

$$\text{Constraint } 2: 4y_1 + 7y_2 + 10y_3 \geq 40$$

$$y_1, y_2, y_3 \geq 0$$

Solving of such form and understanding its relationship with dual is analysed below.

Iteration 0:

In the dual problem, inequalities were transformed into greater than or equal to so to fulfil non-negativity criteria artificial variables were added, which has been shown as basic variables (Table 4.15).

Solution:

In the initial solution, according to values in the z_j row of primal $x_1 = x_2 = s_1 = s_2 = s_3 = 0$. This is indicative of the fact that at first stage, no sales have occurred from both areas. In addition, the value of decision variables of dual y_1, y_2 and y_3 observed from corresponding coefficients of slack variables of primal is also zero. Thus, the value of objective function of dual Z* is also zero.

Feasibility: The primal is feasible as all RHS values are positive. Also, putting values of x_1 and $x_2 = 0$ in constraints shows acceptance of all constraints, making the solution feasible. This is shown below:

TABLE 4.15
Iteration 0

		Primal Problem of Maximization							Dual Problem of Minimization					
		x_1	x_2	s_1	s_2	s_3				y_1	y_2	y_3	s_4	s_5	
		80	40	0	0	0	b_i			14	10	20	0	0	b_i
Basic Variables		Coefficients of Constraint Functions						Basic Variables		Coefficients of Constraint Functions					
s_1	0	9	4	1	0	0	14	a_4	0	9	2	5	−1	0	80
s_2	0	2	7	0	1	0	10	a_5	0	4	7	10	0	−1	40
s_3	0	5	10	0	0	1	20	z_j		0	0	0	0	0	
z_j		0	0	0	0	0									

i. $9x_1 + 4x_2 \leq 14$ *would give* $0 \leq 14$
ii. $2x_1 + 7x_2 \leq 10$ *would give* $0 \leq 10$
iii. $5x_1 + 10x_2 \leq 20$ *would give* $0 \leq 20$

Positive surplus variables obtained from corresponding slack variables of primal indicate *feasibility of dual*. As all these values are zero at this stage, fulfilling non-negativity criteria, thus, dual can be said to be feasible. But feasibility also has to be checked by putting values of decision variables of dual in its constraint. All the constraints are of greater than inequality, so putting values of zero in y_1, y_2 and y_3 would lead to violation of constraint.

i. $9y_1 + 2y_2 + 5y_3 \geq 80$ *would give* $0 \geq 80$
ii. $4y_1 + 7y_2 + 10y_3 \geq 40$ *would give* $0 \geq 40$

and $y_1, y_2, y_3 \geq 0$

This implies at iteration 0 stage dual is not feasible which is because of non-optimality of primal.

Optimality: Primal problem is not optimal because a better solution can further be achieved, which is indicated by positive $c_j - z_j$ values.

Rule 1: *Hence, this results in formation of a rule, which says that if a problem is feasible and not optimal, then its counterpart problem would be infeasible.*

This is checked in all further iterations as well.

Iteration 1:

See Table 4.16.

Solution: Primal problem after first iteration gives the value of decision variables as $x_1 = 1.56$ and $x_2 = 0$ which would give $Z = 80x_1 + 40x_2 = 124.8$. Values of decision variables in dual problem (y_1, y_2 and y_3) are indicated by $s_1 = 8.8$, $s_2 = 0$ and $s_3 = 6.6$. Putting these values in the objective function of dual $Z^* = 14y_1 + 10y_2 + 20y_3 = 14*8.8 + 10*0 + 20*0 = 123.2$. Difference between $Z^* - Z = 124.8 - 123.2 = 0.8$. The difference in objective function values indicates a change in economic activity. However at first iteration in this case, the difference is small.

Feasibility: The primal solution values on the RHS are positive, and putting these values of decision variables in constraint equation would give:

i. $9x_1 + 4x_2 \leq 14$
   $9*1.56 + 4*0 = 14.04$ constraint is violated
ii. $2x_1 + 7x_2 \leq 10$
   $2*1.56 + 7*0 = 3.12$ constraint is not violated
iii. $5x_1 + 10x_2 \leq 20$
   $5*1.56 + 10*0 = 7.8$ constraint is not violated

As all constraints are *not met* simultaneously, the first iteration of primal problem is *infeasible*.

Feasibility of *dual* is indicated by positive surplus variable (s_4, s_5) values, which are indicated by $x_1 = 80$ and $x_2 = 35.2$ of primal. Out of these two surplus variables,

TABLE 4.16
Iteration 1

Primal Problem

Decision variables		x_1 80	x_2 40	s_1 0	s_2 0	s_3 0	b_i
Basis Variables		**Coefficients of Constraint Functions**					
x_1	80	1	0.44	0.11	0	0	1.56
s_2	0	0	6.12	−0.22	1	0	6.88
s_3	0	0	7.78	−0.56	0	1	12.22
z_j		80	35.2	8.8	0	0	
$c_j - z_j$		0	4.8	−8.8	0	0	

Dual Problem

		y_1 14	y_2 10	y_3 20	s_4 0	s_5 0	a_4 M	b_i
Basis Variables		**Coefficients of Constraint Functions**						
a_4	M	7	−1.5	0	−1	0.5	1	60
y_3	20	0.4	0.7	1	1	−0.1	0	4
	z_j	7M + 8	−1.5M + 14	20	−M	0.5M − 2	M	
	$c_j - z_j$	6 − 7M	−4 + 1.5M	0	M	2 − 0.5M	0	

the first iteration provides s_4 in the form of artificial variable a_4 to be part of solution. Now s_4 of dual being equivalent to x_1 of primal implies that out of 80 units 60 have been sold and contributed to profit, whereas 20 are left as surplus available. However, second decision variable x_2 represented by s_5 (artificial variable a_5) has left the solution so at this stage sales have occurred only from one store. Finally, value of y_3 is an incoming variable, which is similar to s_3. Thus, in dual, out of three decision variables, only two would remain in the final solution, implying that these variables would contribute to objective function. This was the case for primal as well where only two variables were contributing to profit.

Feasibility is further checked by putting these values in constraint equations of dual problem. So:

i. $9y_1 + 2y_2 + 5y_3 \geq 80$
   $9*8.8 + 2*0 + 5*0 = 79.2$ is not greater than RHS *constraint is violated*

ii. $4y_1 + 7y_2 + 10y_3 \geq 40$
   $4*8.8 + 7*0 + 10*0 = 35.2$ is not greater than RHS *constraint is violated*
   and $y_1, y_2, y_3 \geq 0$

Thus, dual solution after first iteration is not feasible.

Optimality: All values in net evaluation row of primal are either not zero or negative, which implies that solution can be improved further. The $c_j - z_j$ value in the second column of primal problem is 4.8, implying that one unit of sale in area A would result in increase of profit of $4.8. Thus, the current solution of iteration 1 is not optimal, making corresponding dual infeasible.

Rule 2: *Hence, if a problem is infeasible (either primal or dual), then other problem would be either non-optimal or infeasible.*

Iteration 2:

See Table 4.17.

TABLE 4.17
Iteration 2

Primal Problem

Decision variables		x_1	x_2	s_1	s_2	s_3	
		80	40	0	0	0	b_i
Basis Variables		**Coefficients of Constraint Functions**					
x_1	80	1	0	0.127	−0.07	0	1.06
x_2	40	0	1	−0.04	0.16	0	1.12
s_3	0	0	0	−0.25	−1.24	1	3.46
z_j		80	40	8.56	0.8	0	
$c_j - z_j$		0	0	−8.56	−0.8	0	

Dual Problem

		y_1	y_2	y_3	s_4	s_5	
		14	10	20	0	0	b_i
Basis Variables		**Coefficients of Constraint Functions**					
y_1	14	1	−0.21	0	−0.14	0.07	8.55
y_3	20	0	0.78	1	0.005	−0.13	0.58
z_j		14	12.66	20	−1.86	−1.62	
$c_j - z_j$		0	−2.66	0	1.86	1.62	

Solution: With $x_1 = 1.06$ and $x_2 = 1.12$ value of Z = 80*1.06 + 40*1.12 = 129.6, whereas with $s_1 = y_1 = 8.56$, $s_2 = y_2 = 0.8$ and $s_3 = y_3 = 0$ value of Z*= 14*8.56 + 10*0.8 + 20*0 = 127.84. Thus, there is a marked difference between the value of objective function of primal and dual implying sensitivity of these parameters. This only indicates that an optimal solution lying between 129.6 and 127.84 can be achieved by modifying decision variables of a given problem. How much they can be modified indicating their allowable range was discussed in sensitivity analysis section. This can be meaningfully understood by observing the dual simplex table. According to this solution, the value of y_1= 8.55 and y_3= 0.58 which would give Z* = 14*8.56 + 10*0 + 20*0.58 = 131.44. This value lies outside the range of 129.6 and 127.84, making the solution non-optimal.

Feasibility: Primal is feasible because the values of x_1 and x_2 in the constraint equation fulfil criteria of inequalities as shown below:

i. $9x_1 + 4x_2 \leq 14$
   9*1.06 + 4*1.12 = 14 constraint is not violated
ii. $2x_1 + 7x_2 \leq 10$
   2*1.06 + 7*1.12 = 9.96 constraint is not violated
iii. $5x_1 + 10x_2 \leq 20$
   5*1.06 + 10*1.12 = 16.5 constraint is not violated

Also all values in b_i column are positive, making it feasible.

To check feasibility of Dual solution: checking the constraints gave following results:

i. $9y_1 + 2y_2 + 5y_3 \geq 80$
   9*8.56 + 2*0.8 + 5*0 = 78.64 is not greater than RHS *constraint is violated*
ii. $4y_1 + 7y_2 + 10y_3 \geq 40$
   4*8.56 + 7*0.8 + 10*0 = 39.84 is not greater than RHS *constraint is violated*
   and $y_1, y_2, y_3 \geq 0$

So, dual solution is not feasible.

Optimality: All values in net evaluation row of primal are either zero or negative, which implies that the solution cannot be improved further, making the solution of iteration 2 as optimal. However, a dual solution is not optimal. Net evaluation row indicates that the objective of minimization can be further improved by including y_2 into the solution. A value of –2.66 implies inclusion of one unit of y_2 would minimize the objective by 2.66 units.

As a result, in the third iteration, only the dual solution is presented in Table 4.18.

$$Z^* = 14*8.72 + 10*0.74 + 20*0 = 129.48$$

To check feasibility of dual solution: checking the constraints gave following results:

TABLE 4.18
Iteration 3 of Dual Problem

		y_1	y_2	y_3	s_4	s_5	
		14	10	20	0	0	b_i
Basis Variables		**Coefficients of Constraint Functions**					
y_1	14	1.21	0	0.27	−0.14	0.004	8.72
y_2	10	0	1	1.28	0.006	−0.16	0.74
z_j		16.94	10	16.58	−1.9	−1.54	
$c_j - z_j$		−2.94	0	3.42	1.9	1.54	

i. $9y_1 + 2y_2 + 5y_3 \geq 80$
$9*8.72 + 2*0.74 + 5*0 = 8$ is equal to RHS *constraint is not violated*
ii. $4y_1 + 7y_2 + 10y_3 \geq 40$
$4*8.72 + 7*0.74 + 10*0 = 40.06$ is equal to RHS *constraint is not violated*
and $y_1, y_2, y_3 \geq 0$

Negative value pertaining to y_1 implies further scope for improvement of solution making solution in iteration 3 to be non-optimal. However, y_1 is already present in the solution. So, further iterations would not provide any significant improvement and only lead to cyclical incoming and outgoing of basic variables. Moreover, the solution was found to become feasible as compared to iteration 2 solution. Therefore, even if solution is considered to be non-optimal, it is feasible.

According to rule 2, one solution that is infeasible would result in either infeasibility or non-optimality of other solution. But in this case, the primal solution is both optimal and feasible, whereas the dual is both feasible and non-optimal. So what has gone wrong when, according to simplex method, final iteration provides a feasible and optimal solution? The answer lies in comparison of the values of objective function of both solutions, i.e. Z and Z*. As the objective of dual problem is of minimization, the simplex solution of it has shown the least possible value, and as the objective of primal was of maximization, the solution has given a maximum value. As discussed, all values lying within this range would be optimal. They would be feasible, at least for primal problem. This situation is mostly termed as *sub-optimality* and is acceptable. From Chapter 3, the primal solution (Z) was 129.6 and here the dual solution (Z*) is 129.48. The difference of 0.12 or 120 units is termed as shadow change in objective function. Ultimately, it could be concluded that both solutions are feasible and optimal, leading to the formation of another rule which is:

Rule 3: *If one problem is feasible and optimal, then the other problem is mirror image of it, so according to rule of complementary, the other problem would also be feasible and optimal. Also, it substantiates the rule of complementary relationship between two solutions, which implies that each optimal solution in a primal problem has a complementary optimal solution in form of a dual problem.*

4.4.4 Dual Problem of Non-standard LPP

The above procedure can also be applied for minimization primal as its complementary dual would automatically switch to maximization under property of both being symmetrical. Also, inequalities in function equations would undergo a reversal from primal to dual. Thus, a minimization problem should consist of only ≥ constraint functions, whereas a maximization primal should have only ≤ inequalities. Such forms would be standard forms. Till here it has been established in the previous section. But in the case of non-standard form of a LPP (a minimization problem) that includes both a less than and an equality constraint in addition to greater than or equal to, the following operations are performed to convert it into standard form.

Step 1: If constraint function has inequality of less than or equal to (≤), then multiply both right and left sides by −1 so that inequality is converted to greater than or equal to (≥).

Step 2: If inequality in constraint function is of greater than or equal to (≥), then it should remain the same as it is already in standard form.

Step 3: Equality constraint (=) in constraint function of primal problem is treated similar to ≤ inequality. The only change is that in dual problem, non-negativity criteria of corresponding dual variable are deleted. For example, if constraint equation is:

$x_1 + x_2 = 1$, then two inequalities can be created for it: one with ≥ and other with ≤. This is quite simple logic because if both forms are combined it would result in equality (=). So, the equation can be written as:

$$x_1 + x_2 \geq 1$$

$$x_1 + x_2 \leq 1$$

Now equation with ≤ constraint is multiplied by −1 to make it ≥. So, instead of one constraint we have two constraints:

$$x_1 + x_2 \geq 1$$

$-x_1 - x_2 \geq -1$. This can also be written as $-(x_1 + x_2) \geq -1$. This proves that an equality constraint would yield a corresponding dual value which would be negative. *Thus non-negativity for a primal variable with equality constraint in its corresponding dual form does not hold. This makes such variable unconstrained.* To summarize in non-standard form of minimization problem, dual variable (y_i) corresponding to:

- less than or equal to inequality (≤) should follow criteria of it being less than or equal to zero, i.e. $y_i \leq 0$;
- greater than or equal to inequality (≥) should be greater than or equal to zero i.e. $y_i \geq 0$ and finally
- equality constraint (=) it should be unconstrained, i.e. can take any value both positive and negative.

For better understanding of the above explanation, let's take the illustration of the capacity of a restaurant example given in section 2.5.2. For example, the objective was to minimize loss in revenue due to two types (x_1 and x_2) of seating arrangements going vacant. The optimal solution is intended to estimate the number of both types. The original model is shown below with only one change, i.e. instead of the minimum demand of A type of seating arrangement, the manager fixes the demand to be 20 seats. This would convert inequality of greater than or equal (≥) to equality (=) constraint. The other demand constraint of B type remains the same. Thus, *initial model* would be:

$$\textbf{Minimize} \quad Z = 10x_1 + 15x_2$$

Subject to following constraints:

Constraint 1 (Floor space):

$$16x_1 + 30x_2 \leq 1{,}000$$

Constraint 2 (Demand of A type):

$$1x_1 + 0x_2 = 20$$

Constraint 3 (Demand of B type):

$$0x_1 + 1x_2 \geq 15$$

Constraint 4 (Capacity):

$$1x_1 + 1x_2 \leq 40$$

And x_1, $x_2 \geq 0$.

This model is in a non-standard form as it consists of all types of inequalities. To convert it into the standard form of minimization, i.e. constraints with only ≥ inequalities above-mentioned steps are followed. It would yield:

The objective function remains same

$$\textbf{Minimize} \quad Z = 10x_1 + 15x_2$$

Constraints 1 and 4 are multiplied by −1 converting ≤ inequality into ≥. Constraint 3 would remain the same. Finally, as constraint 2 is of the equality type, two equations of it involving ≥ and ≤ inequalities would be created. Inequality of ≤ would be multiplied by −1 to convert it into ≥. Hence, constraints in *modified model* would be:

$$\text{Constraint 1}: -16x_1 - 30x_2 \geq -1{,}000$$

$$\text{Constraint 2}: 1x_1 + 0x_2 \geq 20 \text{ and}$$

$$\text{Constraint 3}: -1x_1 - 0x_2 \geq -20$$

$$\text{Constraint 4}: 0x_1 + 1x_2 \geq 15$$

$$\text{Constraint 5}: -1x_1 - 1x_2 \geq -40$$

To construct *dual problem*, assume y_1, y_2, y_3, y_4 and y_5 to be its decision variables, each associated with each constraint of primal. The following would be the *initial dual model* formed from a modified version of primal.

$$\text{Maximize } Z^* = -1{,}000y_1 + 20y_2 - 20y_3 + 15y_4 - 40y_5$$

$$\text{Constraint 1}: -16y_1 + 1y_2 - 1y_3 + 0y_4 - 1y_5 \leq 10$$

$$\text{Constraint 2}: -30y_1 + 0y_2 - 0y_3 + 1y_4 - 1y_5 \leq 15$$

- y_1 is variable associated with constraint 1 which was modified by multiplying with −1. Thus, $y_1 \leq 0$.
- Constraints 2 and 3 were associated with equality constraint thus variables y_2 and y_3 are unconstrained.
- Constraint 4 as has remain unchanged so $y_4 \geq 0$.
- Constraint 5 for the same reason as in constraint 1 makes $y_5 \leq 0$.

Solution through the Simplex method involves initial version of both primal and dual models. This is done as under wherein feasibility and optimality are checked at each iteration.

Iteration 0:

In the primal model, x_1 and x_2 are decision variables, s_1 and s_4 are slack, a_2 and a_3 are artificial and s_3 is surplus variable (Table 4.19). In dual problem, y_1, y_2, y_3, y_4 and y_5 are formed to be decision and s_5 and s_6 as slack variables. As dual is equivalent of primal, so according to the rule of symmetry, correspondence between primal and dual variables would be in following manner: coefficients of x_1 and x_2 indicated in z_j row would reflect values of s_5 and s_6; $s_1 = y_1$; $a_2 = y_2 - y_3$; $a_3 = y_4$; $s_4 = y_5$. It is important to note that

- as constraint of equality was divided into two equations so it has been represented by two dual variables y_2 and y_3. Thus, in optimal solution of primal, value of a_2 would indicate difference between y_2 and y_3 in optimal solution of dual.
- according to Big M method as a constant value M is added to the coefficient of each artificial variable to adhere to constraint of non-negativity, so in dual problem, M is subtracted from the corresponding dual variable. This is shown below while obtaining dual solution.

TABLE 4.19
Iteration 0

Primal Problem

Decision variable		x_1	x_2	s_1	a_2	s_3	a_3	s_4	
		10	15	0	M	0	M	0	b_i
Basis Variable		**Coefficients of Constraint Functions**							
s_1	0	16	30	1	0	0	0	0	1,000
a_2	M	1	0	0	1	0	0	0	20
a_3	M	0	1	0	0	−1	1	0	15
s_4	0	1	1	0	0	0	0	1	40
z_j		M	M	0	M	−M	M	0	
$c_j - z_j$		10 − M	15 − M	0	0	M	0	0	

Dual Problem

		y_1	y_2	y_3	y_4	y_5	s_5	s_6	
		−1,000	20	−20	15	−350	0	0	b_i
		Coefficients of Constraint Functions							
s_5	0	−16	1	−1	0	−1	1	0	10
s_6	0	−30	0	0	1	−1	0	1	15
z_j		0	0	0	0	0	0	0	
$c_j - z_j$		−1,000	20	−20	15	−40	0	0	

Solution: With $x_1 = 0$ and $x_2 = 0$ value of Z (primal solution) = 0. Following table would give values of dual variables.

Coefficients of Primal Variables in z_j Row	Represents Variable in Dual	Value of Dual Variable
$x_1 = M$	s_5	$s_5 = M - M = 0$
$x_2 = M$	s_6	$s_6 = M - M = 0$
$s_1 = 0$	y_1	$y_1 = 0$
$a_2 = M$	y_2-y_3	$y_2 - y_3 = M - M = 0$
$a_3 = M$	y_4	$y_4 = M - M = 0$
$s_4 = 0$	y_5	$y_5 = 0$

Putting these values in dual objective function $Z^* = -1{,}000y_1 + 20(y_2 - y_3) + 15y_4 - 350y_5 = 0$. Thus both solutions are equal.

Feasibility: RHS values of primal solution are all positive making iteration 0 as feasible. Also putting values of $x_1 = 0$ and $x_2 = 0$ in constraint equations satisfies all inequalities except that of constraint 3 which has greater than or equal to inequality resulting in initial solution as infeasible. Doing the same procedure with dual shows that dual solution is feasible. Thus, in conclusion, the initial primal solution is infeasible, whereas its counterpart is feasible.

Optimality: Both solutions at iteration 0 level are non-optimal. Net evaluation row of primal indicates that objective of minimization can further be improved whereas same row of dual showed that objective of maximization can further be reached with better solution.

Thus in summary primal solution being infeasible makes either or both of solutions non-optimal. Such a situation indicates unboundedness, i.e. there is restriction on optimal value. This is because of addition of very big assumed value in the form of M. As it can have any value, so no fix solution can be achieved at this stage.

Iteration 1:

See Table 4.20.

Solution: With $x_1 = 20$ and $x_2 = 0$ value of Z (primal solution) = 2,000. As discussed above values of dual variables would be:

Coefficients of Primal Variables in z_j Row	Represents Variable in Dual	Value of Dual Variable
$x_1 = 10$	s_5	$s_5 = 10$
$x_2 = M$	s_6	$s_6 = M - M = 0$
$s_1 = 0$	y_1	$y_1 = 0$
$a_3 = M$	y_4	$y_4 = M - M = 0$
$s_4 = 0$	y_5	$y_5 = 0$

TABLE 4.20
Iteration 1

Primal Problem

Decision variable		x_1	x_2	s_1	a_2	s_3	s_4	
		10	15	0	0	M	0	b_i
Basis Variable		**Coefficients of Constraint Functions**						
s_1	0	16	0	1	0	1	0	550
a_2	M	1	0	0	1	0	0	20
x_2	15	0	1	0	0	−1	0	15
s_4	0	1	0	0	0	1	1	25
z_j		M	15	0				
$c_j - z_j$		0	15 − M	0	M	0	0	

Dual Problem

		y_1	y_2	y_3	y_4	y_5	s_5	s_6	
		−1,000	20	−20	15	−350	0	0	b_i
		Coefficients of Constraint Functions							
y_2	20	−16	1	−1	0	−1	1	0	10
s_6	0	−30	0	0	1	−1	0	1	15
	z_j	−320	20	−20	0	−20	20	0	
	$c_j - z_j$	−680	0	0	15	−330	−20	0	

s_5 has been replaced by incoming decision variable of y_2, thus $y_2 = 100$. Putting these values in dual objective function

$$Z^* = -1{,}000y_1 + 20(y_2 - y_3) + 15y_4 - 40y_5$$

$$Z^* = -1{,}000*0 + 20*(10 - 0) + 15*0 - 40*0 = 200$$

Feasibility: Positive RHS values of primal and dual imply feasibility of solution. Putting estimated values of decision variables in constraint equations would further indicate about feasibility of both solutions. For primal constraint, equations were:

$$\text{Constraint 1}: 16x_1 + 30x_2 \leq 1{,}000$$

$$16*20 + 30*0 = 320 < 1{,}000 \text{ constraint is satisfied}$$

$$\text{Constraint 2}: 1x_1 + 0x_2 = 20$$

$$1*20 + 0*0 = 20 \text{ constraint is satisfied}$$

$$\text{Constraint 3}: 0x_1 + 1x_2 \geq 15$$

$$0*20 + 1*0 = 0 < 15 \text{ constraint is violated}$$

$$\text{Constraint 4}: 1x_1 + 1x_2 \leq 40$$

$$1*20 + 1*0 = 20 < 40 \text{ constraint is satisfied}$$

As all the constraints are not met simultaneously, so this solution is infeasible. Doing the same thing with dual would give:

$$\text{Constraint 1}: -16y_1 + 1y_2 - 1y_3 + 0y_4 - 1y_5 \leq 10$$

$$-16*0 + 1*20 - 1*0 + 0*0 - 1*0 = 20 < 10 \text{ constraint is satisfied}$$

$$\text{Constraint 2}: -30y_1 + 0y_2 - 0y_3 + 1y_4 - 1y_5 \leq 15$$

$$-30*0 + 0*100 - 0*0 + 1*0 - 1*0$$

$$= 0 < 15 \text{ constraint is satisfied}$$

So, dual solution is feasible irrespective of infeasibility of primal. Before making final conclusion about the relationship between primal and dual, we should check their optimality.

Optimality: The net evaluation rows of both solutions indicate that the non-optimality of solution obtained at iteration 1 level implying that a better solution can be achieved further, which would utilize remaining resources. In the case of primal, the introduction of x_2 variable can further result in minimization of objective, whereas the introduction of y_4 in dual can result in improvement in profit maximization.

Thus, in conclusion, primal is neither feasible nor optimal, whereas dual is only feasible but not optimal. Solution at this stage fulfils rule 2 that if a solution is infeasible, then the other solution would be either infeasible or non-optimal.

Iteration 2:

See Table 4.21

Solution: For primal problem, decision variables are $x_1 = 20$ and $x_2 = 15$. So, the objective function of Z = 10*20 + 15*15 = 425.

As discussed above, values of dual variables would be:

Coefficients of Primal Variables in z_j Row	Represents Variable in Dual	Value of Dual Variable
$x_1 = 10$	s_5	$s_5 = 10$
$x_2 = 15$	s_6	$s_6 = 15$
$s_1 = 0$	y_1	$y_1 = 0$
$s_4 = 0$	y_5	$y_5 = 0$

To deduce values of remaining decision variables of dual, it is important to reiterate that slack variables represent the resources remaining or can be interpreted as resources consumed by certain activity. In this case, in from iteration 0 to iteration 2, it was observed in dual problem that two slack variables of s_5 and s_6 were replaced by two decision variables of y_2 and y_4 implying that with induction of economic activity y_2 and y_4 resources are consumed. At the end of iteration 2, the above table shows that maximum availability of these resources in the form of slack variables was $s_5 = 10$ and $s_6 = 15$. Thus, making the level of activity $y_2 = 10$ and $y_4 = 15$. This was also

TABLE 4.21
Iteration 2

Primal Problem

Decision variable		x_1	x_2	s_1	s_3	s_4	
		10	15	0	0	0	b_i
Basis Variable		**Coefficients of Constraint Functions**					
s_1	0	0	0	1	1	0	230
x_1	10	1	0	0	0	0	20
x_2	15	0	1	0	−1	0	15
s_4	0	0	0	0	1	1	5
z_j		10	15	0	−15	0	
$c_j - z_j$		0	0	0	15	0	

Dual Problem

Decision variable		y_1	y_2	y_3	y_4	y_5	s_5	s_6	
		−1,000	20	−20	15	−40	0	0	b_i
Basis Variable		**Coefficients of Constraint Functions**							
y_2	20	−16	1	−1	0	−1	1	0	10
y_4	15	−30	0	0	1	−1	0	1	15
z_j		−770	20	−20	15	−35	20	15	
$c_j - z_j$		−230	0	0	0	−5	−20	−15	

indicated by RHS values of dual. The association of coefficient of surplus variable $s_3 = -15$ with y_4 has been ignored as y_4 cannot be negative and as mentioned above that y_4 implies consumption of entire available resource of 15. Also, y_3 is assumed to be zero as it was associated with artificial variable a_3 which has been removed from the solution.

$$\text{So } Z^* = -1{,}000y_1 + 20(y_2 - y_3) + 15y_4 - 40y_5$$

$$Z^* = -1{,}000*0 + 20\ (100 - 0) + 15*15 - 40*0 = 425$$

Thus, the optimal value of Z and Z* is Z = 425 = Z*.

Feasibility*:* Putting the values in constraint equations of primal:

$$\text{Constraint 1}: 16x_1 + 30x_2 \leq 1{,}000$$

$$16*20 + 30*15 = 770 < 1{,}000 \text{ constraint is satisfied}$$

$$\text{Constraint 2}: 1x_1 + 0x_2 = 20$$

$$1*20 + 0*0 = 20 \text{ constraint is satisfied}$$

$$\text{Constraint 3}: 0x_1 + 1x_2 \geq 15$$

$$1*20 + 0*0 = 20 \text{ constraint is satisfied}$$

$$\text{Constraint 4}: 1x_1 + 1x_2 \leq 40$$

$$1*20 + 1*15 = 35 < 40 \text{ constraint is satisfied}$$

Hence, all the constraints are met simultaneously.

Doing the same thing with dual would give:

$$\text{Constraint 1}: -16y_1 + 1y_2 - 1y_3 + 0y_4 - 1y_5 \leq 10$$

$$-16*0 + 1*10 - 1*0 + 0*15 - 1*0 = 10 \text{ constraint is satisfied}$$

$$\text{Constraint 2}: -30y_1 + 0y_2 - 0y_3 + 1y_4 - 1y_5 \leq 15$$

$$-30*0 + 0*10 - 0*0 + 1*15 - 1*0 = 15 \text{ constraint is satisfied}$$

Thus, both solutions are feasible.

Optimality*:* Net evaluation rows of both solutions indicate optimality as further improvement in objective function of both primal and dual cannot be achieved.

This corroborates rule 3 that if one solution is feasible and optimal, then the other solution would also be the same.

4.5 SOLVED ILLUSTRATIONS

4.5.1 Illustration of Sensitivity Analysis

$$\textbf{Maximize } Z = 35x_1 + 40x_2 + 45x_3$$

Subject to:

$$C1: 3x_1 + 1x_2 + 1x_3 \leq 300$$

$$C2: 2x_1 + 2x_2 + 3x_3 \leq 430$$

$$C3: 4x_1 + 5x_2 + 3x_3 \leq 620 \; x_1, x_2 \text{ and } x_3 \geq 0$$

i. Range of feasibility for b_1
ii. Range of feasibility for b_2
iii. Range of feasibility for b_3
iv. Range of optimality for basic variables in objective function

Solution:

Final Simplex tableau is (Table 4.22):
With $x_1 = 0$, $x_2 = 63.36$ and $x_3 = 101.06$ value of Z would be

$$Z = 35*0 + 40*63.36 + 45*101.06 = 7{,}082.1$$

The initial solution is feasible as all values in RHS column indicating values of decision variables are positive and greater than zero. Also, to check the feasibility of solution further, enter these values in constraint equations:

$$C1: 3x_1 + 1x_2 + 1x_3 \leq 300$$

$$3*0 + 1*63.36 + 1*101.06 = 164.42 < 300$$

TABLE 4.22
Final Solution

Decision variable		x_1	x_2	x_3	s_1	s_2	s_3	b_i
		35	40	45	0	0	0	
Basis Variable		**Coefficients of Constraint Functions**						
s_1	0	2.11	0	0	1	−0.22	−0.11	135.58
x_3	45	0.22	0	1	0	0.55	−0.22	101.06
x_2	40	0.66	1	0	0	−0.33	0.33	63.36
z_j		36.66	40	45	0	11.66	3.33	
$c_j - z_j$		−1.66	0	0	0	−11.66	−3.33	

$$C2: 2x_1 + 2x_2 + 3x_3 \leq 430$$

$$2*0 + 2*63.36 + 3*101.06 = 429.9 \leq 430$$

$$C3: 4x_1 + 5x_2 + 3x_3 \leq 620$$

$$4*0 + 5*63.36 + 3*101.06 = 619.98 \leq 620$$

As all constraints are met, so solution is feasible. Finally, solution is optimal as well indicated from either negative or zero values in the net evaluation row.

i. **Range of feasibility for b_1**: Feasibility condition states that values in RHS column should be greater or equal to zero. So, to estimate the allowable range of b_1, which indicates the maximum resource available and is associated with slack variable s_1, the following formula is used.

$$b_1 \text{value in final solution} + (\text{change in } b_1, \Delta b_1)$$

$$* \ (\text{corresponding value of slack variable}) \geq 0$$

Coefficients in the slack variable column indicate change in corresponding basic variable with the introduction of that slack variable in the solution. For instance, in case of s_1, if one unit of it is introduced in the solution, then corresponding change in b_1 would be to the amount of coefficients of basic variables indicated in s_1 column. So, if coefficient in s_1 column is zero, then it would imply that b_1 would change by zero as corresponding basic variable has changed by zero. Hence, in this case b_1 is insensitive parameter and its value is equivalent to slack variable s_1. Any change in its value would keep s_1 same as values of corresponding basic variables in associated s_1 column are zero. For instance, in constraint equation of b_1, the value of s_1 is deduced to be 135.58. Now if it is increased by one unit, then corresponding values of basic variables in the s_1 column, which are s_1, x_2 and x_3, imply a change in values of x_2 and x_3 as zero, and the changed value of b_1 would be equivalent to change in s_1, i.e. one unit. So, b_1 becomes insensitive. This is shown below:

$$C1: 3x_1 + 1x_2 + 1x_3 + 1s_1 = 300$$

$$3*0 + 1*63.36 + 1*101.06 + 1s_1 = 164.42$$

So $s_1 = 135.58$

Thus, putting $x_1 = 0$, $x_2 = 63.36$ and $x_3 = 101.06$ value of s_1 would be 135.58. Now according to coefficients in the s_1 column, x_1 has not entered the solution, so it is assumed to be zero; x_2 and x_3 have *changed* by zero so their values remain the same, which would result in the same value of s_1.This value according to the final solution table represents b_1 value. Therefore, b1 becomes an insensitive parameter as any amount introduction of s_1 would keep b_1 the same.

ii. **Range of feasibility for b_2**: According to the above-discussed formula, the three values of changed b_2 values can be calculated as:

- $135.58 + \Delta b_2 * (-0.22) \geq 0$

$$135.58 - 0.22 * \Delta b_2 \geq 0$$

$$135.58 \geq 0.22 * \Delta b_2$$

$$\Delta b_2 \leq 616.27$$

- $101.06 + \Delta b_2 * 0.55 \geq 0$

$$\Delta b_2 * 0.55 \geq -101.06$$

$$\Delta b_2 \geq -183.74$$

- $63.36 + \Delta b_2 * (-0.33) \geq 0$

$$63.36 - 0.33 * \Delta b_2 \geq 0$$

$$63.36 \geq 0.33 * \Delta b_2$$

$$\Delta b_2 \leq 192$$

Calculation of above three equations helps to decide lower and upper bound of b_2 value. It was found that b_2 can take a minimum of –183.74 and a maximum of 616.27. So, the range of b_2 values was found to be:

$$-183.74 \leq \Delta b_2 \leq 616.27$$

Initial constraint was a maximum of 430. By adding this 430 to the minimum and maximum values of Δb_2 range of feasibility can be found. Thus,

$$430 - 183.74 \leq b_2 \leq 430 + 616.27$$

$$246.26 \leq b_2 \leq 1{,}046.27$$

So, solution would remain feasible within this range.

iii. **Range of feasibility for b_3**: According to the above discussed formula, the three values of changed b_3 values can be calculated as:

- $135.58 + \Delta b_3 * (-0.11) \geq 0$

$$135.58 - 0.11 * \Delta b_3 \geq 0$$

$$135.58 \geq 0.11 * \Delta b_3$$

$$\Delta b_3 \leq 1{,}232.55$$

- $101.06 + \Delta b_3 * (-0.22) \geq 0$

$$101.06 - 0.22 * \Delta b_3 \geq 0$$

$$101.06 \geq 0.22 * \Delta b_3$$

$$\Delta b_3 \leq 459.36$$

- $63.36 + \Delta b_3 * (0.33) \geq 0$

$$0.33 * \Delta b_3 \geq -63.36$$

$$\Delta b_3 \geq -192$$

Calculation of the above three equations helps to decide lower and upper bound of b_2 value. It was found that b_3 can take minimum of −192 and maximum of 1,232.55. So, the range of b_3 values was found to be:

$$-192 \leq \Delta b_3 \leq 1{,}232.55$$

Initial constraint was a maximum of 620. By adding this 620 to the minimum and maximum values of Δb_3 range of feasibility can be found. Thus,

$$620 - 192 \leq b_3 \leq 620 + 1{,}232.55$$

$$428 \leq b_3 \leq 1{,}852.55$$

So, solution would remain feasible within this range.

iv. **Range of optimality**: Final solution shown in Table 4.23 indicates x_2 and x_3 as basic variables. Range of optimality for each is found individually. In first case, profit of $40 is replaced by c_1, resulting in final solution tables shown as under (Table 4.23).

For maximization, the LPP solution would be optimal if values in net evaluation row are either zero or less than zero. This would lead to:

- $25.1 - 0.66c_1 \leq 0$

$$25.1 \leq 0.66c_1 \text{ or}$$

$$c_1 \geq 38.03$$

TABLE 4.23
Final Solution

Decision variable		x_1	x_2	x_3	s_1	s_2	s_3	b_i
		35	c_1	45	0	0	0	
Basis Variable		**Coefficients of Constraint Functions**						
s_1	0	2.11	0	0	1	−0.22	−0.11	135.58
x_3	45	0.22	0	1	0	0.55	−0.22	101.06
x_2	c_1	0.66	1	0	0	−0.33	0.33	63.36
z_j		$9.9 + 0.66c_1$	c_1	45	0	$24.75 - 0.33c_1$	$0.33c_1 - 9.9$	
$c_j - z_j$		$25.1 - 0.66c_1$	0	0	0	$-24.75 + 0.33c_1$	$-0.33c_1 + 9.9$	

- $-24.75 + 0.33c_1 \leq 0$

$$0.33c_1 \leq 24.75 \text{ or}$$

$$c_1 \leq 75$$

- $9.9 - 0.33c_1 \leq 0$

$$9.9 \leq 0.33c_1 \text{ or}$$

$$c_1 \geq 30$$

Thus, from the above inferences, the value of c_1 would lie between

$$38.03 \leq c_1 \leq 75$$

Hence, this illustration entails that range of objective function coefficient of a basic variable over which the current solution would remain optimal is a minimum of 38.03 and a maximum of 75.

In the second case, profit of $45 is replaced by c_2 resulting in final solution tables shown as under (Table 4.24).

For maximization, the LPP solution would be optimal if values in net evaluation row are either zero or less than zero. This would lead to:

- $-0.22c_2 + 8.6 \leq 0$

$$8.6 \leq 0.22c_2 \text{ or}$$

$$c_2 \geq 39.1$$

TABLE 4.24
Final Solution

Decision variable		x_1 35	x_2 40	x_3 c_2	s_1 0	s_2 0	s_3 0	b_i
Basis Variable		**Coefficients of Constraint Functions**						
s_1	0	2.11	0	0	1	−0.22	−0.11	135.58
x_3	c_2	0.22	0	1	0	0.55	−0.22	101.06
x_2	40	0.66	1	0	0	−0.33	0.33	63.36
z_j		$0.22c_2 + 26.4$	40	c_2	0	$0.55c_2 - 13.2$	$-0.22\ c_2 + 13.2$	
$c_j - z_j$		$-0.22c_2 + 8.6$	0	0	0	$13.2 - 0.55c_2$	$0.22c_2 - 13.2$	

- $13.2 - 0.55c_2 \le 0$

$$13.2 \le 0.55c_2 \text{ or}$$

$$c_2 \ge 24$$

- $0.22c_2 - 13.2 \le 0$

$$0.22c_2 \le 13.2 \text{ or}$$

$$c_2 \le 60$$

Thus, from the above inferences, the value of c_2 would lie between

$$39.1 \le c_2 \le 60$$

Hence, this illustration entails that the range of objective function coefficient of a basic variable over which current solution would remain optimal is a minimum of 39.1 and a maximum of 60.

Excel Solution

The final solution of the above illustration by using Solver is shown in Figure 4.4.

Sensitivity analysis report is shown in Figure 4.5. The results show that only dependent variables x_2 and x_3 contribute to the objective function. Moreover, any further change in their values would not alter the objective function, making them insensitive parameters. Also, the allowable range shows that the solution remains feasible with a maximum value of x_2 to be 40 + 35 = 75 and a minimum value of 40 – 2.5 = 37.5. Similarly, the allowable range of x_3 would be a maximum of 45 + 15 = 60 and a minimum of 45 – 7.5 = 37.5.

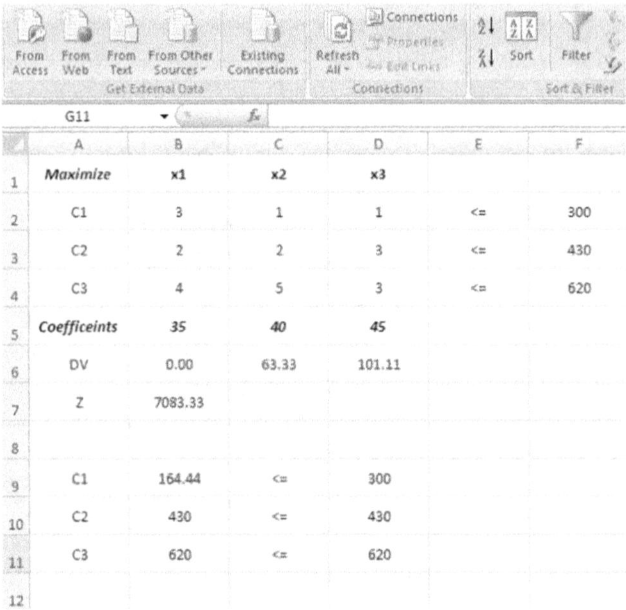

	A	B	C	D	E	F
1	*Maximize*	x1	x2	x3		
2	C1	3	1	1	<=	300
3	C2	2	2	3	<=	430
4	C3	4	5	3	<=	620
5	*Coefficeints*	*35*	*40*	*45*		
6	DV	0.00	63.33	101.11		
7	Z	7083.33				
8						
9	C1	164.44	<=	300		
10	C2	430	<=	430		
11	C3	620	<=	620		
12						

FIGURE 4.4 Excel solution.

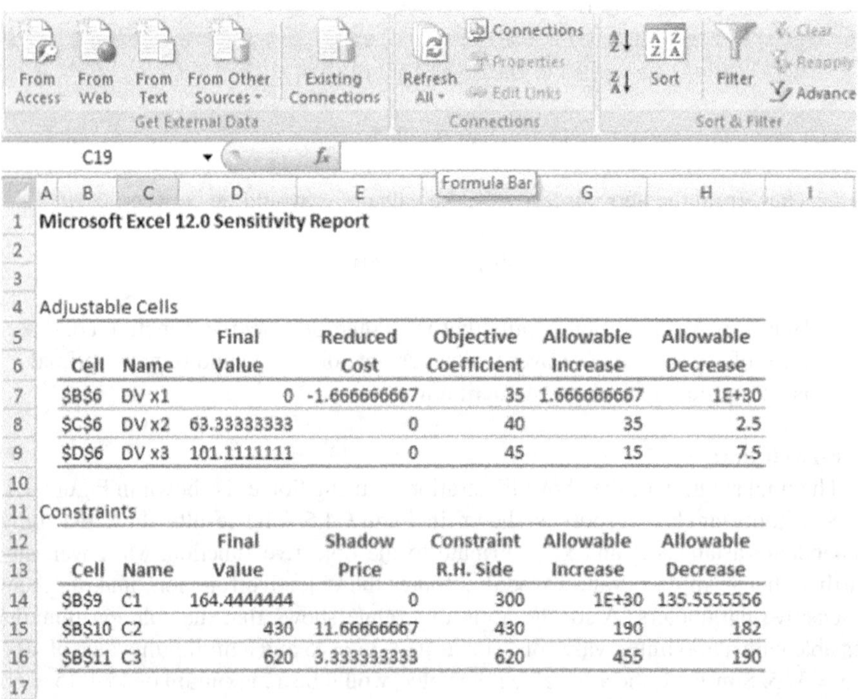

Microsoft Excel 12.0 Sensitivity Report

Adjustable Cells

Cell	Name	Final Value	Reduced Cost	Objective Coefficient	Allowable Increase	Allowable Decrease
$B$6	DV x1	0	-1.666666667	35	1.666666667	1E+30
$C$6	DV x2	63.33333333	0	40	35	2.5
$D$6	DV x3	101.1111111	0	45	15	7.5

Constraints

Cell	Name	Final Value	Shadow Price	Constraint R.H. Side	Allowable Increase	Allowable Decrease
$B$9	C1	164.4444444	0	300	1E+30	135.5555556
$B$10	C2	430	11.66666667	430	190	182
$B$11	C3	620	3.333333333	620	455	190

FIGURE 4.5 Sensitivity analysis.

Sensitivity analysis of constraint equations show that an increase of one unit in resource availability of the second constraint (C2) objective function would change by 11.66 as indicated by shadow price column. In addition, this change would keep solution feasible if RHS of C2 ranges between 430 + 190 = 620 and 430 − 182 = 248. Similarly, an increase in one unit of C3 would have a positive change in objective function by 3.33. Feasible range values would be 620 + 455 = 1,075 and 620 − 190 = 430.

4.5.2 Illustration of Duality

$$\textbf{Maximize } Z = 2x_1 + 6x_2 + 5x_3$$

Subject to:

$$x_1 + 2x_2 + 2x_3 \leq 10$$

$$3x_1 + 3x_2 + 2x_3 \leq 12 \text{ with } x_1, x_2 \geq 0$$

i. Construct the dual form of given primal problem.
ii. Solve only primal by simplex method and identify basic solution at each iteration. Also, identify basic and non-basic variables of dual problem.

Solution:

i. Dual form would be:

$$\textbf{Minimize} \quad Z^* = 10y_1 + 12y_2$$

Subject to:

$$1y_1 + 3y_2 \geq 2$$

$$2y_1 + 3y_2 \geq 6$$

$$2y_1 + 2y_2 \geq 5 \text{ with } y_1, y_2 \geq 0$$

ii. Following would be initial models of both primal and dual:

$$\textbf{Maximize} \quad Z = 2x_1 + 6x_2 + 5x_3 + 0s_1 + 0s_2$$

Subject to:

$$x_1 + 2x_2 + 2x_3 + 1s_1 + 0s_2 = 10$$

$$3x_1 + 3x_2 + 2x_3 + 0s_1 + 1s_2 = 12 \text{ with } x_1, x_2 \geq 0$$

where s_1, s_2 are slack variables.

$$\textbf{Minimize}\quad Z^* = 10y_1 + 12y_2 + 0s_3 + 0s_4 + 0s_5 + Ma_3 + Ma_4 + Ma_5$$

Subject to:

$$1y_1 + 3y_2 - 1s_3 + 0s_4 + 0s_5 + 1a_3 + 0a_4 + 0a_5 = 2$$

$$2y_1 + 3y_2 + 0s_3 - 1s_4 + 0a_3 + 1a_4 + 0a_5 = 6$$

$$2y_1 + 2y_2 + 0s_3 + 0s_4 - 1s_5 + 0a_3 + 0a_4 + 1a_5 = 5 \text{ with } y_1, y_2 \geq 0$$

where s_3, s_4 and s_5 are surplus variables and a_3, a_4 and a_5 are corresponding artificial variables.

Taking above model into consideration following iterations shows solution only for primal problem (Tables 4.25–4.27).

Iteration 0:
See Table 4.25

Iteration 1:
See Table 4.26

TABLE 4.25
Iteration 0

Decision variable		x_1	x_2	x_3	s_1	s_2	b_i
		2	6	5	0	0	
Basis Variable			**Coefficients of Constraint Functions**				
s_1	0	1	2	2	1	0	10
s_2	0	3	3	2	0	1	12
z_j		0	0	0	0	0	
$c_j - z_j$		2	6	5	0	0	

TABLE 4.26
Iteration 1

Decision variable		x_1	x_2	x_3	s_1	s_2	b_i
		2	6	5	0	0	
Basis Variable			**Coefficients of Constraint Functions**				
s_1	0	−1	0	2/3	1	−2/3	2
x_2	6	1	1	2/3	0	1/3	4
z_j		6	6	4	0	2	
$c_j - z_j$		−4	0	1	0	−2	

TABLE 4.27
Iteration 2

Decision variable		x_1	x_2	x_3	s_1	s_2	b_i
		2	6	5	0	0	
Basis Variable		**Coefficients Of Constraint Functions**					
x_3	5	−3/2	0	1	3/2	−1	3
x_2	6	2	1	0	−1	1	2
z_j		9/2	6	5	3/2	1	
$c_j - z_j$		−5/2	0	0	−3/2	−1	

Iteration 2:

See Table 4.27

The following table provides solutions for both primal and dual:

	Primal						Dual						
	Coefficients of Decision Variables			Coefficients of Slack Variables		Value of Objective Function Z	Value of Decision Variables		Value of Surplus Variables			Value of Objective Function Z*
	x_1	x_2	x_3	s_1	s_2		y_1	y_2	s_3	s_4	s_5	
Iteration 0	0	0	0	0	0	$x_1 = x_2 = x_3 = 0$ so $Z = 0$	0	0	0	0	0	0
Iteration 1	6	6	4	0	2	$x_1 = 0$; $x_2 = 4$; $x_3 = 0$ so $Z = 24$	0	2	6	6	4	24
Iteration 2	9/2	6	5	3/2	1	$x_1 = 0$; $x_2 = 2$; $x_3 = 3$ so $Z = 27$	3/2	1	9/2	6	5	27

Form iteration 2 $x_1 = 0$; $x_2 = 2$; $x_3 = 3$. Putting these values in constraint equations

$$x_1 + 2x_2 + 2x_3 + 1s_1 + 0s_2 = 10$$

$$3x_1 + 3x_2 + 2x_3 + 0s_1 + 1s_2 = 12$$

Would give $s_1 = s_2 = 0$ satisfying non-negativity criteria, making solution feasible. Out of these five variables, non-zero variables x_2 and x_3 are basic variables as these are non-zero. Associated dual variables deduced from the above given solutions are s_4 and s_5. According to the concept of primal and dual being complementary to each other, these variables would be non-basic in dual solution. From definition, non-basic variables are variables with zero value, making s_4 and s_5 equal to zero.

Putting values of s_4 and s_5 in constraint equations of dual:

$$2y_1 + 3y_2 + 0s_3 - 1s_4 = 6$$

$$2y_1 + 2y_2 + 0s_3 + 0s_4 - 1s_5 = 5$$

would give y_1 as 3/2 and y_2 as 1 which is same as derived from iteration 2, thus corroborating the solution.

4.6 SUMMARY

This chapter discussed concepts of sensitivity analysis and duality by illustrating linear programming problems. It was inferred that coefficients and parameters used in formulation of a linear programming model are only estimates, and it is important to apply sensitivity analysis to identify which of these are sensitive and which are insensitive to change. Sensitivity analysis was performed by finding an allowable feasible range of RHS values, which indicate resource availability and also a range of optimality was identified for objective function coefficients. It is important to note that in this chapter such deduction was done by changing only one coefficient, keeping all others as unchanged, though limited analysis is possible by adopting multiple changes.

Another section was devoted to understanding the concept of duality that was helpful in implementing and interpreting sensitivity analysis. A mirror image called dual can be constructed of original problem that is complementary to primal. Solving any one would provide the solution for the other. The benefit of creating and solving dual is that any variation in solution identifies economic contribution by variable that has caused such variation. The formulation of dual problem and its relationship with primal was discussed in detail.

4.7 CASE STUDY: MODERN FOODS INDIA LIMITED

Modern Foods India Limited (MFIL) is the second largest bread manufacturer in the country. Bread as a food has seen increased adoption among urban Indians. The company is involved in producing three different varieties of bread: wheat based (A), wholegrain (B) and multi-grain (C). To increase its market share and become a leading player, the company intends to increase the volume of each variety and also keep the cost down. The present process applies batch production system where each machine produces in bulk of one type of bread. For instance, the first process is of mixing ingredients. As some of the ingredients are particular to a particular type of bread, the producer mixes certain quantity of them for the production of a batch of breads. Then this batch moves to second machine carrying out process of moulding and, finally, to oven. The batch type of production system results in building of lot of work-in-process (WIP) inventory which adds to cost of production and thus effecting profitability. Application of the current system allows producer to earn profit of $1.0, $1.2 and $1.1 on per kilogram of bread.

With increase in demand for all types of breads, producers intend to shift operations to a combination of batch and line production known as hybrid system. Such a shift results in a reduction of non-value added activities or wastes. Management decided to further reduce wastes by applying various lean manufacturing techniques.

Three of such techniques, namely (i) modification of push system (batch) to pull system (hybrid); (ii) reduction in change over time and (iii) improvement in preventive maintenance method, were identified after thorough brainstorming with shop floor managers and workers. It was also estimated that at least 10% improvement in profits be achieved across all three types of breads as a result of implementation of selecting lean manufacturing techniques. To find whether objective of obtaining potential gains from implementing such techniques was achieved, a model was constructed which compared estimated results of modified system with the current system. Details of three techniques with regard to current and modified system are described as under. The system operated in three shifts of 8 hours each for 30 days a month. The data was recorded for a production schedule which was formed for 1 week and then created or modified again depending on customer demand.

The performance of push and pull system was evaluated through analysis of WIP inventory. The pull system is considered to be more effective in reducing inventory as in this system production is initiated only when a customer's order is received, whereas in the push system, products are manufactured without estimation of demand. Current system at the plant suggested an average inventory equivalent of 0.58, 0.2 and 0.36 days for A, B and C, respectively. Past data showed piling of stocks at the end of the month equivalent to inventory of 15 days. With modification to the hybrid system, optimistic but realistic inventory values were estimated to reduce to 0.45, 0.12 and 0.25 days of three types of products. Also, maximum inventory should be reduced to 09 days at the end of month. Collecting weekly data indicated change over time under the current system to be on an average of 15, 120 and 90 minutes, respectively, for A, B and C. The maximum allowed change over time for such a system was 800 minutes. Application of lean manufacturing technique resulting in reduction of such time should result in reduction of maximum change over time to 200 minutes and 5, 20 and 20 minutes for three products. Finally, production schedule data indicated that out of a total 1,440 minutes allocated for preventive maintenance of entire process, an average of 90, 70 and 50 minutes is attributed to production of A, B and C type of bread. Suggested values for the modified model were 36, 20 and 15 minutes out of a total of 240 minutes.

i. Is modified model feasible? Infer the difference with original model.
ii. Estimate the maximum and minimum values of profits that could be earned from production of three types of breads.
iii. What would be a feasible range of inventory level, change over time and preventive maintenance time within which optimal profits could be achieved?
iv. Construct dual model for both original and modified models and compare the results. Identify shadow prices by comparing the dual solution of the modified model with its primal form.
v. Average combined demand of three types of bread per month was calculated as 48,000 units. Infer changes to both models that would occur due to addition of this constraint.

4.8 GLOSSARY

- **Sensitivity analysis:** The technique of identifying sensitive parameters out of numerous parameters effecting objective function.
- **Sensitive parameters:** The parameters whose value cannot be changed without causing a change in optimal solution.
- **Post-optimality analysis:** A sequence of solutions where each solution is better than the previous one.
- **Primal problem:** A linear programming problem in its original form.
- **Dual problem:** A linear programming problem which is mirror image of primal.
- **Shadow price:** It is the difference between value of 'Z' from original model and new value obtained from marginal increase (i.e. increase of one unit) of 'b'.
- **Insensitive parameters:** Any realistic changes in parameters pertaining to non-basic variables does not alter optimality, feasibility and profit function of initial solution are known as insensitive parameters.
- **Standard LPP:** A maximization problem involving functional constraints with only less than or equal to (≤) inequalities and a minimization problem with only greater than or equal to constraints is considered to be a standard form.
- **Feasibility of dual:** Positive surplus variables obtained from corresponding slack variables of primal indicate feasibility of dual. It provides range of values over which values of 'b' may vary but solution still remains feasible.
- **Optimality of dual:** It provides a range of values over which value of objective function might change, but solution still remains optimal.

4.9 MODEL QUESTIONS

For all of the following linear programming problems:

i. Construct the dual form.
ii. Solve only primal by simplex method and identify basic solution at each iteration. Also identify complementary dual solution from primal.
iii. Identify basic and non-basic variables of dual problem.
iv. Determine whether primal and dual has a feasible and optimal solution.

4.9.1 **Maximize** $Z = 2x_1 + 4x_2$

Subject to:

$$1x_1 - 2x_2 \geq 2$$

$$4x_1 + 3x_2 \leq 12 \quad x_1, x_2 \geq 0$$

4.9.2 **Maximize** $Z = 2x_1 + 3x_2 + 4x_3$

Subject to:

$$4x_1 + 3x_2 + 5x_3 \leq 8$$

$$0x_1 - 2x_2 + 4x_3 \leq 5x_1, x_2 \geq 0$$

In this case, solve dual problem and identify shadow prices if any. What would be effect on basic solution if coefficients of x_3 in both constraints are changed to 4 and 2, respectively?

4.9.3 **Maximize** $Z = 3x_1 + 7x_2 + 9x_3$

Subject to:

$$4x_1 + 0x_2 + 4x_3 \leq 12$$

$$0x_1 + 2x_2 + 8x_3 \leq 10x_1, x_2 \geq 0$$

4.9.4 **Maximize** $Z = 4x_1 + 4x_2$

Subject to:

$$5x_1 + 10x_2 = 20$$

$$10x_1 + 1x_2 \geq 10x_1, x_2 \geq 0$$

Convert primal to standard form and construct its dual. Also show that this dual is equivalent to the dual in non-standard form.

4.9.5 **Minimize** $Z = 8x_1 + 7x_2$

Subject to:

$$5x_1 + 4x_2 \geq 5$$

$$2x_1 + 1x_2 \leq 3$$

$$1x_1 + 1x_2 = 2x_1, x_2 \geq 0$$

4.9.6 For the problem illustrated in Section 2.4.5, find:

i. Feasible range of all three resources available for both vendors. Which vendor uses available resources more effectively?
ii. Identify basic and non-basic variables for both models. Construct their dual and find dual solution from primal.
iii. Optimal range of profit generated from each type of seat. Which seat can contribute maximum to profit function?

4.9.7 For the problem illustrated in 2.10.4 under section 2.10, solve independently following questions:

i. Feasible range for maximum availability of raw material. What happens if for same quantity ordered now available raw material is 3,000 kg.
ii. Feasible range of hours of production. Interpret changes if any to objective function if available production time is now 5,200 hours.

iii. Feasible range of cost of production. What happens if such cost gets increased to $500?
iv. Optimal range of profit earned from T1.
v. Optimal range of profit earned from T2.
vi. Optimal range of profit earned from T3.

4.9.8 For the problem illustrated in 2.10.7 under section 2.10, solve independently following questions:

i. What would be feasibility range of available time for braiding process?
ii. What would be feasibility range of available time for tube extrusion process?
iii. What would be feasibility range of available time for cover extrusion process?
iv. What would happen to objective function if demand of hose H1 is increased to 50,000 units?
v. What would happen to objective function if demand of hose H2 is decrease to 10,000 units?
vi. Find optimal range of cost incurred in manufacturing of H2. Also interpret the changes if due efficiency in production processes it gets reduced to $100.
vii. Similarly interpret changes to objective function if cost of buying H1 gets increased to $200. Also find its optimality range.
viii. Finally, identify insensitive parameter in the objective function.

5 Network Model I
Transportation Model

5.1 INTRODUCTION

Linear programming models discussed in previous chapters were primarily formulated to solve the problems of resource allocation in various disciplines of management. However, these models can be further applied to solve more complex problems by creating special structures of linear programming, which otherwise could only be solved by using computer technology. One of such problems was concerned with those of distribution involving the supply of material from various sources to different destinations. Such problems are categorized as network flow problems consisting of nodes and arcs. Nodes represent supply and demand centres, whereas arcs imply direction of flow and level of activity that could be time, distance or cost. The models could be constructed and solved using the rules of simplex method as discussed in previous chapters. This and two subsequent chapters are devoted to explanation and solution by using algorithms and Excel of certain network models.

Network flow problems primarily arise in industrial logistics concerning the distribution of a single homogeneous product from one or more origins (supply centres) to one or more destinations (demand centres). Total units to be supplied from each centre and demand of each are known with certainty. The products supplied can move either directly from supply centres to demand centres or through intermediaries of warehouses, for instance. In other variations of network flow models, the flow might involve routing a vehicle or salesperson back to origin after fulfilling the demand of each destination. Furthermore, problems become more complex though more realistic when capacity constraints of delivery vehicle are brought into the model. However, in all cases, the objective is to reduce shipping cost, travel time and distance.

This chapter on network flow models has discussed a commonly faced problem in logistics of finding the minimum total transportation cost of distributing products from multiple origins to multiple destinations with each source being able to supply to any of the demand centre due to homogeneity of product. Such model is termed as transportation model. However, such a model assumes the absence of any intermediaries. Flow models with intermediaries have been discussed under the transshipment section that has been formulated as a special case of transportation models.

This model could be applied to understanding the assignment of jobs to employees where the skill set of all employees is homogeneous. The measure of performance in such cases could be assignment done in a manner that would minimize the job completion time. The assignment model has been constructed as an integer linear programming model where either a specific job has been allocated to a particular employee or not. Therefore, decision variable in this case is either zero or one as compared to the number of units in the transportation model.

DOI: 10.1201/9781003212966-5

The last chapter on network flow models discussed more complex and special forms of network flow models. The travelling salesman problem involved finding a route involving multiple destinations in such a way that the minimum performance metric is achieved by visiting each node only once and then returning to the origin. A variant of it in the form of vehicle routing problem was discussed further, which involved capacity constraints and multiple vehicles. This is a form of salesman problem with not one but multiple salesmen. Finally, in another variant of flow problem, the shortest route from a particular source to a destination was deduced in the possibility of multiple alternative routes. These special cases provide insight into the solution of more practical industrial problems through well-established algorithms.

5.2 STRUCTURE OF TRANSPORTATION MODEL

The transportation problem is a special kind of linear programming problem that is used to determine the number of units to be distributed from limited supply centres to known demand centres. In the majority of cases, there are more than one supply and demand centres. This makes the problem of allocating the number of units from one supply centre to a particular demand centre peculiar. With the increase in the number of supply and demand centres, the problem becomes more complex. Typically, the objective of a transportation model is to find the minimum cost required to fulfill the demand of various destinations when material is sourced from various suppliers. The transportation model determines the number of units to be sourced from one supplier and to be supplied to a particular distributor at the minimum or most optimal transportation cost.

A transportation model specifically includes the following parameters:

- Limited number of suppliers and capacity of each supplier.
- Known number of demand centres with known demand of each centre.
- Known cost of transporting one unit from each supply to each demand centre.

These three elements are shown in Table 5.1 indicating generalized structure of a transportation model.

TABLE 5.1
Demand centres

Supply Centres	D1	D2	D3	D4	Total Supply of Each Center
S1	a_{11}	a_{12}	a_{13}	a_{14}	b_1
S2	a_{21}	a_{22}	a_{23}	a_{24}	b_2
S3	a_{31}	a_{32}	a_{33}	a_{34}	b_3
Total demand of each center	c_1	c_2	c_3	c_4	

Table 5.1 illustrates:

- Four demand centres: D1, D2, D3 and D4.
- Three supply centres: S1, S2 and S3.
- Total supply of each supply centre: Supply of S1 = b_1, supply of S2 = b_2 and supply of S3 = b_3
- Total demand of each centre: Demand of D1 = c_1, demand of D2 = c_2, demand of D3 = c_3 and demand of D4 = c_4.
- Cost of transporting one unit from each supply to each demand centre. a_{11} is the cost of transporting one unit from S1 to D1 and so forth.

This structure would be used for solving all transportation problems in this chapter.

The same model can also be formulated as a linear programming model. The construction of such a model discussed in previous chapters involves knowledge of objective function, decision variables and constraint equations. In the case of the transportation model, the objective is to minimize transportation cost incurred in distributing a fixed number of units from a limited number of supply centres to a known number of distribution centres. This objective depends on finding out such number of units relating to each supply and demand centre. In Table 5.1, there are three supply centres denoted by coefficient i (= 1, 2 and 3) and four demand centres denoted by coefficient j (= 1, 2, 3 and 4). For instance, x_{11} would imply the number of units to be transported from S1 to D1. Similarly, other decision variables can be read. The values in cells indicate the cost of transportation of one unit of product. Therefore, if the cost of transporting one unit is a_{11}, then that of x_{11} units would be $a_{11}x_{11}$. This information can be used to create the following objective function.

$$\textbf{Minimize}\quad Z = a_{11}x_{11} + a_{12}x_{12} + a_{13}x_{13} + a_{14}x_{14} + a_{21}x_{21} + a_{22}x_{22} + a_{23}x_{23}$$

$$+ a_{24}x_{24} + a_{31}x_{31} + a_{32}x_{32} + a_{33}x_{33} + a_{34}x_{34}$$

Fulfillment of objective depends on certain constraints. In this case, constraints relate to the supply of *fixed* units from a particular supply centre and *fixed* demand from a particular demand centre. The important thing to note here is that, at least for now, we are focusing only on when supply and demand are fixed. However, in cases discussed in later sections, supply and demand might not be fixed value. For instance, a supply centre due to manufacturing limitations might be able to produce only a certain limited and not fixed quantity, whereas a demand centre might ask for a certain minimum amount of units. Such cases result in formulation of constraint equation with inequality constraints. However, for the time being for the above case, we formulate constraints for fixed quantity. This leads to two set of constraints; one pertaining to supply and the other to demand. Constraints pertaining to supply would be:

$$x_{11} + x_{12} + x_{13} + x_{14} \le b_1$$

$$x_{21} + x_{22} + x_{23} + x_{24} \le b_2$$

$$x_{31} + x_{32} + x_{33} + x_{34} \le b_3$$

Constraints pertaining to demand would be:

$$x_{11} + x_{21} + x_{31} = c_1$$

$$x_{12} + x_{22} + x_{32} = c_2$$

$$x_{13} + x_{23} + x_{33} = c_3$$

$$x_{14} + x_{24} + x_{34} = c_4$$

where $x_{11}, x_{12}, x_{13}, x_{14}, x_{21}, x_{22}, x_{23}, x_{24}, x_{31}, x_{32}, x_{33}, x_{34} \leq 0$

Formulation of transportation model should follow certain assumptions that are derived from understanding of model as discussed above.

5.3 ASSUMPTIONS OF TRANSPORTATION PROBLEMS

- **Requirement assumption**: Each supplier has a fixed number of units to be supplied, i.e. it has a fixed output that has to be supplied to various destinations. Each demand centre has a fixed demand, which is to be met by the number of units supplied from various and limited suppliers.
- **Feasibility**: A transportation problem is considered to be feasible if total output from various supply sources equals total demand of various destinations. From Table 5.1, this can be illustrated as:

  $$b_1 + b_2 + b_3 = c_1 + c_2 + c_3 + c_4$$

  $$\text{i.e.} \sum b_i = \sum c_j$$

  In some cases, this assumption gets violated as practically each source indicates its maximum capacity to produce and each demand centre indicates its maximum demand. However, such a situation can be resolved, which will be discussed later in detail.
- **Cost assumption**: the cost of transporting units from one source to a particular destination is directly proportional to the number of units to be transported. For example, if the transportation cost per unit from one source to a destination is $5, then transporting ten units would be $50.
- **Integer values**: A transportation problem would have every demand and supply value as an integer value. The number of units allocated to each destination and cost of transporting are also integer values.

5.4 TRANSPORTATION PROBLEM

5.4.1 Case: Musashi Auto Parts Michigan, Inc.

Let us illustrate the above-discussed concepts by considering a transportation problem faced by Musashi Auto Parts Michigan, Inc. The company provides gears, ball joints, camshafts and other motor vehicle components to various car manufacturers in

North America. Musashi Auto Parts Michigan operates worldwide. It is a vital supplier to Honda car manufacturing plants in the USA. The company has even been honoured by award of excellence in quality by Honda motors. It has three major plants in Battle Creek, Michigan; Southfield, Michigan and Ontario Canada catering to North American market of Honda. These three supply centres provide auto parts and components to three major Honda motor plants in Marysville Auto Plant Ohio, East Liberty Ohio and Greensburg Indiana. Importantly, each plant can supply to all three plants. The transportation cost of supplying components from a supply centre to a demand centre is directly affected by the distance between them. Taking into consideration the distance transportation cost per unit of each supply centre to each demand centre is shown. For illustration, the transportation model is constructed only for three supply centres S1, S2 and S3 and for three demand centres D1, D2 and D3. These destinations are catered by the mentioned supply centres as they are within their scope. The annual average total demand for parts for these three demand centres is 120,000 units. By using the following data of demand and production capacity of each centre and transportation cost per unit, optimal routes were estimated through which fixed quantity of units would be transported in such a way that total transportation cost is minimum.

Production capacities of all three manufacturing plants are as follows:

Origin	Manufacturing Plant (Supply Centre)	Production Capacity (units)
1.	Ontario (S1)	55,000
2.	Southfield (S2)	40,000
3.	Battle Creek (S3)	25,000
	Total	120,000

The average annual demand enumerated from past data of three distribution centres is:

Destination	Demand Centre	Demand (units)
1.	Marysville (D1)	65,000
2.	East Liberty (D2)	45,000
3.	Greensburg (D3)	10,000
	Total	120,000

Cost data in unit of thousands indicating per unit transportation cost from each supply centre to each distribution centre is shown in the following table.

	Demand Centre		
Origin	D1	D2	D3
S1	5	11	6
S2	4	6	5
S3	4	6	9

TABLE 5.2
Network Diagram

Production Capacity (units)	Manufacturing Plant (Supply Centre)	Transportation Cost	Routes	Demand Centre	Demand (Units)
55,000	S1	5			65,000
		11		D1	
		6			
40,000	S2	4			45,000
		6		D2	
		5			
25,000	S3	4			10,000
		6		D3	
		9			

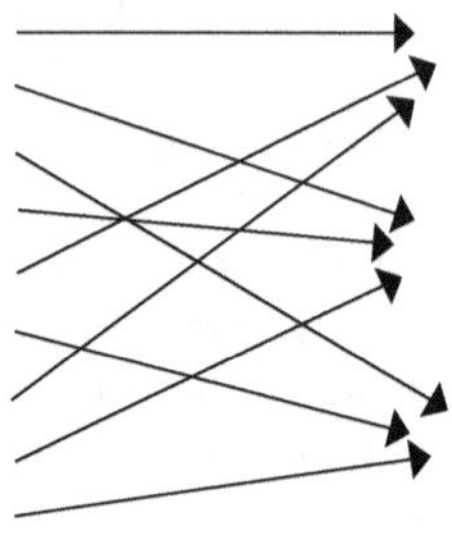

By using the above-mentioned data management, we would like to determine how many units should be shipped from each manufacturing plant to each demand centre. The above data indicates each production facility is capable of meeting the demand of each destination but with varying costs. This results in the number of possible routes that are shown in Table 5.2. There are a total of nine possible routes. The purpose is to find which of them is the most feasible route and how many units would be transported through these viable routes.

From Table 5.2, nine feasible and viable routes are estimated by applying a transportation model. The solution would calculate the number of units to be transported from feasible routes from which total cost would be calculated.

It can also be formulated as LPP, which is shown below.

$$\textbf{Minimize} \quad Z = 5x_{11} + 11x_{12} + 6x_{13} + 4x_{21} + 7x_{22} + 5x_{23} + 4x_{31} + 6x_{32} + 9x_{33}$$

Subject to:

Constraints pertaining to supply would be:

$$x_{11} + x_{12} + x_{13} \leq 55{,}000$$

$$x_{21} + x_{22} + x_{23} \leq 40{,}000$$

$$x_{31} + x_{32} + x_{33} \leq 25{,}000$$

Constraints pertaining to demand would be:

$$x_{11} + x_{21} + x_{31} = 65{,}000$$

$$x_{12} + x_{22} + x_{32} = 45{,}000$$

$$x_{13} + x_{23} + x_{33} = 10{,}000$$

where $x_{11}, x_{12}, x_{13}, x_{21}, x_{22}, x_{23}, x_{31}, x_{32}, x_{33} \leq 0$

5.5 TRANSPORTATION SOLUTION METHODS

The above discussion shows that a transportation problem can be solved through its formulation by using a linear programming technique. Such an LPP model can further be reached to a final solution by solving it through the simplex method. However, the above example involving only a few resources in terms of supply and demand resulted in the creation of quite a tedious LPP. It would become more complex if the problem consisted of many more supply and distribution centres, which happen in many industrial problems. Such an issue can be resolved by applying the transportation version of the simplex method. Solving of such a version is detailed below.

5.5.1 Formulation of Model

The first step is to check the feasibility of transportation problem. Feasibility is tested by comparing total demand with total supply. The transportation model demands them to be equal. In this case:

$$\text{Total supply} = 55{,}000 + 40{,}000 + 25{,}000 = 120{,}000$$

$$\text{Total demand} = 65{,}000 + 45{,}000 + 10{,}000 = 120{,}000$$

As both are equal, we can proceed with further steps. Thus, the model would be as shown in Table 5.3.

The simplex form of the above model shows that the problem of minimization of transportation cost is limited by six constraints – three of which indicate supply constraints and three demand constraints. All the constraints are equality constraints and so solving it by the simplex method would involve the introduction of six artificial variables. These artificial variables in the form of basic variables would result in an initial feasible solution satisfying non-negativity criteria. However, a final feasible solution would be achieved through numerous iterations, making the entire procedure cumbersome. The transportation model solves the same problem by using a simpler procedure where directly a basic feasible model is constructed in the form of the transportation initial table shown in Table 5.3.

TABLE 5.3
Initial Model

	Destination			
Origin	**Marysville (D1)**	**East Liberty (D2)**	**Greensburg (D3)**	**Supply (Thousand Units)**
Ontario (S1)	5	11	6	55
Southfield (S2)	4	6	5	40
Battle Creek (S3)	4	6	9	25
Demand (thousand units)	65	45	10	

To check its feasibility, the relationship between the number of basic variables and functional constraints needs to be understood. Ordinarily, there is one basic variable for each constraint. Thus, for transportation problems with 'm' sources and 'n' destinations, the number of constraints and thus the number of basic variables would be m + n. However, a basic feasible solution would always or most of the time would give m + n – 1 allocation. These allocations are considered as basic variables if interpreted in terms of the simplex method. This implies that a solution through the transportation model would require solving for one less constraint, whereas in the simplex procedure, all m + n functions needed to be solved. Thus, any one of the constraint functions is considered redundant and estimating the value of m + n – 1 equation would automatically lead to solving of remaining extra constraint. Hence,

Number of basic variables = m + n – 1.

So, if three supply and demand centres leading to six constraint equations, then solving for five equations (could be a combination of supply and demand sources) would be sufficient to estimate total transportation cost. This solution of five equations is represented by number of allocations in solution of transportation model. Thus, feasibility condition of a transportation model is tested at two stages. First, when model is formulated where total supply is equal to total demand, it would indicate the feasibility of model. Second, after finding the initial solution, it is important to check the equality between *number of allocations = m + n – 1.* The second step to check feasibility is tested after finding the initial solution.

5.5.2 Initial Solution

Any of the following mentioned methods can be used to find an initial solution for given transportation problem. It is termed as initial because this is the first solution that is being applied so for without any improvements. Such a solution satisfies all the demand and supply constraints in the sense that no supply centre produces more than the mentioned capacity and no demand centre receives more than its requirement. These constraints are met in such a fashion that the flow of goods from one destination to another incurs the minimum possible cost. The three discussed methods of finding an initial solution are:

- Least cost method
- North-West Corner method
- Vogel's approximation method (VAM)

Initial solution from these methods has been illustrated one by one.

5.5.2.1 Least Cost Method

This method involves the following steps:

i. Find the least cost cell in entire cost matrix. In Table 5.3, routes S2 – D1 and S3 – D1 each qualifies for minimum cost condition. Both have a transportation cost of $4 per unit. In case of situation of tie, the cell with the maximum allocation is selected. For route S2 – D1, the allocation of number of units

TABLE 5.4
Least Cost Method after Final Allocation

Origin	Marysville (D1)	East Liberty (D2)	Greensburg (D3)	Supply (Thousand Units)
	Destination			
Ontario (S1)	5	11 (*iii* = *45*)	6 (*iv* = *10*)	***55/30/0***
Southfield (S2)	4 (*i* = *40*)	6	5	***40/0***
Battle Creek (S3)	4 (*ii* = *25*)	6	9	***25/0***
Demand (thousand units)	***65/25/0***	***45/0***	***10/0***	

would be 40, whereas that of S3 – D1 would be 25. The concept is that the maximum number of units should be transported at least possible cost. Thus, cell a_{21} implying cost per unit of route S2 – D1 is selected. Allocate the number of units required or available. If demand is more than supply, assign the number of available units and if supply is more than demand, assign the number of units required. Supply centre S2 can produce a fixed quantity of 40 units whereas demand of D1 centre is fixed quantity of 65 units. So Southfield manufacturing facility can provide only this much of products to Marysville plant. The remaining demand of 25 units of D1 is yet to be fulfilled. But production capacity is fully utilized. So any resource that is fully utilized and becomes zero should be struck off as shown in Table 5.4.

ii. The next minimum cost cell is a_{31} indicating transportation cost of Rs.4 per unit through route S3 – D1. As demand is of only 25 units from available 25 units so 25 units would be allocated and column D1 as well as row S3 would be struck off.
iii. Cell a_{12} indicating route S1 – D2 is allocated 45 units.
iv. Cell a_{13} indicating route S1 – D3 is allocated 10 units.

Summary of least cost method findings in sequential form and calculation of total cost is shown in Table 5.5.

TABLE 5.5
Total Cost from Least Cost Method

Supply Centre	Demand Centre	Units Shipped	Cost Per Unit ($)	Total Cost
Southfield (S2)	Marysville (D1)	40	4	4*40 = 160
Battle Creek (S3)	Marysville (D1)	25	4	4*25 = 100
Ontario (S1)	East Liberty (D2)	45	11	45*11 = 495
Ontario (S1)	Greensburg (D3)	10	6	10*6 = 60
Total				$815

5.5.2.2 North-West Corner Method

This method involves the following steps:

i. As the name suggests begin allocation of units from North-West cell of cost matrix, which is a_{11} cell.
ii. Allocate the number of units required or available, i.e.

- If demand is more than supply set, the cell equal to supply and proceed *vertically*.
- If demand is less than supply set, the cell equal to demand and proceed *horizontally*.
- If demand is equal to supply, proceed *diagonally*.

These steps are detailed as under (Table 5.6) by using same example shown in Table 5.3.

iii. Starting from first cell i.e. a_{11} indicating route S1 – D1 supply is 55 units and demand is 65 units. As requirement is more than available so only 55 units are allocated to this cell. Remaining demand of 10 units needs to be satisfied from other sources. This allocation finishes supply of S1 so row 1 is eliminated shown by dashed line. Next allocation would be done to cell following rule of that if demand is more than supply so proceed vertically. So allocation would be done to cell a_{21}.
iv. The remaining demand of D1 of 10 units is allocated to cell a_{21} indicating route S2 – D1. S2 is left with 30 units. As demand is less than supply so next allocation would be done to the cell by moving horizontally. Therefore, the allocation would be done to cell a_{22}.
v. Demand of demand centre D2 is 45 units and available are 30 from S2. So 30 units are allocated to cell a_{22}. After this allocation, demand of 15 units of

TABLE 5.6
North-West Corner Solution

	Destination			
Origin	**Marysville (D1)**	**East Liberty (D2)**	**Greensburg (D3)**	**Supply (Thousand Units)**
Ontario (S1)	5	11	6	***55/0***
	(i = 55)			
Southfield (S2)	4	6		***40/30/0***
	(ii = 10)	*(iii = 30)*	5	
Battle Creek (S3)	4	6	9	***25/10/0***
		(iv = 15)	*(10)*	
Demand (thousand units)	***65/10/0***	***45/15/0***	***10/0***	

TABLE 5.7
Total Cost from North-West Corner Method

Supply Centre	Demand Centre	Units Shipped	Cost Per Unit ($)	Total Cost
Ontario (S1)	Marysville (D1)	55	5	55*5 = 275
Southfield (S2)	Marysville (D1)	10	4	10*4 = 40
Southfield (S2)	East Liberty (D2)	30	6	30*6 = 180
Battle Creek(S3)	East Liberty (D2)	15	6	15*6 = 90
Battle Creek(S3)	Greensburg (D3)	10	9	10*9 = 90
Total				$675

East Liberty (D2) centre is still to be fulfilled. Following the rule of demand greater than supply, next allocation would be done to cell a_{32}.

vi. So 15 units are allocated to route S3 – D2.
vii. Remaining 10 units to cell a_{33} indicating route S3 – D3.

Summary of North-West Corner method findings in sequential form and calculation of total cost is shown in Table 5.7.

5.5.2.3 Vogel's Approximation Method (VAM)

This method involves the following steps:

- Find the difference between the smallest and second smallest cost cell for each row and column.
- Select the row or column with the greatest difference.
- In the selected row or column assign the number of units to the least cost cell. If demand is more than supply, assign the number of available units and if supply is more than demand, assign the number of units required.
- If there is a tie in the difference between smallest and second smallest cost cell, then select row or column with minimum cost cell. If there is further tie between minimum cost cells, then select that row or column where maximum units can be allocated to that tied cell.

These steps are detailed as under by using the same example shown in Table 5.3 and the final solution is shown in Table 5.8.

viii. In the first step, the difference between the smallest and second smallest value row-wise indicated by r1 as well as column-wise indicated by c1 is calculated. Out of these values select the maximum value. In this case, it is 2 (shown in bold italics). In the selected row, allocation should be done to minimum cost cell, which is cell a_{31} with cost of $4 per unit. Twenty-five units are allocated to a_{31} and S3 is eliminated with demand for 40 units of D1 yet to be fulfilled.
ix. Repeat the process of finding out differences between the smallest and the next smallest value. These are indicated in r2 and c2. The maximum value

TABLE 5.8
VAM Method

	Destination					
Origin	**Marysville (D1)**	**East Liberty (D2)**	**Greensburg (D3)**	**Supply (Thousand Units)**	**Difference**	
Ontario (S1)	5 (*40*)	11 (*5*)	6 (*10*)	***55***	1	1
Southfield (S2)	4	6 (*ii = 40*)	5	***40/0***	1	1
Battle Creek(S3)	4 (*i = 25*)	6	9	***25/0***	**2**	
Demand (thousand units)	***65/40/0***	***45/5/0***	***10/0***			
Difference	0	0	1			
	1	**2**	1			

was obtained from D2. In this row, the minimum cost cell is a_{22} with cost of \$6. An allocation of 40 units is done to route S2 – D2 exhausting all capacity of S2, whereas 5 units of demand are left unfulfilled.

x. Finally, only cells a_{11}, a_{12} and a_{13} are left, which are allocated 40, 5 and 10 units, respectively.

From Table 5.8, transport routes and total cost can be calculated as shown in Table 5.9.

Importantly, after finding the initial solution, a second step involved in testing the feasibility of the solution should be applied. It has been discussed that this involves the following rule:

$$\text{Number of allocations} = m + n - 1$$

$$5 = 3 + 3 - 1 = 5$$

The equality of both right- and left-hand side indicates the feasibility of a solution.

The above-discussed methods provide an initial basic solution. As discussed earlier, the allocated or occupied cells represent basic variables. These variables directly contribute to the objective function of minimization. Thus, these methods can also be termed as methods of identifying basic variables. A comparison of these three methods used in the identification of basic variables is important in their selection

TABLE 5.9
Total Cost from VAM Method

Supply Centre	Demand Centre	Units Shipped	Cost Per Unit ($)	Total Cost
Battle Creek (S3)	Marysville (D1)	25	4	25*4 = 100
Southfield (S2)	East Liberty (D2)	40	6	40*6 = 240
Ontario (S1)	Marysville (D1)	40	5	40*5 = 200
Ontario (S1)	East Liberty (D2)	5	11	5*11 = 55
Ontario (S1)	Greensburg (D3)	10	6	10*6 = 60
Total				$655

of finding an initial solution. The North-West Corner method is easy to apply, but it ignores the importance of unit costs a_{ij}. These costs indicate the transportation cost of one unit. Thus, one should always select the minimum cost cell. This deficiency is removed to a good extent by the least cost method. In the minimum cost method, allocations are made based on minimum unit costs. The method is further advanced by Vogel's Approximation method. This method has been quite popular in its role of finding initial cost effectively. The criteria of considering the difference between the smallest and next smallest value implies extra cost incurred if a cell with minimum cost for some reason cannot be used for allocation, making this method more effective than the least cost method. This was aptly illustrated in the above-discussed example, where by using the least cost method, the total transportation cost was \$660, which is marginally higher than the cost produced by the VAM method (\$640). Hence, predominantly, the VAM method is more frequently used.

The next step is to check whether or not this initial solution is optimal by applying the optimality test.

5.5.3 Optimality Test

Optimality test implies conversion of identified basic variables into non-basic variables by including them in the solution. It is important to reiterate that each cell in the transportation table represents the cost per unit of a specific route. Therefore, if another such route is incorporated in the solution, it would imply that a certain number of units would be transported from a source to destination through this route. In addition, such a change in route would affect the existing routes and their capacity of transportation. Thus, an optimality test would involve:

- identifying a new route that is unused in the current solution and where allocating a fixed number of units will result in the largest per unit reduction in total cost. This new cost cell is termed as *incoming cell.*
- allocating flow denotes the number of units to be transported through a new route by adjusting the flow of previous routes in such a way that a solution remains feasible.
- inclusion of another basic variable would make the solution infeasible as the number of allocations, i.e. routes would become m + n rather than m + n − 1. Thus, the existing non-basic variable has to be converted into basic variable to keep the next solution feasible. Such cost cell or route is termed as *outgoing cell.*

We start with *identifying incoming cell.* The criteria for selecting a new route are that it should reduce total per unit transportation cost to the maximum in relation to other unoccupied cells (basic variables). Thus, to identify such a route, a procedure should be applied to compute cost for each unoccupied cell by which per unit total cost would be reduced. Such a method referring to modified distribution of flow or units to be transported is succinctly called as *MODI method.*

The method requires the introduction of certain coefficients that are symbolized as u_i for each row and v_j for each column. Calculation of these coefficients requires the building of equations in such a way that cost per unit of occupied cell is equal to

TABLE 5.10
VAM Method

	v_1	v_2	v_3	Supply (Thousand Units)
u_1	5 *(40)*	11 *(5)*	6 *(10)*	*55*
u_2	4	6 *(40)*	5	*40*
u_3	4 *(25)*	6	9	*25*
Demand (thousand units)	*65*	*45*	*10*	

the sum of corresponding u_i and v_j. The following is a set of such equations created from the solution of the above-discussed question through the VAM method. The solution with u_i and v_j coefficients is reproduced in Table 5.10.

The set of equations formulated from the requirement of $u_i + v_j = a_{ij}$ for all occupied cells in the initial solution given above would be:

$$u_1 + v_1 = 5;\ u_1 + v_2 = 11;\ u_1 + v_3 = 6;\ u_2 + v_2 = 6;\ u_3 + v_1 = 4$$

These equations give six variables and five equations. Hence, to solve them, any one variable can be assumed to take value of 0 (zero). There is no fixed rule regarding putting a particular variable as zero. We always assume $v_1 = 0$. Doing so and solving the above given set of equations would give the following values of other variables.

$$v_1 = 0;\ u_1 = 5;\ u_2 = 0;\ v_2 = 6;\ u_3 = 4 \text{ and } v_3 = 1$$

From these values, net cost implying a change in total transportation cost due to allocating one unit to unoccupied cells is calculated. Thus, the next step is to find net cost of all unoccupied cells, which is calculated by using formula: Net cost $e_{ij} = a_{ij} - u_i - v_j$

For instance, in initial solution Table 5.10 $a_{21} = 4$. If this unoccupied cell is converted to a non-basic variable i.e. a unit is allocated to it for transportation, then the change in total cost would be $e_{21} = 4 - 3 - 0 = 1$. Cells shown as (a) indicate occupied cells.

Recall that purpose of the above exercise is to identify incoming cell from all unoccupied cells. The criterion for it was that it should reduce cost to the maximum. Net evaluation cost (Table 5.11) of all unoccupied cells except a_{32} is positive, implying

TABLE 5.11
Net Cost of Unoccupied Cells

	$v_1 = 0$	$v_2 = 6$	$v_3 = 1$	Supply (Thousand Units)
$u_1 = 5$	a	a	a	*55*
$u_2 = 0$	$4 - 0 - 0 = 4$	a	$5 - 0 - 1 = 4$	*40*
$u_3 = 4$	a	$6 - 4 - 6 = \mathbf{-4}$	$9 - 4 - 1 = 4$	*25*
Demand (thousand units)	*65*	*45*	*10*	

that if any of it is introduced into a new solution, then it would result in an increase in cost rather than a decrease. Net cost of −4 indicates that transportation cost would reduce by \$4 if one unit is transported from source S3 to destination D2. Thus, cell a_{32} would become *incoming cell*. Here, it is important to mention two scenarios: first, if there are two cells with net cost as negative, then select the cell that reduces the cost to the maximum, as that is the purpose; second, if there is a tie among net costs, then select the cell with the maximum allocation, as the logic would be to transport maximum units with the least possible cost.

The next step of deciding the amount of flow or number of units that would be transported from route S3 – D2 is decided by using the stepping stone method.

Stepping stone method: Till now, it has been decided that a_{32} as the new route in the next solution would reduce cost by one unit. We would like to allocate as many units to this route as possible so that cost is reduced by maximum. However, this would ask for adjustments among allocations on other routes. The stepping stone method helps in identifying these adjustments. In addition, another result of this method is that during the adjustment process, it would also identify the *outgoing cell.*

Suppose one unit is allocated to an incoming cell, i.e. route S3 – D2. This would result in total allocated units to supply centre as 26, whereas its production capacity is 25. So, to maintain feasibility, one unit less would be allocated to an allocated cell in the same row, which in this case is a_{31} (route S3 – D1). Now, the allocation of one unit to a_{32} also increases the demand for D2 to 46 units, whereas the demand of D2 destination should not be more than 45. Again, to maintain feasibility, one unit has to be subtracted from an unallocated cell in the D2 column. However, in the same column, there are two such cells a_{22} with flow of 40 units and a_{12} with a flow of 5 units. If one unit is subtracted from a_{22}, the supply of S2 centre falls to 39, when it should be 40. But if a unit is subtracted from a_{12} making flow to 4, then the same amount of units has to be added other allocations in row S1. It cannot be done to a_{13} as demand D3 would become 11 but required is 10. So this unit is added to cell a_{11} making allocation to 41. Similarly, to maintain feasibility, the allocation of a_{31} has to be reduced to 24. This adjustment is shown in Table 5.12.

To generalize the above procedure, the following points should be underscored, which would result in the formation of a closed loop.

TABLE 5.12
Cycle of Adjustment to Decide Number of Units Allocated at Incoming Cell

Origin	Destination: Marysville (D1)	East Liberty (D2)	Greensburg (D3)	Supply (Thousand Units)
Ontario (S1)	5 (~~*40*~~ ***41***)	*8* (~~*5*~~ ***4***)	*6* (*10*)	***55***
Southfield (S2)	4	6 (*40*)	5	***40***
Battle Creek(S3)	4 (~~*25*~~ ***24***)	6 (***1***)	9	***25***
Demand (thousand units)	***65***	***45***	***10***	

TABLE 5.13
Stepping Stone Path

	Destination			
Origin	**Marysville (D1)**	**East Liberty (D2)**	**Greensburg (D3)**	**Supply (Thousand Units)**
Ontario (S1)	5 *(40)* (+)	11 *(5)* (−)	6 *(10)*	***55***
Southfield (S2)	4	6 *(40)*	5	***40***
Battle Creek(S3)	4 *(25)* (−)	6 (+)	9	***25***
Demand (thousand units)	***65***	***45***	***10***	

- Start a flow from incoming cell.
- Move either horizontally or vertically to the next allocation. But never move diagonally.
- The flow has to stop only at an allocated cell and not at unoccupied cell.
- Keep on traversing the path till flow reaches back to the starting point. Thus, at the end, a closed loop would be formed. The loop formed is $a_{32}-a_{12}-a_{11}-a_{31}-a_{32}$ (shown in Table 5.13).

The addition of another cell in the solution would violate the feasibility condition, that is, number of allocations = m + n − 1. As shown in Table 5.12, the number of allocations has increased to six, making the solution infeasible. Thus, the next step would be to identify an *outgoing cell.* For this, the following steps should be followed:

- Mark the starting point, i.e. incoming cell as (+) and then the next corner, i.e. occupied cell as (–). Repeat the process alternatively. The point is to consider each corner as a stone from which a feasible solution can step to the next corner. It was shown quite illustratively in Table 5.13 that allocations are added and subtracted alternatively. So, the procedure of marking positive and negative alternatively to occupied cells in the selected loop. (+) sign indicates that allocation will increase and (–) sign indicates that allocation will decrease. This adjustment would allocate certain units to incoming cell and also decide outgoing cell.
- To determine the maximum units allocated to the incoming cell, identify the *minus sign cell with the smallest amount allocated* (Table 5.13). In this case, the smallest such value would be 5, which are number of units allocated to cell a_{12}. This amount should be added to cells with (+) sign and subtracted from cells with (–) sign. This would lead to a new solution with allocations as shown in Table 5.14. Value in bold indicates allocations of incoming cells.

From Table 5.14 transport routes and total cost can be calculated as shown in Table 5.15.

TABLE 5.14
New Solution

Origin	Marysville (D1)	East Liberty (D2)	Greensburg (D3)	Supply (Thousand Units)
Ontario (S1)	5 (*45*)	11 (*0*)	6 (*10*)	***55***
Southfield (S2)	4	6 (*40*)	5	***40***
Battle Creek(S3)	4 (*20*)	6 (*5*)	9	***25***
Demand (thousand units)	***65***	***45***	***10***	

TABLE 5.15
Total Cost from New Solution

Supply Centre	Demand Centre	Units Shipped	Cost Per Unit ($)	Total Cost
Ontario (S1)	Marysville (D1)	45	5	45*5 = 225
Ontario (S1)	Greensburg (D3)	10	6	10*6 = 60
Southfield (S2)	East Liberty (D2)	40	6	40*6 = 240
Battle Creek(S3)	Marysville (D1)	20	4	20*4 = 80
Battle Creek(S3)	East Liberty (D2)	5	6	5*6 = 30
Total				$635

TABLE 5.16
Optimality Test of New Solution

Origin	v_1	v_2	v_3	Supply (Thousand Units)
u_1	5 (*45*)	11	6 (*10*)	***55***
u_2	4	6 (*40*)	5	***40***
u_3	4 (*20*)	6 (*5*)	9	***25***
Demand (thousand units)	***65***	***45***	***10***	

The cost has decreased from $655 to $635. This new solution has to be checked for its optimality to infer whether cost can be reduced further or not. The solution would be incomplete without checking the optimality of the new solution. This is done by again applying the MODI method as discussed before (Table 5.16).

The set of equations from occupied cells would be:

$$u_1 + v_1 = 5;\ u_1 + v_3 = 6;\ u_2 + v_2 = 6;\ u_3 + v_1 = 4 \text{ and } u_3 + v_2 = 6$$

Putting $v_1 = 0$ and solving for other equations, we get:

$$v_1 = 0;\ u_1 = 5;\ v_3 = 1;\ u_3 = 4;\ v_2 = 2;\ u_2 = 4$$

TABLE 5.17
Net Cost of Unoccupied Cells

	$v_1 = 0$	$v_2 = 2$	$v_3 = 1$	Supply (Thousand Units)
$u_1 = 5$	[a]	$11 - 5 - 2 = 4$	[a]	*55*
$u_2 = 4$	$4 - 4 - 0 = 0$	[a]	$5 - 4 - 1 = 0$	*40*
$u_3 = 4$	[a]	[a]	$9 - 4 - 1 = 4$	*25*
Demand (thousand units)	*65*	*45*	*10*	

The following table shows the net evaluation cost of unoccupied cells by using the calculated value of u_i and v_j variables (Table 5.17).

As the net cost of all unoccupied cells is positive or zero thus, it can be concluded that the new solution achieved is the most optimal. Transportation cost of $635,000 would be incurred in distributing 1,20,000 units to three destinations from three sources.

Excel Solution

The final Excel solution of the above-explained transportation problem is shown below.

Step 1: The first step is to create a transportation table in an Excel spreadsheet with origins, destinations, cost per unit from each supply centre to demand centre, total supply capacity and demand of each centre. This is shown in the Excel matrix indicated by cells A6:E6.

Step 2: At the bottom of this transportation matrix, a linear model is constructed with decision variables, objective function and both supply and demand constraints. This is shown in cells A10:G17.

Step 3: As previously discussed in the linear programming formulation of the above-discussed problem, the decision variables are the number of units to be transported from each supply to each demand centre. These are to be calculated. They are shown in cells B12:D14.

Step 4: *Supply constraints:* Cells B12:D12 represent supply constraint of S1; cells B13:D13 represent supply constraint of S2; cells B14:D14 of S3. The sum of each supply centre should not be more than its maximum capacity. So, the first supply constraint is the sum of cells B12:D12. This is shown in cell E12=sum(B12:D12). The supply constraint of S2 is shown in cell E13=sum(B13:D13). Finally, supply constraint of S3 is shown in cell E14=sum(B14:D14).

Step 5: *Demand constraints:* Cells B12:B14 represent demand constraint of D1; cells C12:C14 of D2 and D12:D14 of D3. The sum of each demand centre is equivalent to the total number of units required as shown in cells B6:B8. So, the first demand constraint of D1 is shown in cell B15=sum(B12:B14). Demand constraint of D2 is C15=sum(C12:C14). Finally, the demand constraint of D3 is D15=sum(D12:D14).

Step 6: Right-hand side value of constraints is copied from the original transportation matrix. For instance, for S1, right-hand side value is shown in cell G12=E3. This is copied to G13 and G14. For D1, the right-hand side value is shown in cell B17=B6. This is copied to B18 and B19.

Step 7: *Objective function Z:* In the transportation model, the objective function is to minimize cost of transportation, which is function of number of units to be transported from each centre and corresponding per unit cost. The cost per unit is shown in cells B3:D5. The number of units transported would be calculated in matrix B12:D14. So, objective function Z is shown in cell B8=sumproduct (B3:D5, B12:D14).

The solution solved by using the Solver application is shown in Figure 5.1

Set target cell = B8

By changing cells = B12:D14

Subject to the constraints:

Demand constraints: B15=B17; C15=C17; D15=D17

Supply constraints: E12 ≤ G12; E13 ≤ G13; E14 ≤ G14

The solution indicates the total cost of transportation to be $635. That is the same as solved by using the transportation model.

B8 =SUMPRODUCT(B3:D5,B12:D14)

	A	B	C	D	E	F	G
1			**Destination**		***Supply (thousand units)***		
2	**Origin**	Marysville (D1)	East Liberty (D2)	Greensburg (D3)			
3	Ontario (S1)	5	11	6	55		
4	Southfield (S2)	4	6	5	40		
5	Battle Creek (S3)	4	6	9	25		
6	***Demand (thousand units)***	65	45	10			
7							
8	***Minimize***	635					
9							
10			**Destination**		***Supply (thousand units)***		
11	**Origin**	Marysville (D1)	East Liberty (D2)	Greensburg (D3)			
12	Ontario (S1)	45	0	10	55	<=	55
13	Southfield (S2)	0	40	0	40	<=	40
14	Battle Creek (S3)	20	5	0	25	<=	25
15	***Demand (thousand units)***	65	45	10			
16		=	=	=			
17		65	45	10			

FIGURE 5.1 Excel solution.

5.6 UNBALANCED TRANSPORTATION MODEL

In realistic situations, majority of times supply or demand is not fixed, and they vary depending on change in production constraints or customer requirements. In such cases of inequality between supply and demand, the transportation model problem becomes an unbalanced one. This can be resolved by adding a *dummy supply centre if total supply is less than total demand or add a dummy demand centre if total supply is more than total demand.* As such a centre is a dummy, the cost of transporting each unit would be least and assumed to be zero. After matching the supply and demand requirements, the problem can be solved as basic transportation problem which has been discussed in detail in previous sections. For illustration above, the discussed case of Musashi India Ltd. is advanced further in two ways. The first case discusses when production capacity is maintaining demand at fixed levels. The second case discusses the scenario of demand being more than production capacity which is kept at fixed levels.

5.6.1 Scenario 1: Supply is More than Demand

Due to production constraints such as maintenance, breakdowns, variation in the availability of raw material, etc., the manufacturer, in this case, Musashi Auto Parts Michigan, tends to produce and stock finished auto parts even before the order is placed. This is done to avoid any bottlenecks in production when the process of fulfillment of an order begins. The amount of overproduction is based on previous demand of three consumers. Thus, considering both production and demand constraints, three supply centres were able to produce a maximum of 65, 50 and 35 (in thousands) units. The demand for all three destinations was at a fixed level. Estimate the most viable route to supply and minimum total transportation cost.

Step 1: Formulation of model:

As discussed in the previous section, the notation for different variables would be:

x_{ij} = number of units to be transported from supply centre i to demand centre j.

b_i = maximum production capacity of supply centre *i*.

c_j = fixed demand of destination j.

Constraints regarding both supply and demand constraints are formulated by using a linear programming method. Supply of each source is not fixed and is given at maximum production capacity. This would involve construction of supply constraints with $\leq$ inequality. Demand for each destination is kept at a fixed level, keeping demand constraints with equality constraint function. Thus,

Constraints pertaining to demand would be:

$$x_{11}+x_{21}+x_{31}=c_1$$

$$x_{12}+x_{22}+x_{32}=c_2$$

$$x_{13}+x_{23}+x_{33}=c_3$$

Constraints pertaining to supply would be:

$$x_{11}+x_{12}+x_{13}\leq b_1$$

$$x_{21}+x_{22}+x_{23} \leq b_2$$

$$x_{31}+x_{32}+x_{33} \leq b_3$$

where $x_{11}, x_{12}, x_{13}, x_{21}, x_{22}, x_{23}, x_{31}, x_{32}, x_{33} \geq 0$

In order to fit the transportation model, these inequalities need to be converted into equalities, which was done by adding slack variables in each constraint function was done in the chapter of linear programming. These slack variables indicate unused capacity. In the transportation model, these slack variables are allocations done to a dummy destination. The demand for dummy destination representing unused or spare production capacity is calculated quite understandably as:

$$(b_1+b_2+b_3) - (c_1+c_2+c_3)$$

For this case, it would be: $(65 + 50 + 35) - (65 + 45 + 10) = 30$

Costs associated with these dummy destination routes are zero as these are fictitious allocations and no cost can be attributed to them. The resulting final transportation model is shown in Table 5.18.

With the addition of a dummy destination (D4), the initial transportation model becomes feasible as it fulfils the criteria of total supply being equal to total requirement, i.e.

Total supply = 65 + 50 + 35 = Total demand = 65 + 45 + 10 + 30 = 150 units.

Step 2: *Initial solution:* The first feasible solution is calculated by using the VAM method shown below (Table 5.19).

Table 5.20 shows the calculation of total transportation cost.

The second step involved in testing the feasibility of solution is to check:

$$\text{Number of allocations} = m + n - 1$$

$$6 = 3 + 4 - 1 = 6$$

The equality of both right- and left-hand side indicates the feasibility of solution.

TABLE 5.18
Model

	Destination				
Origin	**Marysville (D1)**	**East Liberty (D2)**	**Greensburg (D3)**	**Dummy destination (D4)**	**Supply (Thousand Units)**
Ontario (S1)	5	11	6	0	***65***
Southfield (S2)	4	6	5	0	***50***
Battle Creek(S3)	4	6	9	0	***35***
Demand (thousand units)	***65***	***45***	***10***	***30***	

TABLE 5.19
Initial Solution by VAM Method

Origin	Marysville (D1)	East Liberty (D2)	Greensburg (D3)	D4	Supply (Thousand Units)	Difference			
Ontario (S1)	5 (*25*)	11	6 (*10*)	0 (*30*)	***65/35***	***5***	1	1	1
Southfield (S2)	4 (*5*)	6 (*45*)	5	0	***50/5/0***	4	1	1	***1***
Battle Creek(S3)	4 (*35*)	6	9	0	***35/0***	4	**2**		
Demand (thousand units)	***65/30/25***	***45/0***	***10***	***30/0***					
Difference	0	0	1	0					
	0	0	1						
	1	***5***	1						
	1		1						

TABLE 5.20
Total Cost from Initial Solution

Supply Centre	Demand Centre	Units Shipped	Cost Per Unit ($)	Total Cost
Ontario (S1)	D4	30	0	30*0 = 30
Battle Creek(S3)	Marysville (D1)	35	4	35*4 = 140
Southfield (S2)	East Liberty (D2)	45	6	45*6 = 270
Southfield (S2)	Marysville (D1)	5	4	5*4 = 20
Ontario (S1)	Marysville (D1)	25	5	25*5 = 125
Ontario (S1)	Greensburg (D3)	10	6	6*10 = 60
Total				$615

TABLE 5.21
Optimality Test

Origin	v_1	v_2	v_3	v_4	Supply (Thousand Units)
u_1	5 (*25*)	11	6 (*10*)	0 (*30*)	***65***
u_2	4 (*5*)	6 (*45*)	5	0	***50***
u_3	4 (*35*)	6	9	0	***35***
Demand (thousand units)	***65***	***45***	***10***	***30***	

Step 3: *Optimality test:* Applying the MODI method, also known as u – v method, will give Table 5.21

The set of equations formulated from the requirement of $u_i + v_j = a_{ij}$ for all occupied cells in the initial solution given above would be:

TABLE 5.22
Net Evaluation Cost of Unoccupied Cells

	Destination				
Origin	$v_1 = 0$	$v_2 = 2$	$v_3 = 1$	$v_4 = -5$	**Supply (Thousand Units)**
$u_1 = 5$	a	$11 - 5 - 2 = 4$	a	a	*65*
$u_2 = 4$	a	a	$5 - 4 - 1 = 0$	$0 - 4 + 5 = 1$	*50*
$u_3 = 4$	a	$6 - 4 - 2 = 0$	$9 - 4 - 1 = 4$	$0 - 4 + 5 = 1$	*35*
Demand (thousand units)	*65*	*45*	*10*	*30*	

$$u_1 + v_1 = 5;\ u_1 + v_3 = 6;\ u_1 + v_4 = 0;\ u_2 + v_1 = 4;\ u_2 + v_2 = 6;\ u_3 + v_1 = 4$$

The solution of these equations would give:

$v_1 = 0$; $u_1 = 5$; $v_3 = 1$; $v_4 = -5$; $u_2 = 4$; $v_2 = 2$ and $u_3 = 4$. Using these values, the net evaluation cost for unoccupied cells is calculated as shown in Table 5.22.

As all net evaluation cost of unoccupied cells is either zero or positive, so the initial solution is most optimal. It is interesting to note that this case provided the scenario in which, due to fluctuating demand, all supply centres tend to produce more than required, leading to mismatch between supply and demand. This was resolved by adding a dummy or fictional demand centre to make the model feasible. In reality, this demand centre does not exist. Now the question is are all supply centres are left with unused production capacity or only one of them faces such a situation. This inference was drawn from the final solution, which indicated that only supply centre S1 was left with idle units of 30. So, out of produced 65 units, 35 were transported to various demand centres and 35 were left idle. Increased production capacity of other sources was fully utilized. This is an important interpretation of unbalanced transportation model when production is more than requirement. The next illustration would help in reaching a similar conclusion when demand or requirement is higher than production capacity.

Excel Solution

See Figure 5.2.

5.6.2 Scenario 2: Demand is More than Supply

This case takes into consideration the scenario of fixed production capacity and estimated demand of three destinations. Performing under normal conditions three supply centres were producing a fixed quantity of 55, 40 and 25 (in thousands) units. The data corresponded with requirement data indicated the difference between capacity and increased demand of three destinations. To offset this Musashi mostly resort to producing in advance resulting in creation of stock and increase in inventory cost. The varying demand data of past 6 months show that minimum requirement of auto parts from D1 was 75,000 units whereas that D2 was found to be 60,000 units. Finally, data showed a marginal decrease in demand from D3 for last 6 months. But due to approaching festive season, plant at D3 estimated to sell at least 15,000 units.

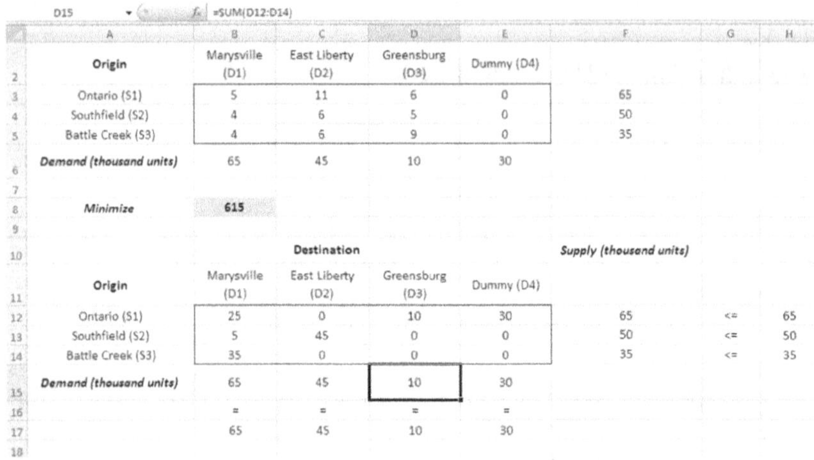

D15 =SUM(D12:D14)

	A	B	C	D	E	F	G	H
2	Origin	Marysville (D1)	East Liberty (D2)	Greensburg (D3)	Dummy (D4)			
3	Ontario (S1)	5	11	6	0	65		
4	Southfield (S2)	4	6	5	0	50		
5	Battle Creek (S3)	4	6	9	0	35		
6	Demand (thousand units)	65	45	10	30			
7								
8	Minimize	615						
9								
10			Destination			Supply (thousand units)		
11	Origin	Marysville (D1)	East Liberty (D2)	Greensburg (D3)	Dummy (D4)			
12	Ontario (S1)	25	0	10	30	65	<=	65
13	Southfield (S2)	5	45	0	0	50	<=	50
14	Battle Creek (S3)	35	0	0	0	35	<=	35
15	Demand (thousand units)	65	45	10	30			
16		=	=	=	=			
17		65	45	10	30			
18								

FIGURE 5.2 Excel solution.

Estimate number of units to be transported from each source in such a way that total transportation cost is minimum.

Step 1: *Formulation of model:*

Constructing the model in a linear programming method would result in the following constraints. It is important to note that in this case, demand constraints are formulated with inequality constraint functions, whereas supply constraints are formed with equality functions. It is important to note that in this case of demand outstripping supply, inequalities would be of $\geq$ type as demand centres would like Musashi to fulfil at least a certain limit of demand and suppliers are allowed to supply more than this limit.

Constraints pertaining to demand would be:

$$x_{11}+x_{21}+x_{31} \geq c_1$$

$$x_{12}+x_{22}+x_{32} \geq c_2$$

$$x_{13}+x_{23}+x_{33} \geq c_3$$

Constraints pertaining to supply would be:

$$x_{11}+x_{12}+x_{13}=b_1$$

$$x_{21}+x_{22}+x_{23}=b_2$$

$$x_{31}+x_{32}+x_{33}=b_3$$

where $x_{11}, x_{12}, x_{13}, x_{21}, x_{22}, x_{23}, x_{31}, x_{32}, x_{33} \geq 0$

These inequalities are converted into equalities by introducing surplus and artificial variables in the equations in such a way that a criterion of non-negativity is fulfilled. This was discussed in detail in the chapter of linear programming. These surplus variables are represented by allocations assigned to cost cells of newly incorporated dummy supply centre. Thus, the final solution would provide an inference regarding the number of units allocated (surplus variable) to which demand centre implying unfulfilled demand. For this, first, the transportation model is formulated by understanding what the unmet demand is. This is shown below:

Total supply = 55 + 40 + 25 = 120 whereas

Total demand = 75 + 60 + 15 = 150

Thus, demand is more than supply by 20 units, so to make transportation model feasible, another fictitious or dummy supply centre needs to be added – the production capacity for which would be 150 – 120 = 30 units. This implies that the extra demand of 30 units cannot be met by existing supply centres and needs to be satisfied by another source that, in this case, is a dummy centre. This results in formulation of the following transportation model (Table 5.23):

With addition of a dummy supply centre (S4), the initial transportation model as shown in Table 5.22 becomes feasible as it fulfils the criteria of total supply being equal to total requirement, i.e.

Total supply = 55 + 40 + 25 + 30 = Total demand = 75 + 60 + 15 = 150 units.

Step 2: *Initial solution*: The first feasible solution is calculated by using the VAM method shown below (Table 5.24).

Total cost would be (Table 5.25):

The second step involved in testing the feasibility of solution is to check:

$$\text{Number of allocations} = m + n - 1$$

$$6 = 4 + 3 - 1 = 6$$

The equality of both right- and left-hand side indicates the feasibility of solution.

Step 3: *Optimality test*: The application of the MODI method, also known as u – v method, is given in Table 5.26.

So,

$u_1 + v_1 = 5$; $u_1 + v_3 = 6$; $u_2 + v_1 = 4$; $u_2 + v_2 = 6$; $u_3 + v_1 = 4$ and $u_4 + v_2 = 0$. The solution of these equations would give values of various four 'u' and three 'v' variables shown in Table 5.27 along with net evaluation costs of unoccupied cells.

TABLE 5.23
Model

	Destination			
Origin	**Marysville (D1)**	**East Liberty (D2)**	**Greensburg (D3)**	**Supply (Thousand Units)**
Ontario (S1)	5	11	6	***55***
Southfield (S2)	4	6	5	***40***
Battle Creek (S3)	4	6	9	***25***
Dummy supply centre (S4)	0	0	0	***30***
Demand (thousand units)	***75***	***60***	***15***	

TABLE 5.24
Initial Solution by VAM Method

Origin	Destination: Marysville (D1)	East Liberty (D2)	Greensburg (D3)	Supply (thousand units)	Difference
Ontario (S1)	5 (*40*)	11	6 (*15*)	***55/0***	1 1 1 1
Southfield (S2)	4 (*10*)	6 (*30*)	5	***40/10/0***	1 1 1 ***1***
Battle Creek(S3)	4 (*25*)	6	9	***25/0***	2 ***2***
Dummy supply centre (S4)	0	0 (*30*)	0	***30/0***	0
Demand (thousand units)	***75/50/40/0***	***60/30/0***	***15***		
Difference	4	***6***	5		
	0	0	1		
	1	***5***	1		
	1		1		

TABLE 5.25
Total Cost from Initial Solution

Supply Centre	Demand Centre	Units Shipped	Cost Per Unit (Rs.)	Total Cost
S4	East Liberty (D2)	30	0	30*0 = 0
Battle Creek(S3)	Marysville (D1)	25	4	25*4 = 100
Southfield (S2)	East Liberty (D2)	30	6	30*6 = 180
Southfield (S2)	Marysville (D1)	10	4	10*4 = 40
Ontario (S1)	Marysville (D1)	40	5	40*5 = 200
Ontario (S1)	Greensburg (D3)	15	6	15*6 = 90
Total				$610

TABLE 5.26
Optimality Test

Origin	Destination: v_1	v_2	v_3	Supply (Thousand Units)
u_1	5 (*40*)	11	6 (*15*)	***65***
u_2	4 (*10*)	6 (*30*)	5	***50***
u_3	4 (*25*)	6	9	***35***
u_4	0	0 (*30*)	0	30
Demand (thousand units)	***65***	***45***	***10***	

TABLE 5.27
Net Evaluation Cost of Unoccupied Cells

	Destination			
Origin	$v_1 = 0$	$v_2 = 2$	$v_3 = 1$	**Supply (Thousand Units)**
$u_1 = 5$	a	$11 - 5 - 2 = 4$	a	*65*
$u_2 = 4$	a	a	$5 - 4 - 1 = 0$	*50*
$u_3 = 4$	a	$6 - 4 - 2 = 0$	$9 - 4 - 1 = 4$	*35*
$u_4 = -2$	0	$0 + 2 - 2 = 0$	0	30
Demand (thousand units)	*65*	*45*	*10*	

As all net evaluation cost of unoccupied cells is either zero or positive, so initial solution is most optimal. In addition to the purpose of finding the most optimal route for transporting a certain number of units from a particular source to a destination at the most optimal cost, an unbalanced transportation problem also indicates unmet demand of a destination. According to Table 5.23, 30 units should be supplied from dummy source centre S4 to destination D2. As this source is a fictitious one, it implies that 30 unit demand for D2 is unmet, whereas the increased demand from other demand centres has been met. Thus, supplier in this case Musashi, should first fulfil the demand of D2 if they decide to increase their production capacity.

Excel Solution

See Figure 5.3.

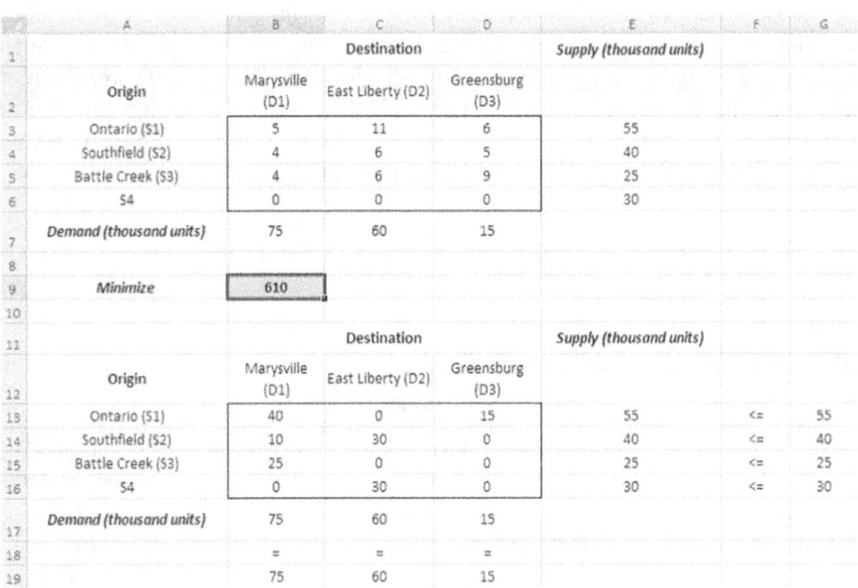

FIGURE 5.3 Excel solution.

5.7 DEGENERACY

The chapter on simplex explained that if in a solution there is a tie in the right-hand side values leading to making one or more basic variables as zero in a subsequent iteration, then the solution is termed as degenerate. In the transportation model, degeneracy is identified by comparing the number of allocated or occupied cells after initial solution with one less number of constraint functions, i.e. number of allocations = m + n – 1. Basic variables in the transportation model are represented by occupied cells. Now if the above-stated condition is not satisfied, then a transportation solution would be termed as degenerate. For reaching a basic feasible solution, out of remaining unoccupied cells one of the cost cells would be part of the solution, making it a basic variable that would satisfy the condition of the number of allocated cells. These remaining cells are considered to have allocation of zero units, substantiating the concept of basic variable becoming zero in the case of degeneracy. Thus, any of these cells which would be incoming cell would have an allocation of zero units. It is important to remember that this cell is artificially created with a flow of zero units.

Now the question is which of the cell from all of the unoccupied cells would be an artificial cell. For this, two considerations should be met. First, as the purpose is to minimize total transportation costs, so cell with least per unit transportation cost from unoccupied cells should be selected. In the case of a tie, then the cell with the maximum allocation is selected. Second and more importantly, the selected cell should be able to compute the remaining row (u) and column (v) variables.

To illustrate the solution procedure, Table 5.4 has been reproduced here in the form of Table 5.28.

Number of allocations = 4, whereas m + n – 1 = 5. Thus, the condition of feasibility is violated, and the initial solution is degenerate. Out of the remaining unoccupied cells, a_{11} and a_{23} have least per unit cost of $5. However, there is a tie so, which of them would be considered as artificial variable. The second condition says that it should be able to solve for all 'u' and 'v' variables. So we would solve first by taking a_{11} and then considering a_{23} as an artificial (basic) variable.

TABLE 5.28
Initial Solution from Least Cost Method

	Destination			
Origin	**Marysville (D1)**	**East Liberty (D2)**	**Greensburg (D3)**	**Supply (Thousand Units)**
Ontario (S1)	5	11 (*45*)	6 (*10*)	***55/30/0***
Southfield (S2)	4 (*40*)	6	5	***40/0***
Battle Creek (S3)	4 (*25*)	6	9	***25/0***
Demand (thousand units)	***65/25/0***	***45/0***	***10/0***	

5.7.1 Scenario 1: With A_{11} as New Basic Variable

An allocation of zero units (shown in bold italics) is done to cell a_{11} implying that initially zero units are transported from S1 – D1 (Tables 5.29–5.34).

5.7.2 Scenario 2: With A_{23} as New Basic Variable

Comparing two cases clearly indicates that making a_{11} as the new basic variable is better as it results in less total transportation cost per unit ($635) than with cell a_{23} ($815) which is significantly lower (Tables 5.35–5.38). Other important inferences can be drawn regarding the routes of transportation. For instance, in case 1, supply centre S1 was supplied to D1 and D3, whereas in case 2, the same is supplied to D2

TABLE 5.29
Initial Solution from Least Cost Method (Iteration 0)

	Destination			
Origin	v_1	v_2	v_3	**Supply (Thousand Units)**
u_1	5 (***0***)	11 (*45*)	6 (*10*)	***55***
u_2	4 (*40*)	6	5	***40***
u_3	4 (*25*)	6	9	***25***
Demand (thousand units)	***65***	***45***	***10***	

TABLE 5.30
Net Evaluation Cost of Unoccupied Cells

	Destination			Supply
Origin	$v_1 = 0$	$v_2 = 6$	$v_3 = 1$	**(Thousand Units)**
$u_1 = 5$	5 (***0***)	11 (*45*)	6 (*10*)	***55***
$u_2 = 4$	4 (*40*)	6 − 4 − 6 = −4	5 − 4 − 1 = 0	***40***
$u_3 = 4$	4 (*25*)	6 − 4 − 6 = −4	9 − 4 − 1 = 4	***25***
Demand (thousand units)	***65***	***45***	***10***	

TABLE 5.31
Stepping Stone Method

	D1	**D2**	**D3**	
S1	5 (***0***) (+)	11 (*45*) (−)	6 (*10*)	***55***
S2	4 (*40*) (−)	6 (+)	5	***40***
S3	4 (*25*)	6	9	***25***
	65	***45***	***10***	

TABLE 5.32
New Solution (Iteration 1)

	D1	D2	D3	
S1	5 (*40*)	11 (*5*)	6 (*10*)	***55***
S2	4	6 (40)	5	***40***
S3	4 (*25*)	6	9	***25***
	65	***45***	***10***	

TABLE 5.33
Final Solution (Iteration 2)

	D1	D2	D3	
S1	5 (*45*)	11	6 (*10*)	***55***
S2	4	6 (40)	5	***40***
S3	4 (*20*)	6 (5)	9	***25***
	65	***45***	***10***	

TABLE 5.34
Total Cost from Final Solution

Supply Centre	Demand Centre	Units Shipped	Cost Per Unit (Rs.)	Total Cost
S1	D1	45	5	45*5 = 225
S1	D3	10	6	10*6 = 60
S2	D2	40	6	40*6 = 240
S3	D1	20	4	20*4 = 80
S3	D2	5	6	5*6 = 30
Total cost				$635

TABLE 5.35
Iteration 0

	Destination			
Origin	v_1	v_2	v_3	**Supply (Thousand Units)**
u_1	5	11 (*45*)	6 (*10*)	***55***
u_2	4 (*40*)	6	5 (*0*)	***40***
u_3	4 (*25*)	6	9	***25***
Demand (thousand units)	***65***	***45***	***10***	

TABLE 5.36
New Solution (Iteration 1)

	D1	D2	D3	Supply (Thousand Units)
S1	5	11 *(45)*	6 *(10)*	***55***
S2	4 *(40)*	6 *(0)*	5	***40***
S3	4 *(25)*	6	9	***25***
	65	***45***	***10***	

TABLE 5.37
Final Solution (Iteration 2)

	D1	D2	D3	Supply (Thousand Units)
S1	5	11 *(45)*	6 *(10)*	***55***
S2	4 *(40)*	6	5	***40***
S3	4 *(25)*	6 *(0)*	9	***25***
	65	***45***	***10***	

TABLE 5.38
Total Cost from Final Solution

Supply Centre	Demand Centre	Units Shipped	Cost Per Unit (Rs.)	Total Cost
S1	D2	45	11	45*11 = 495
S1	D3	10	6	10*6 = 60
S2	D1	40	4	40*4 = 160
S3	D1	25	4	25*4 = 100
S3	D2	0	6	0*6 = 0
Total cost				$815

and D3. Similarly, S2 was used to source material to fulfil demand of centre D2 in case 1, whereas S2 is fulfilling the demand of D1 in case 2. Finally, S3 in both cases is supplying to the same demand centres in both but different amounts of units. D2 is being shipped zero units from S3 in case 2, whereas 5 units were supplied in case 1. Thus, though both cases would provide feasible solution, but introduction of particular cost cell as basic variable has important practical applications. This realistic picture provided by the application of transportation model helps managers to decide among alternatives and reach a better decision.

5.8 MAXIMIZATION

The transportation model can also be used for maximization objectives. In the transportation tableau, instead of cost cells, the routes from various sources to various destinations would represent per unit profit. The purpose would be to transport a certain number of units at maximum possible profit after giving due consideration to identifying routes with minimum transportation cost. As profit is a function of cost, so reduction in cost would in all likelihood increase profit. Thus, finding out routes in a transportation model that would result in maximum profit is linked with the objective of cost minimization. There are two methods to finding an initial solution for maximization transportation problems.

Method 1: Identify the maximum value from the profit matrix and subtract it from all other values in the matrix. The resultant matrix would be a cost matrix, and it can be solved in a similar fashion to other minimization problems.

Method 2: First, while finding an initial solution, for instance by the VAM method, instead of allocating to the least cell, allocate a certain number of units to the maximum cell. Second, when using the MODI method to check for optimality, instead of selecting the most negative cell while deciding incoming cell, select cell with the highest positive e_{ij} value. In the case of minimization, the most negative e_{ij} used to become the incoming variable.

Both the methods are illustrated by considering the same model shown in Table 5.4. The model will have the same number of suppliers and demand centres. After deducting shipping costs, the values in cells imply profit earned by supplying thousands of units from each supply centre to other demand centre. The model is given in Table 5.39.

Method 1:

To convert profit matrix into cost matrix, subtract maximum value i.e. 70 from all other values. The following is cost matrix (Table 5.40):

The initial feasible solution is found by using the VAM method shown in Table 5.41.

Number of allocations = 5 = m + n – 1 = 5

Further optimality is checked by applying the MODI method. Values of u and v variables and net evaluation costs of unoccupied cells are shown in Tables 5.42–5.48.

Thus, the solution is optimal.

$$\text{Total profit} = 55*55 + 67*40 + 50*10 + 66*5 + 65*10 = \$7,185$$

TABLE 5.39
Initial Model (Profit Matrix)

	Destination			
Origin	**Marysville (D1)**	**East Liberty (D2)**	**Greensburg (D3)**	**Supply (Thousand Units)**
Ontario (S1)	55	70	62	55
Southfield (S2)	50	67	60	40
Battle Creek (S3)	50	66	65	25
Demand (thousand units)	65	45	10	

TABLE 5.40
Initial Model (Cost Matrix)

Origin	Destination: Marysville (D1)	East Liberty (D2)	Greensburg (D3)	Supply (Thousand Units)
Ontario (S1)	15	0	8	*55*
Southfield (S2)	20	3	10	*40*
Battle Creek(S3)	20	4	5	*25*
Demand (thousand units)	*65*	*45*	*10*	

TABLE 5.41
Initial Solution (Iteration 0)

Origin	Destination: D1	D2	D3	Supply (thousand units)	Difference	
S1	15 (*10*)	0 (*45*)	8	*55/10/0*	*8*	7
S2	20 (*40*)	3	10	*40/0*	7	10
S3	20 (*15*)	4	5 (*10*)	*25/15/0*	1	*15*
Demand (thousand units)	*65/55/40/15*	*45/0*	*10/0*			
Difference	5	3	3			
	5		3			

TABLE 5.42
Net Evaluation Cost

Origin	Destination: $v_1 = 0$	$v_2 = -15$	$v_3 = -15$	Supply (Thousand Units)
$u_1 = 15$	15 (*10*)	0 (*45*)	$8 - 15 + 15 = 0$	*55*
$u_2 = 20$	20 (*40*)	$3 - 20 + 15 = -2$	$10 - 20 + 15 = 5$	*40*
$u_3 = 20$	20 (*15*)	$4 - 20 + 15 = -1$	5 (*10*)	*25*
Demand (thousand units)	*65*	*45*	*10*	

Method 2:

Another method to solving maximization problem is to start finding initial solution by allocating to maximum profit cell. Initial solution and iterations involved in entire process of obtaining optimal and feasible solution are shown in Tables 5.49–5.56.

$$\text{Total profit} = 55*55 + 50*10 + 67*20 + 60*10 + 66*25 = \$7{,}115.$$

TABLE 5.43
Stepping Stone Method

Origin	**D1**	**D2**	**D3**	**Supply (Thousand Units)**
	Destination			
S1	15 (*10*)	0 (*45*)	8	
	(+)	(−)		
S2	20 (*40*)	3	10	***55***
	(−)	(+)		
S3	20 (*15*)	4	5 (*10*)	***40***
Demand (thousand units)	***65***	***45***	***10***	***25***

TABLE 5.44
Iteration 1

	D1	**D2**	**D3**	**Supply (Thousand Units)**
S1	15 (50)	0 (5)	8	***55***
S2	20	3 (40)	10	***40***
S3	20 (15)	4	5 (10)	***25***
Demand (thousand units)	***65***	***45***	***10***	

TABLE 5.45
Net Evaluation Cost

Origin	**Destination** $v_1 = 0$	$v_2 = -15$	$v_3 = -15$	**Supply (Thousand Units)**
$u_1 = 15$	15 (50)	0 (5)	$8 - 15 + 15 = 8$	***55***
$u_2 = 18$	$20 - 18 - 0 = 2$	3 (40)	$10 - 18 + 15 = 7$	***40***
$u_3 = 20$	20 (15)	$4 - 20 + 15 = -1$	5 (10)	***25***
Demand (thousand units)	***65***	***45***	***10***	

TABLE 5.46
Stepping Stone Method

Origin	**D1**	**D2**	**D3**	**Supply (Thousand Units)**
	Destination			
S1	15 (50)	0 (5)	8	
	(+)	(−)		
S2	20	3 (40)	10	***55***
S3	20 (15)	4	5 (10)	***40***
	(−)	(+)		
Demand (thousand units)	***65***	***45***	***10***	***25***

TABLE 5.47
Iteration 2

	D1	D2	D3	Supply (Thousand Units)
S1	15 (55)	0	8	*55*
S2	20	3 (40)	10	*40*
S3	20 (10)	4 (5)	5 (10)	*25*
Demand (thousand units)	*65*	*45*	*10*	

TABLE 5.48
Net Evaluation Cost

	Destination			
Origin	$v_1 = 0$	$v_2 = -16$	$v_3 = -15$	**Supply (Thousand Units)**
$u_1 = 15$	15 (55)	0 – 15 + 16 = 1	8 – 15 + 15 = 8	*55*
$u_2 = 19$	20 – 19 – 0 = 1	3 (40)	10 – 19 + 15 = 6	*40*
$u_3 = 20$	20 (10)	4 (5)	5 (10)	*25*
Demand (thousand units)	*65*	*45*	*10*	

TABLE 5.49
Initial Solution (Iteration 0)

	Destination					
Origin	**D1**	**D2**	**D3**	**Supply (Thousand Units)**	**Difference**	
S1	55 (*55*)	70	62	*55/0*	7	7
S2	50 (*10*)	67 (*20*)	60 (*10*)	*40/20/10/0*	10	10
S3	50	66 (*25*)	65	*25/0*	*15*	
Demand (thousand units)	*65/10/0*	*45/20/0*	*10/0*			
Difference	0	1	2			
	5	3	2			

TABLE 5.50
Net Evaluation Cost

	Destination			
Origin	$v_1 = 0$	$v_2 = 17$	$v_3 = 10$	**Supply (thousand units)**
$u_1 = 55$	55 (*55*)	70 – 55 – 17 = –2	62 – 55 – 10 = –3	*55*
$u_2 = 50$	50 (*10*)	67 (*20*)	60 (*10*)	*40*
$u_3 = 49$	50 – 49 – 0 = 1	66 (*25*)	65 – 49 – 10 = *6*	*25*
Demand (thousand units)	*65*	*45*	*10*	

TABLE 5.51
Stepping Stone Method

Origin	D1	D2	D3	Supply (Thousand Units)
S1	55 (*55*)	70	62	***55***
S2	50 (*10*)	67 (*20*)	60 (*10*)	***40***
		(+)	(−)	
S3	50	66 (*25*)	65	***25***
		(−)	(+)	
Demand (thousand units)	***65***	***45***	***10***	

TABLE 5.52
Iteration 1

Origin	D1	D2	D3	Supply (Thousand Units)
S1	55 (*55*)	70	62	***55***
S2	50 (*10*)	67 (*30*)	60	***40***
S3	50	66 (*15*)	65 (*10*)	***25***
Demand (thousand units)	***65***	***45***	***10***	

TABLE 5.53
Net Evaluation Cost

	Destination			Supply
Origin	$v_1 = 0$	$v_2 = 17$	$v_3 = 16$	**(Thousand Units)**
$u_1 = 55$	55 (*55*)	70 − 55 − 17 = −2	62 − 55 − 16 = −9	***55***
$u_2 = 50$	50 (*10*)	67 (*30*)	60 − 50 − 16 = −6	***40***
$u_3 = 49$	50 − 49 − 0 = *1*	66 (*15*)	65 (*10*)	***25***
Demand (thousand units)	***65***	***45***	***10***	

TABLE 5.54
Stepping Stone Method

Origin	D1	D2	D3	Supply (Thousand Units)
S1	55 (*55*)	70	62	***55***
S2	50 (*10*)	67 (*30*)	60	***40***
	(−)	(+)		
S3	50	66 (*15*)	65 (*10*)	***25***
	(+)	(−)		
Demand (thousand units)	***65***	***45***	***10***	

TABLE 5.55
Iteration 2

Origin	D1	D2	D3	Supply (thousand units)
S1	55 (*55*)	70	62	***55***
S2	50	67 (*40*)	60	***40***
S3	50 (*10*)	66 (*5*)	65 (*10*)	***25***
Demand (thousand units)	***65***	***45***	***10***	

TABLE 5.56
Net Evaluation Cost

	Destination			Supply
Origin	$v_1 = 0$	$v_2 = 16$	$v_3 = 15$	**(Thousand Units)**
$u_1 = 55$	55 (*55*)	70 – 55 – 16 = –1	62 – 55 – 15 = –8	***55***
$u_2 = 51$	50 – 51 – 0 = –1	67 (*40*)	60	***40***
$u_3 = 50$	50 (*10*)	66 (*5*)	65 (*10*)	***25***
Demand (thousand units)	***65***	***45***	***10***	

As all net values of unoccupied cells are negative so profit cannot be increased further, resulting in optimal solution.

$$\text{Total profit} = 55*55 + 67*40 + 50*10 + 66*5 + 65*10 = \$7,185$$

As the final answer from both methods is exactly the same, so any of the methods can be applied.

5.8.1 Linear Programming Formulation

It can also be formulated as LPP, which is shown below. It is worth noting two variations from minimization problem. First, objective changes to maximization with coefficients in objective function indicate profit of transporting thousand units from a particular supply centre to a demand centre. Second, demand constraint inequalities changes from '=' to '≤'.

Maximize $Z = 55x_{11} + 70x_{12} + 62x_{13} + 50x_{21} + 67x_{22} + 60x_{23} + 50x_{31} + 66x_{32} + 65x_{33}$

Subject to:

Constraints pertaining to supply would be:

$$x_{11} + x_{12} + x_{13} \leq 55{,}000$$

$$x_{21} + x_{22} + x_{23} \leq 40{,}000$$

	A	B	C	D	E	F	G
2	**Origin**	Marysville (D1)	East Liberty (D2)	Greensburg (D3)			
3	Ontario (S1)	55	70	62	55		
4	Southfield (S2)	50	67	60	40		
5	Battle Creek (S3)	50	66	65	25		
6	***Demand (thousand units)***	65	45	10			
7							
8	***Minimize***	**7185**					
9							
10			**Destination**		***Supply (thousand units)***		
11	**Origin**	Marysville (D1)	East Liberty (D2)	Greensburg (D3)			
12	Ontario (S1)	55	0	0	55	<=	55
13	Southfield (S2)	0	40	0	40	<=	40
14	Battle Creek (S3)	10	5	10	25	<=	25
15	***Demand (thousand units)***	65	45	10			
16		<=	<=	<=			
17		65	45	10			
18							

FIGURE 5.4 Excel solution.

$$x_{31}+x_{32}+x_{33}<= 25000$$

Constraints pertaining to demand would be:

$$x_{11}+ x_{21}+ x_{31} \leq 65{,}000$$

$$x_{12}+ x_{22}+ x_{32} \leq 45{,}000$$

$$x_{13}+ x_{23}+ x_{33} \leq 10{,}000$$

where $x_{11}, x_{12}, x_{13}, x_{21}, x_{22}, x_{23}, x_{31}, x_{32}, x_{33} \geq 0$

Excel Solution
See Figure 5.4.

5.9 UNACCEPTABLE ROUTES

Sometimes certain routes are infeasible due to high costs or are not possible due to lack of accessibility. Such situations are handled by imposing very high cost almost close to infinity denoted by 'M' so that the cost cell does not enter the solution at any stage. The above example expanded in the following way explains the method.

Due to increased demand for the Honda CRV and Accura series, Honda motors decided to increase production in their Lincoln, Alabama plant. The production from the plant is planned to cater to the needs of both the US and foreign customers. The number of cars exported from this plant outnumbers those from the Marysville plant.

Earlier, 10,000 cars were intended for export from the Marysville plant each month, but this demand has now been decided to be met by the Lincoln plant, increasing total production of export-oriented cars from Lincoln to 20,000 units. Musashi was selected as supplier of auto component parts as was for the other three plants. However, this increased demand can be met by only the Ontario (S1) plant due to its capacity to increase production at a transportation cost of $15 per unit. Also, S1 can supply more than 5,000 units. Demand cannot be met by S2 and S3 as they have exhausted their production capacity.

The transportation model for the above illustration would be (Table 5.57):

$$\text{Total supply} = 60 + 40 + 25 = 125$$

$$\text{Total demand} = 55 + 45 + 10 + 20 = 130$$

As supply is not equal to demand, so the model is an unbalanced one. To fulfil the remaining demand of 5 (in thousands) units, another source has to be added which is considered as dummy or fictitious. This would result in (Table 5.58):

Solving it by the VAM method would give the following initial basic feasible solution (Table 5.59).

TABLE 5.57
Transportation Model

	Destination				
Origin	**Marysville (D1)**	**East Liberty (D2)**	**Greensburg (D3)**	**Lincoln (D4)**	**Supply (Thousand Units)**
Ontario (S1)	5	11	6	15	***60***
Southfield (S2)	4	6	5	***M***	***40***
Battle Creek(S3)	4	6	9	***M***	***25***
Demand (thousand units)	***55***	***45***	***10***	***20***	

TABLE 5.58
Balanced Transportation Model

	Destination				
Origin	**Marysville (D1)**	**East Liberty (D2)**	**Greensburg (D3)**	**Lincoln (D4)**	**Supply (Thousand Units)**
Ontario (S1)	5	11	6	15	***60***
Southfield (S2)	4	6	5	***M***	***40***
Battle Creek(S3)	4	6	9	***M***	***25***
S4 (dummy)	0	0	0	0	***5***
Demand (thousand units)	***55***	***45***	***10***	***20***	

TABLE 5.59
Initial Solution by VAM (Iteration 0)

Origin	Marysville (D1)	East Liberty (D2)	Greensburg (D3)	Lincoln (D4)	Supply (thousand units)	Difference
	Destination					
Ontario (S1)	5 *(30)*	11 (5)	6 *(10)*	15 *(15)*	***60/30/25/15/0***	1 1 1
Southfield (S2)	4	6 *(40)*	5	*M*	***40/0***	1 1 1
Battle Creek(S3)	4 *(25)*	6	9	*M*	***25/0***	2 ***2***
S4 (dummy)	0	0	0	0 *(5)*	***5/0***	0
Demand (thousand units)	***55/30/0***	***45/5/0***	***10/0***	***20/15/0***		
Difference	4	***6***	5	15		
	0	0	1			
	1	***5***	1			

TABLE 5.60
Total Cost from Initial Solution

Supply Centre	Demand Centre	Units Shipped	Cost Per Unit ($)	Total Cost
S1	D1	30	5	30*5 = 150
S1	D2	11	5	11*5 = 55
S1	D3	10	6	10*6 = 60
S1	D4	15	15	15*15 = 225
S2	D2	40	6	40*6 = 240
S3	D1	25	4	25*4 = 100
S4	D4	5	0	5*0 = 0
Total cost				$830

The total cost would be (Table 5.60):
Optimality test:
Number of allocations = 7 = m + n – 1 = 7
Thus, the MODI method can be applied as shown in Table 5.61.
So,

$$u_1 + v_1 = 5;\ u_1 + v_2 = 11;\ u_1 + v_3 = 6;\ u_1 + v_4 = 15;$$

$$u_2 + v_2 = 6;\ u_3 + v_1 = 4 \text{ and } u_4 + v_4 = 0.$$

TABLE 5.61
Net Evaluation Cost of Unoccupied Cells

Origin	Destination $v_1 = 0$	$v_2 = 6$	$v_3 = 1$	$v_4 = 10$	Supply (Thousand Units)
$u_1 = 5$	a	a	a	a	*60*
$u_2 = 0$	$4 - 0 - 0 = 4$	a	$5 - 0 - 1 = 4$	M	*40*
$u_3 = 4$	a	$6 - 4 - 6 = \mathbf{-4}$	$9 - 4 - 1 = 4$	M	*25*
$u_4 = -10$	$0 - (-10) - 0 = 10$	$0 + 10 - 6 = 4$	$0 + 10 - 1 = 9$	a	*5*
Demand (thousand units)	*55*	*45*	*10*	*20*	

TABLE 5.62
Stepping stone method

Origin	Destination D1	D2	D3	D4	Supply (Thousand Units)
S1	5 (*30*) (+)	11 (5) (−)	6 (*10*)	15 (*15*)	*60*
S2	4	6 (*40*)	5	*M*	*40*
S3	4 (*25*) (−)	6 (+)	9	*M*	*25*
S4	0	0	0	0 (*5*)	*5*
Demand (thousand units)	*55*	*45*	*10*	*20*	

Solution of these equations would give values of various four 'u' and three 'v' variables shown in Table 5.61 along with net evaluation costs of unoccupied cells.

Table 5.62 shows incoming cell and new allocations

Iteration 1:

See Table 5.63.

Optimality test:

Number of allocations = 7 = m + n − 1 = 7

Thus, the MODI method can be applied.

So, $u_1 + v_1 = 5$; $u_1 + v_3 = 6$; $u_1 + v_4 = 15$; $u_2 + v_2 = 6$; $u_3 + v_1 = 4$; $u_3 + v_2 = 6$ and $u_4 + v_4 = 0$. Solution of these equations would give values of various four 'u' and three 'v' variables shown in Table 5.64 along with net evaluation costs of unoccupied cells.

As all values are positive or zero, the deduced solution in iteration 1 is most optimal. The total cost under this solution is given in Table 5.65:

Thus, the cost has decreased from \$830 to \$810.

TABLE 5.63
Initial solution

Origin	D1	D2	D3	D4	Supply (Thousand Units)
	Destination				
S1	5 (*35*)	11	6 (*10*)	15 (*15*)	***60***
S2	4	6 (*40*)	5	*M*	***40***
S3	4 (*20*)	6 (*5*)	9	*M*	***25***
S4	0	0	0	0 (*5*)	***5***
Demand (thousand units)	***55***	***45***	***10***	***20***	

TABLE 5.64
Net Evaluation Cost of Unoccupied Cells

Origin	$v_1 = 0$	$v_2 = 2$	$v_3 = 1$	$v_4 = 10$	Supply (Thousand Units)
	Destination				
$u_1 = 5$	a	11 −5−2 = 4	a	a	***60***
$u_2 = 4$	4−0−0 = 4	a	5 −4−1 = 0	M	***40***
$u_3 = 4$	a	a	9 −4−1 = 4	M	***25***
$u_4 = -10$	0−(−10) −0 = 10	0 + 10 −2 = 8	0 + 10 − 1 = 9	a	***5***
Demand (thousand units)	***55***	***45***	***10***	***20***	

TABLE 5.65
Total Cost from Iteration 1

Supply Centre	Demand Centre	Units Shipped	Cost Per Unit ($)	Total Cost
S1	D1	35	5	35*5 = 175
S1	D3	10	6	10*6 = 60
S1	D4	15	15	15*15 = 225
S2	D2	40	6	40*6 = 240
S3	D1	20	4	20*4 = 80
S3	D2	5	6	5*6 = 30
S4	D4	5	0	5*0 = 0
Total cost				$810

5.10 TRANSSHIPMENT PROBLEM: THEORY

The transshipment problem is a more general form of transportation problem reflecting more practical situations of transportation problem. In general, transportation occurs not directly between manufacturer and retailer. An intermediary in the form of warehouse sources finished goods from various producers, stores them at one place and then supplies to final demand centres. Thus, transportation occurs between producer to warehouse and then from warehouse to retailer. This kind of problem is represented by the transshipment model. The model involves nodes and arcs where nodes represent supply, transshipment and demand centres. On the other hand, arcs represent routes between various centres. This is represented in Figure 5.5. It is important to note that the purpose of both transportation and transshipment problem is exactly the same of minimization of transportation cost.

5.10.1 Case: Food Corporation of India

Food Corporation of India is one of the largest corporations in India to procure and distribute food grains from across the country. Its operations are humungous as more than half of India's population is involved in farming. The organization is considered the largest organization in Asia involved in distributing grains throughout the country. The operations of the organization are a perfect case study for managing supply chains. One of the effective examples of such efficient management of supply chain of food grains is the case of the state of Punjab. The state is considered to supply more than 50% of food grains to FCI's requirement, thus aptly called as granary of India. Punjab has the most structured and effective way of procuring grains from farmers and fulfilling the requirement of FCI. The process of procuring and distributing is divided into three steps. First, after harvest, farmers bring their produce to various

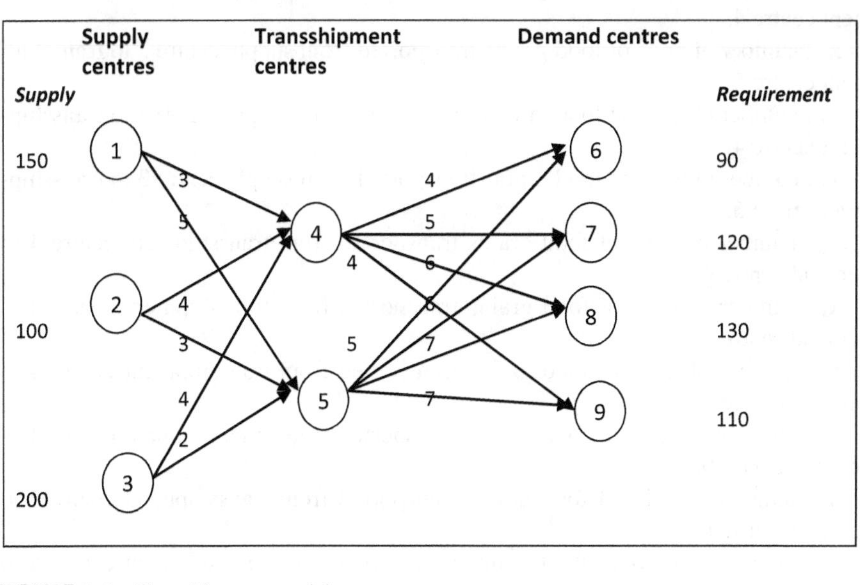

FIGURE 5.5 Transhipment model.

grain markets spread all over the state. Second, from these markets, FCI procures and collects them for various large warehouses. Finally, FCI distributes from its warehouses depending on demand to different demand centres of the state. The top three grain markets in the state, which procures approximately 70% of grains, are located in the cities of Khanna, which is considered Asia's largest grain market followed by Rajpura and Sirhind. From these three markets, grains are collected at two biggest warehouses in the cities of Patiala and Bhatinda. Finally, grain is supplied to three big demand centres of three most populous cities, namely Amritsar, Ludhiana and Jalandhar. The supply centres (S1, S2 and S3) are called as origins, two warehouses are called as W1 and W2 (transshipment centres) and four demand centres are D1, D2, D3 and D4. These are represented as nodes 1, 2 and 3 (supply centres) 4 and 5 (transshipment centres) and 6, 7, 8 and 9 as demand centres. S1 can supply a maximum quantity of 150 million tonnes (mt), whereas S2 and S3 can supply to the tune of 100 and 200 mt, respectively. Demand centre D1, D2, D3 and D4 have fixed requirements of 90, 120, 130 and 110 mt of food grains. Their demand is met through supply from two warehouses. Each origin (1, 2 and 3) can supply to each warehouse (4 and 5), and each warehouse can supply to each demand centre (6, 7, 8 and 9). The transportation cost (Rs.) of each route along with capacity and requirement is shown in the figure.

To understand the constraints and objective of transshipment model, the above-stated problem is constructed by using a linear programming method. The following notation is used.

x_{ij} represents number of units shipped from centre i to centre j. Thus,

x_{14}: number of units of food grains transported from supply centre 1 to transshipment centre 4.

x_{15}: number of units of food grains transported from supply centre 1 to transshipment centre 5.

x_{24}: number of units of food grains transported from supply centre 2 to transshipment centre 4.

x_{25}: number of units of food grains transported from supply centre 2 to transshipment centre 5.

x_{34}: number of units of food grains transported from supply centre 3 to transshipment centre 4.

x_{35}: number of units of food grains transported from supply centre 3 to transshipment centre 5.

x_{46}: number of units of food grains transported from transshipment centre 4 to demand centre 6.

x_{47}: number of units of food grains transported from transshipment centre 4 to demand centre 7.

x_{48}: number of units of food grains transported from transshipment centre 4 to demand centre 8.

x_{49}: number of units of food grains transported from transshipment centre 4 to demand centre 9.

x_{56}: number of units of food grains transported from transshipment centre 5 to demand centre 6.

x_{57}: number of units of food grains transported from transshipment centre 5 to demand centre 7.

x_{58}: number of units of food grains transported from transshipment centre 5 to demand centre 8.

x_{59}: number of units of food grains transported from transshipment centre 5 to demand centre 9.

Supply constraints:

Supply from S1 is maximum 150 mt and is supplying to transshipment centres 4 and 5, so

$$x_{14} + x_{15} \leq 150 \quad (5.1)$$

Similarly

$$x_{24} + x_{25} \leq 100 \quad (5.2)$$

$$x_{34} + x_{35} \leq 200 \quad (5.3)$$

Transshipment constraints:

Each transshipment centre supplies a number of units to each distribution centre equivalent to the number of units received by it from different supply centres. This implies that both centres 4 and 5 receive from centers 1, 2 and 3 and supplies to centers 6, 7, 8 and 9. This leads to the formation of the following constraints:

Number of units received by centre 4 = $x_{14} + x_{24} + x_{34}$

Number of units shipped out from centre 4 = $x_{46} + x_{47} + x_{48} + x_{49}$

These should be equal. So:

$$x_{14} + x_{24} + x_{34} = x_{46} + x_{47} + x_{48} + x_{49}$$

$$\text{or } x_{14} + x_{24} + x_{34} - x_{46} - x_{47} - x_{48} - x_{49} = 0 \quad (5.4)$$

Similarly for node 5:

Number of units received by centre 5 = $x_{15} + x_{25} + x_{35}$

Number of units shipped out from centre 5 = $x_{56} + x_{57} + x_{58} + x_{59}$

These should be equal. So:

$$x_{15} + x_{25} + x_{35} = x_{56} + x_{57} + x_{58} + x_{59}$$

$$\text{or } x_{15} + x_{25} + x_{35} - x_{56} - x_{57} - x_{58} - x_{59} = 0 \quad (5.5)$$

Demand centre constraints:

Each demand centre has a certain fixed requirement fulfilled by supply from various transshipment centres. Thus:

$$x_{46} + x_{56} = 90 \quad (5.6)$$

$$x_{47} + x_{57} = 120 \quad (5.7)$$

$$x_{48} + x_{58} = 130 \tag{5.8}$$

$$x_{49} + x_{59} = 110 \tag{5.9}$$

Objective function:

The objective is to find a certain number of units of food grains transported through certain routes in such a way so as to minimize total transportation cost. Shipping cost associated with each route is shown in figure, so the objective function would be:

$$\textbf{Minimize} \quad Z = 3x_{14} + 5x_{15} + 4x_{24} + 3x_{25} + 4x_{34} + 2x_{35} + 4x_{46}$$
$$+ 5x_{47} + 6x_{48} + 4x_{49} + 5x_{56} + 6x_{57} + 7x_{58} + 7x_{59}$$

Assumptions:

The above illustration helps in the formation of the following assumptions regarding the transshipment problem:

- Each node represents either an origin, transshipment or demand centre.
- There can be 'n' number of nodes but it should be finite.
- There can be only one constraint per node. For instance, node 1 is associated with only supply constraint and so on.
- Each arc represents the route from one center to other.
- Transportation cost per unit is the variable shown on each route. This yields the information of objective of minimization.

The supply centres can also directly supply to demand centres without going through transshipment centres. The only variation would be the addition of more variables in the constraint equations. For instance, if supply centre 3 supplies directly to destination 9 in addition to supplying to warehouses as well, then another variable x_{39} implying the number of units transporting from 3 to 9 would be added to demand constraint (5.9) and to objective function. So demand constraint (5.9) would be

$$x_{39} + x_{49} + x_{59} = 110$$

and objective function would be if per unit shipping cost from 3 to 9 is \$10

$$\textbf{Minimize} \quad Z = 3x_{14} + 5x_{15} + 4x_{24} + 3x_{25} + 4x_{34} + 2x_{35} + 10x_{39} + 4x_{46}$$
$$+ 5x_{47} + 6x_{48} + 4x_{49} + 5x_{56} + 6x_{57} + 7x_{58} + 7x_{59}$$

The solution of transshipment problem through discussed methods of transportation models is tedious and thus not discussed here. However, the Excel solution is provided below.

Excel Solution

Step 1: Formulate transshipment model: This is done as shown on the left side of Excel worksheet. It involves a list of all supply centres termed as origins (A5:A18). In the next column, a list of destinations is created involving warehouses and demand centres (B5:B18). The third column shows data of per unit cost from each supply to destination centre (C5:C18). The fourth column is represented by decision variables indicating the number of units that need to be transported so as to minimize total transportation cost (D5:D18).

Step 2: The right side of Excel sheet shows the formulation of linear programming model involving calculation of objective functions and constraints. First, a list of all nodes, i.e. three supply centres; two warehouses and three demand centres are created (G5:G13). The next two columns show the quantity of incoming and outgoing units from each centre. S1, S2 and S3 are outgoing centres, i.e. units are shipped out of these centres and received by W1 and W2 warehouses. Further, W1 and W2 are outgoing centres as well as grain collected at these centres is shipped to three demand centres. D1, D2 and D3 are only incoming centres receiving final shipment from two warehouses. Now as S1, S2 and S3 ships to W1 and W2, so units going out of S1, for instance, is the sum of units received by W1 and W2. So formulas to be used in the incoming and outgoing column would be:

Incoming units:
H8: = D5 + D7 + D9
H9: = D6 + D8 + D10
H10: = D11 + D15
H11: = D12 + D16
H12: = D13 + D17
H13: = D14 + D18

Outgoing units:
I5: = D5 + D6
I6: = D7 + D8
I7: = D9 + D10
I8: = D11 + D12 + D13 + D14
I9: = D15 + D16 + D17 + D18

The next column of net flow of units (J5:J13) indicates the left-hand side of supply constraints, which is the difference between outgoing and incoming units. Finally, the right-hand side of supply constraints indicating the maximum resource available is listed in cells L5:L13. D1, D2, D3 and D4 are demand centres, so minus signs indicate net inflow rather than outflow of units.

Step 3: Objective function is calculated as =Sumproduct(C5:C18,D5:D18). The final solution is shown in Figure 5.6.

Transshipment problem

Food corporation of India

			Decision variables		Units shipped				
Origin	Destination	Transportation cost per unit	No. of Units to be shipped	Node	In	Out	Net		Supply at each node
S1	W1	3	150	S1		150	150	<=	150
S1	W2	5	0	S2		100	100	<=	100
S2	W1	4	50	S3		200	200	<=	200
S2	W2	3	50	W1	200	200	0	=	0
S3	W1	4	0	W2	250	250	0	=	0
S3	W2	2	200	D1	90		-90	=	-90
W1	D1	4	90	D2	120		-120	=	-120
W1	D2	5	0	D3	130		-130	=	-130
W1	D3	6	0	D4	110		-110	=	-110
W1	D4	4	110						
W2	D1	5	0						
W2	D2	6	120	Z	3630				
W2	D3	7	130						
W2	D4	7	0						

FIGURE 5.6 Excel solution to transhipment problem.

5.11 SUMMARY

This chapter has discussed two special cases of network flow problems. The first model explained is the transportation model that involves a network of nodes and arcs. Nodes represent origins and destinations, whereas arcs represent linkage between these nodes. The transportation model is a kind of many-to-many network problem wherein resources can be supplied from various origins, also termed as supply centres to various destinations, also known as demand centres. The objective of this model is to determine the number of units that can be transported from a particular supply centre to a particular demand centre in a manner that minimizes total transportation cost. Specifically, the model fulfils two aims: first, it provides information about the number of units to be transported from each supply centre and the number of units required by each demand centre; second, what would be the cost of such flow. Thus, the model optimizes the allocation of resources. That is why the chapter also illustrates the transportation model as a linear programming problem. Different variants of transportation model such as maximization, unbalanced and degeneracy have also been discussed. Each variant has also been solved by using Excel.

Another kind of network flow problem discussed in the chapter is the transshipment model. This model is an advanced version of the transportation model wherein in addition to origins and destination, there are transfer nodes between them. These problems are primarily applied to distribution problems. However, this also causes many-to-many network flow problems. The chapter has explained the linear programming construction of the network with an explanation of constraints and the objective function. Using a case on procurement and distribution of food grains, the chapter has illustrated the Excel solution of such a complicated model.

5.12 CASE STUDY 1: NORLAND PLASTICS CO.

Norland Plastics is a premier supplier in a variety of markets serving the needs of many large global manufacturing companies. Among them, OEMs and Tier 1 clients in the automobile industry are two of the main axes of company's business development strategy. Norland Plastics makes quality automotive injection moulded parts for various international car manufacturers. One of the products supplied to various OEMs is emblems for Ford, Nissan and GM. The supplier uses a single production line to manufacture three different emblems for these three players. Management wanted to improve productivity of production line by eliminating or reducing non-value added activities (NVA). One example of such NVA is transportation cost incurred between transporting finished emblems from assembly section to inspection section. The process of manufacturing involved a sequence of moulding, chrome plating, painting, assembly and final inspection. With a change in type emblem, moulds have to be changed in the moulding section. Thus, the management of supplier decided to produce emblems in batches. One production run produces 1,500 emblems of Ford (E1); 5,400 of Nissan (E2) and 4,300 of GM (E3) model. Each batch of emblem in transported to one of three inspection lines (I1, I2 and I3) for testing the quality of emblem. Each line is capable of inspecting any type of emblem depending on the line's availability. The lines were designed to inspect a maximum of 3,000 units per shift. As any line can inspect any of the emblem, so capacity of inspection lines was not considered inadequate. Transportation cost per unit between assembly and inspection line of each type of emblem is shown in the following table.

	I1	I2	I3
E1	1	1.5	1.3
E2	2	1.7	1.1
E3	1.4	1.2	1.9

The management wants to calculate the minimum possible transportation cost by estimating the number of emblems allocated to different inspection lines. Formulate and solve a transportation model.

5.13 CASE STUDY 2: HONDA MOTORS

The production of motor vehicle bodies requires welding of numerous metal parts stamped from sheet metal. These stampings are produced by running a large roll of sheet steel into an automated press to produce a number of blanks slightly larger than the required size. These blanks are then inserted into large stamping presses containing matched upper and lower dies. After pushing these dies together, the required metal shape is obtained, whether that is of chassis, door, or something else. Honda decided to launch for new variants of mid-size cars targeted at different demographics. They were two-door coupe, station wagon, four-door sedan and a four-door hardtop. Each car required different kinds of stampings and in varying quantities. The management decided to launch all variants simultaneously, so production was also done at the same time. The number of dies required to manufacture metal parts for one variant to another was quite large. Thus, it was decided to dedicate a pressing machine to the needs of one particular car model. The capacity of each pressing line was: Machine A produced 70,000 stampings for two-door coupe; machine B produced 60,000 for station wagon; machine C 90,000 for four-door sedan and machine D produced 50,000 for a four-door hardtop. Three assembly plants P1, P2 and P3 assembled the stampings of different cars with other components to produce a finished car. Each plant was assigned to manufacture a particular type of cars depending on their capability. P1 was assigned to production of two-door coupe and station wagons; P2 can manufacture a four-door sedan and four-door hardtop; and finally, P3 can manufacture a station wagon and four-door hardtop. These final products are shipped to three demand centres (D1, D2 and D3) located in three different geographies. Each demand centre was allocated a fixed quantity of 90,000 cars. Though due to the push system, the total quantity to be sold was the same at all three demand centres, but the combination was different. According to extensive market analysis, D1 could cater to demand for coupe, station wagon and four-door sedan. D2 sold only coupe, four-door sedan and four-door hardtop, whereas D3 distributed coupe, station wagon and four-door hardtop. JIT philosophy applied by the company made supply of stampings as and when required by assembly plants. This required their establishment close to all plants. Transportation cost per unit of stamping to a particular assembly plant as discussed in the plan was: $1.1 from A to P1; $1.4 and $1.2 from B to P1 and P3; $1.3 from C to P2 and $1.4 and $1.5 from D to P2 and P3. Similarly, transportation cost per unit of finished car from three assembly plants to three customers was: $25, $27 and $40 from P1 to C1, C2 and C3; $30, $42 and $35 from P2 to C1, C2 and C3; and $39, $37 and $41 from P3 to C1, C2 and C3. Given this data:

- Draw network diagram indicating various nodes and routes.
- Formulate the problem in the LPP form.

5.14 GLOSSARY

- **Transportation problem:** It is a special kind of linear programming problem, which is used to determine the number of units to be distributed from limited supply centres to known demand centres.
- **Network model:** A model involving nodes and arcs. Nodes represent origins and destinations while arcs represent linkage between different nodes.
- **Transportation cost:** Total cost incurred in transporting a fixed number of units from known centres to fulfil fixed demand of known destinations.
- **Least cost method:** A method of finding initial solution to a transportation problem by allocating units for transportation to minimum cost cells.
- **North-West Corner method:** Another method of finding an initial solution by allocating first to a cost cell located at the North-West Corner of the cost matrix and then moving according to the availability of resources.

- **Vogel's approximation method (VAM):** It is considered to be the most efficient though a complex method of finding an initial solution.
- **Feasible solution:** Number of allocations equal to m + n − 1 where 'm' is number of rows and 'n' being number of columns makes a transportation problem feasible.
- **MODI method**: An acronym for *modified distribution.* The method is applied to compute cost for each unoccupied cell by which per unit total cost would be reduced. This method is used in identifying incoming cell of a new solution so as to check the optimality of the solution.
- **Stepping stone method:** it identifies outgoing cost cell in the process of finding an optimal solution.
- **Unbalanced transportation model:** inequality between supply and demand transportation model problem makes it an unbalanced one.
- **Dummy centre:** could be either supply or demand with zero cost incorporated in the transportation model to make supply and demand equivalent so that the initial solution becomes feasible.
- **Degeneracy:** In the transportation model, if the condition of feasibility i.e. number of allocations = m + n – 1 is violated, then model is said to be degenerate.
- **Transshipment problem:** In these kinds of models, transportation occurs between producer to warehouse and then from warehouse to retailer.

5.15 MODEL QUESTIONS

5.15.1 A LPG distributor has four supply centres to cater to the demand of three demand areas in a particular city. These four supply centres (S1, S2, S3 and S4) can supply on a daily basis 1,000, 1,200, 750 and 900 cylinders of LPG to fulfil the demand of 1,800, 1,300 and 750 units of areas A1, A2 and A3. The distributor wants to find appropriate routes to supply a certain number of units in a manner which would minimize transportation cost. The cost of transporting one unit from each supply centre to each demand centre is shown below:

	A1	A2	A3
S1	10	8	11
S2	9	12	7
S3	11	9	8
S4	6	10	14

5.15.2 Electrical batteries in automobiles are imported by four luxury car makers (C1, C2, C3 and C4) at three ports (P1, P2 and P3). As similar type of battery is installed in each type of car, so distributor of batteries can supply to all four customers. The number of units supplied from each port, requirement of each car maker and shipping cost of per unit are given as:

	C1	C2	C3	C4	Supply
P1	3	7	7	3	40
P2	2	3	6	8	60
P3	4	5	3	6	50
Demand	50	35	45	40	

i. Find an initial solution by NWCM and least cost method. Explain the reasons for difference in total cost if any.
ii. Check the optimality of solution by both methods. Is solution unique? If not, find an alternative solution.

5.15.3 An LED manufacturer from its three warehouses distributes a fixed number of units to three retail outlets. The following table shows per unit shipping cost, demand of each retail store and capacity of each distributor to transport per month. Calculate optimum transportation cost indicating the allocation of number of units that needs to be transported from particular distributor to particular store.

	C1	C2	C3	Supply
D1	10.5	11.4	10.9	20,000
D2	12.4	15.7	12.7	10,000
D3	9.7	9.2	10.1	12,000
Demand	14,000	13,000	10,000	

How the solution changes if C3 asks for 4000 units more?

5.15.4 A text book of operations management is supplied by four distributors (D1, D2, D3 and D4) to book shops of three universities (S1, S2 and S3). Due to variation in distance between distributor and each university, the selling price was found to be different. S1 was selling one unit at $75; S2 at $73 and S3 at $76. The production cost of book including printing and packaging was $50 per unit. The fixed number of units that each distributor can supply to each book shop, requirement of each shop and transportation cost per unit are shown in the following table. Estimate the appropriate number of units that each distributor should supply so as to maximize profit.

	S1	S2	S3	Supply
D1	5	7	6	700
D2	4	3	5	750
D3	8	7	4	650
D4	7	5	3	500
Demand	800	850	950	

5.15.5 Each of three wholesales (W1, W2 and W3) can supply cartons of apples to five retailers (R1, R2, R3, R4 and R5) and earn different profits. This difference is because of the difference in shipping cost associated with transportation from one wholesaler to a particular retailer. The supply is done once to meet weekly demand. W1 can supply maximum of 450; W2 300 and W3 475 cartons, whereas demand of each retailer is 350, 250, 200, 200 and 225 units respectively. Profits per supply route are shown below. Solve by using NWCM.

	R1	R2	R3	R4	R5
W1	40	35	32	30	37
W2	28	27	30	31	29
W3	24	23	27	28	25

5.15.6 Determine initial feasible solution of following transportation problem by using VAM and then test its optimality.

	D1	D2	D3	D4	D5	Supply
S1	4	13	12	5	9	6
S2	3	6	9	4	3	10
S3	5	11	6	10	14	11
Demand	5	5	6	7	8	

5.15.7 Determine the initial feasible solution of following transportation problem and then test its optimality.

	Shipping cost ($ per unit)				
	R1	R2	R3	R4	Supply
P1	40	50	30	10	15
P2	10	80	5	20	25
P3	20	30	10	5	25
P4	10	5	20	10	15
Demand	25	15	15	25	

5.15.8 For the following problem:

	Shipping cost ($ per unit)			
	P	Q	R	Supply
A	70	40	10	50
B	60	30	20	100
Demand	40	30	50	

iii. Apply VAM to find initial solution and test its optimality.
iv. Apply NWCM to find initial solution and test its optimality.

5.15.9 Following table give profit values associated with shipping a unit from a source to a particular destination. Find the number of units to be transported from each source so as to maximize profits.

	Profits ($ per unit)		
	P	**Q**	**Supply**
A	13	10	6
B	11	9	4
C	12	8	6
Demand	7	9	

5.15.10 Following transportation problem shows profit values associated with shipping one unit.

	Profit matrix ($ per unit)			
	1	**2**	**3**	**Supply**
1	13	11	10	300
2	10	10	10	300
3	14	12	8	300
4	13	11	9	300
Demand	400	400	400	

i. Find number of units transported by each supply centre in a manner which maximizes profit.
ii. Solve for maximization of profit if for some reason supply of centre '4' is restricted to 200 units.

6 Network Model II
Assignment Model

6.1 INTRODUCTION

Assignment model is another form of network flow problem that deals with the problem of assigning *n* persons to *n* jobs. Assignment is done based on certain criteria. The model can be constructed as a network consisting of nodes and arcs as a transportation model. Nodes again imply supply and demand centres, whereas arcs represent flow of resource from one centre to another. However, as specific purpose of assignment model is assigning or allocating for instance one worker to only one job or one store to only one retailer so demand and supply capacity of each node is equal to one. The assignment model due to these aspects is also considered as a special case of linear programming problem.

In the assignment model, the purpose is not of distribution of resources as was in the case of transportation but assigning of tasks like assigning a project to a particular team or jobs to machines, etc. Such assignments are governed by certain criteria that could be, for instance, least completion time or the highest performance. Furthermore, if there are 'n' projects to be dealt with by n teams, in the case of time criteria, the objective of the assignment model would be to allocate each project to a particular time in a manner which would minimize total project completion time. However, the objective would be of maximization if a criterion of allocation is performance. Thus, an assignment model consists of three elements: an *assignee* who/which needs to be allocated a *task* based on certain decided *criteria*. The assignee can be a person, a machine, a store, etc. that requires allocation to a particular task. The following applications illustrate this.

- Foreman needs to assign three employees to three machines. Each of them can work on three machines but with different capabilities, resulting in different costs. Foreman needs to decide on the optimal assignment so as to minimize total cost. In this case, assignee was a person who needed to be assigned to a particular machine.
- In a batch production system, four machines doing the same process are to be allocated four locations on the shop floor. Criteria of allocation were minimum material handling cost between machine and packaging section. In this case, the assignee was a machine that was allocated to a particular location.
- A shopping mall has ten stores to be allocated to ten different apparel companies. The criteria for assignment were rental cost. The store paying the

DOI: 10.1201/9781003212966-6

highest would be assigned a store. In this case, the assignee was a store that was to be assigned to a particular company in a manner which would maximize rent earned by the mall manager.

Thus, there are numerous applications of the assignment model. The above-discussed examples facilitate the formulation of the model of the assignment problem, which is discussed in the next section.

6.2 ASSIGNMENT PROBLEM: CONSTRUCTION OF MODEL

6.2.1 Case: MarketOne International LLP

The first step in product design and development is to conduct large-scale market surveys in order to understand the needs and desires of targeted consumers. The new car development process is, on an average, a 4-year process. General Motors (GM) one of the largest car manufacturers in the world, faces stiff competition from especially Japanese car manufacturers like Honda and Toyota. However, Toyota, with its successful application of lean methods of production, mostly beats GM in new product launches. GM decided to snub this trend by expediting the launch of its new models for which it first intends to identify customer expectations regarding four models of its cars. Four different models were categorized as compact, sub-compact, intermediate and full size. It was important to know the differentiating characteristics regarding appearance, features, mileage, price and power of each model so that car manufacturers could accurately incorporate them while designing the product for each segment. The effective conduct of market survey was given to MarketOne International consultancy, which provides services of market surveys which result in quantifiable sales outcomes. The company provides accurate and reliable data about prospective buyers in the highly competitive USA car market. Such information becomes very important for companies like GM, whose product category is capital-intensive and customer expectations are very high. To understand customers' tastes and preferences regarding GMs four offerings, the manager of consultancy organized its employees into four teams (Team 1, Team 2, Team 3 and Team 4). Each team could conduct market analysis for each model of car. But the manager had to fulfil four projects, each pertaining to a particular type of car model at the earliest. Thus, assignments had to be made with the condition of least completion time. The times in months by each team for each project are shown in Table 6.1.

TABLE 6.1
Estimated Completion Time in Months

	Project 1 (Compact)	Project 2 (Sub-compact)	Project 3 (Intermediate)	Project 4 (Full Size)
Team 1	9	8	10	10
Team 2	6	3	8	9
Team 3	7	2	5	10
Team 4	3	4	3	7

The above model includes certain decision variables that are expressed in the objective function as:

$$\text{Min}\sum_{i=1}^{n}\sum_{j=1}^{n} C_{ij}X_{ij}$$

where

'i' represents a particular person and goes from 1, 2….., n;

'j' represents a particular job and goes from 1, 2…., n;

c_{ij} = criteria per unit, i.e. cost/time or profit per unit in assigning an assignee to a particular task; for instance, c_{11} indicates time incurred when person 1 is assigned to job 1, which is 9 months.

x_{ij} = indicates a particular assignment, i.e. $x_{11} = 1$ implies that person 1 has been assigned to carry out job 1 and if $x_{11} = 0$ means person 1 is not qualified to do job 1.

The objective function is fulfilled under certain constraints that are:

$$\sum_{i=1}^{n} X_{ij}=1; \text{ and}$$

$$\sum_{j=1}^{n} X_{ij}=1;$$

x_{ij} = 1 or 0 for all i = 1, 2…, n and j = 1, 2…, n

Using the above notation and understanding, the objective function and constraint equations are developed as:

Minimize $Z = 9x_{11} + 8x_{12} + 10x_{13} + 10x_{14} + 6x_{21} + 3x_{22} + 8x_{23} + 9x_{24} + 7x_{31} + 2x_{32} + 5x_{33} + 10x_{34} + 3x_{41} + 4x_{42} + 3x_{43} + 7x_{44}$

Subject to:

Assignment of team 1: $x_{11} + x_{12} + x_{13} + x_{14} \leq 1$

Assignment of team 2: $x_{21} + x_{22} + x_{23} + x_{24} \leq 1$

Assignment of team 3: $x_{31} + x_{32} + x_{33} + x_{34} \leq 1$

Assignment of team 4: $x_{41} + x_{42} + x_{43} + x_{44} \leq 1$

Assignment of project 1: $x_{11} + x_{21} + x_{31} + x_{41}= 1$

Assignment of project 2: $x_{12} + x_{22} + x_{32} + x_{42}= 1$

Assignment of project 3: $x_{13} + x_{23} + x_{33} + x_{43}= 1$

Assignment of project 4: $x_{14} + x_{24} + x_{34} + x_{44}= 1$

The constraints are of equality types as the number of projects are equal to the number of teams, i.e. 4. Constraints could be written as of inequality type if the number of assignments is unequal. For instance, if a number of assignees are more than tasks, then the constraint changes from exactly one assignment to each assignee being allocated at least one task, making the inequality to less than or equal to (≤). The network diagram of the model is shown in Figure 6.1.

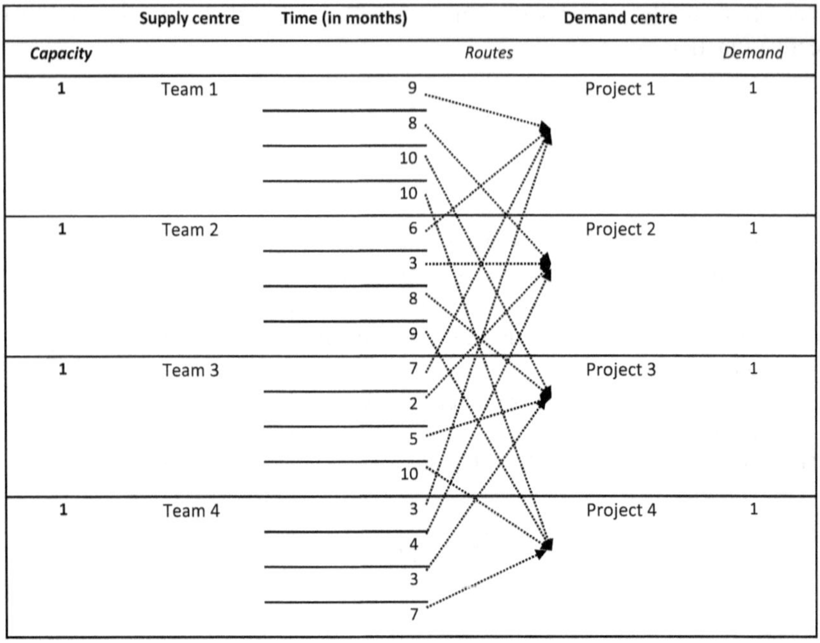

FIGURE 6.1 Network diagram.

TABLE 6.2
Structure of Assignment Model

	Task (j = 1, 2…., n)				
Employees (i = 1, 2…, n)	t_1	t_2	…	t_n	**Demand**
e_1	c_{11}	c_{12}	…	c_{14}	1
e_2	c_{21}	c_{22}	…	c_{24}	1
…	…	…	…	…	1
e_n	c_{n1}	c_{n2}	…	c_{nn}	1
Supply	1	1	1	1	

The above illustration helps us to create a generalized structure of an assignment model as shown in Table 6.2.

6.3 ASSUMPTIONS

The solution of the above assignment problem, formulated as shown in Table 6.2, follows certain *assumptions* which are:

1. The number of employees to be assigned a job is equal to the number of jobs available. Thus, there would be n assignments.

2. Each employee, though qualified to do all jobs, will be assigned only one job.
3. Similarly, though each job can be performed by any of the employees, each task has to be performed by only one employee.
4. Performing a job incurs cost. Cell elements indicate cost associated with each possible assignment. As each employee can perform each task, the matrix indicates cost for each possible combination of employee and task.
5. As pointed by 2 and 3, any employee can be assigned to any job incurring a certain cost. Thus, the purpose of the assignment model is to identify that combination which results in minimization of total cost.

However, assumptions 1, 2 and 3 are mostly not applicable in realistic situations. For instance, fewer people are available to work on more machines; much more applications are received for a limited number of positions; limited classrooms are available for assigning a large number of periods, etc. Thus, certain variations are required in the assignment model, which will be discussed in later sections.

6.4 COMPARISON WITH TRANSPORTATION MODEL

The assignment problem is a special case of the transportation model. This is better understood by comparing the assignment model shown in Table 6.1 with transportation model shown in Table 5.1 of the chapter on Transportation model. This is reproduced here as Table 6.3.

First, let's discuss similarities between the two models.

Similarities:

- First, both models have a limited and fixed number of suppliers or employees and a limited and known number of demand centres or jobs.
- Second, both models deals with aspect of combining resources with demand and tasks. In the case of transportation, the number of units from different supply centres is allocated to fulfil demand of various destinations in a manner to minimize transportation cost. Whereas, in the assignment model, the same kind of allocation is done based on certain criteria of either minimizing cost or minimizing time to do a specific job.

Difference:

- But both models have one major difference. In the case of transportation problem, different supply and demand centres can have different capacities of output and requirements respectively, shown by b_i and c_j in Table 6.3. Whereas in the case of assignment problem, all supplies and demands equal to one. This is because of fundamental of assignment problem, i.e. one worker can be assigned to do only job at a time, one geographic area can be assigned only one store or department and so on. Thus, the cost incurred in transporting 'x' units from supply centre 1 to demand centre 1 would be $c_{11}x_{11}$. For the assignment model, it would be the same but the value of x_{11} would be either 1 or 0.

TABLE 6.3
Transportation Model

Supply Centers	D1	D2	D3	D4	Total Supply of Each Center
	Demand Centres				
S1	a_{11}	a_{12}	a_{13}	a_{14}	b_1
S2	a_{21}	a_{22}	a_{23}	a_{24}	b_2
S3	a_{31}	a_{32}	a_{33}	a_{34}	b_3
Total demand of each center	c_1	c_2	c_3	c_4	

- The optimality of the transportation model is checked by equating total available units with total requirement. Whereas in the case of assignment, it is checked by equating the number of persons with the number of tasks. Thus, the condition of suppliers/employees to be equal to demand centres/tasks in assignment is essential, though not a necessary condition.
- Finally, feasibility of transportation is checked by equating $m + n - 1$ with the number of basic variables (allocated cells). If it is not fulfilled then deficit variables are termed as degenerate. Thus in the case of assignment where $m = n$ number of allocations/assignments should be equal to $2n - 1$. However, because the assignment problem involves assigning *n* employees to n jobs so, there would be only n assignments. So deficit, i.e. $n - 1$ variables would degenerate. This is depicted by $x_{ij} = 0$, implying no assignment.

6.5 HUNGARIAN ALGORITHM

Transportation problem is a general-purpose algorithm involving allocation of units available from certain sources to certain destinations. However, it does not provide sufficient resolution for a special structure of transportation problem where all demand and supplies equal to one unit. Hungarian algorithm is a special purpose method to resolve such problems termed as assignment problem. Steps involved in the algorithm have been discussed by using the illustration shown in Table 6.1.

Step 1: Feasibility check: In the assignment problem, the feasibility of model is checked by finding equality between the number of rows and columns. Such problem is called as balanced problem. The example considered is shown in Table 6.4. The number of rows representing teams to be assigned is four, which is equal to the number of columns representing the number of projects.

The next two steps are based on the concept of matrix reduction that implies that if a constant value is subtracted or added from the entire set of values in the matrix, it would yield an optimal solution. The reduced values indicate the opportunity cost (time in this case) indicating the next best opportunity. For instance, for team 1, the best possible assignment would be project 2 as it entails the least cost/time, whereas the next best would be project 1.

TABLE 6.4
Estimated Completion Time in Months

	Project 1 (Compact)	Project 2 (Sub-compact)	Project 3 (Intermediate)	Project 4 (Full Size)
Team 1	9	8	10	10
Team 2	6	3	8	9
Team 3	7	2	5	10
Team 4	3	4	3	7

Step 2: Row reduction: From each row, subtract the minimum cell value (in this case, time) from all other values of that row. In row 1, represented by Team 1, the minimum value is 8, which is subtracted from other values. This would give us the following result (Table 6.5):

Step 3: Column reduction: From each column, subtract the minimum cell value (in this case, time) from all other values of that column. In column 1 represented by Project 1, the minimum value is 0, which is subtracted from other values. This would give values as shown in Table 6.6.

Step 4: Make assignments:

i. Starting from the first row, make assignment or allocations in the row which has only one zero. This would imply assigning an employee to a particular job. The first row has two zeros, so assignment cannot be done as the most optimal assignment for team 1 would be both projects 2 and 4.

TABLE 6.5
Row Reduction

	Project 1 (Compact)	Project 2 (Sub-compact)	Project 3 (Intermediate)	Project 4 (Full Size)
Team 1	1	0	2	2
Team 2	3	0	5	6
Team 3	5	0	3	8
Team 4	0	1	0	4

TABLE 6.6
Column Reduction

	Project 1 (Compact)	Project 2 (Sub-compact)	Project 3 (Intermediate)	Project 4 (Full Size)
Team 1	1	0	2	0
Team 2	3	0	5	4
Team 3	5	0	3	6
Team 4	0	1	0	2

However, according to restrictive assumptions of the model, one employee/team can be assigned only one job, so in the first step, no assignment is made in the first row. Moving to second row it has only one zero so assign team 2 to project 2 *(indicated by zeroes in brackets)*. This implies that team 2 would perform on project 2 or project 2 would be conducted by team 2. So, this team should not handle any other project or this project should not be conducted by any other team. Thus, any other zeros in the column of assigned cell would be struck off as shown in Table 6.7. In this case, it would be cell x_{12} and x_{32}. Row 3 has no zeros left as zero in cell x_{32} has been struck off. Row 4 again has two zeros so no assignment can be made.

ii. The same procedure is adopted column-wise. In the first column, there is only one zero (cell x_{41}), so assignment can be made. Thus, project 4 is assigned to team 4. Team 4 cannot be assigned any other project so zero in cell x_{43} is struck off. Columns 2 and 3 have no zeros left. In the fourth column, one unassigned zero in cell x_{14} allows us to assign team 1 to conduct project 4 (Table 6.8).

Step 5: Check for optimality:

An assignment solution is said to be optimal if all jobs are assigned. In this case, only three jobs, i.e. projects 1, 2 and 4 have been assigned to team 4, team 2 and team 1, respectively. Thus, the solution is not optimal. Moving towards the next solution, i.e. iteration 1, involves the following steps:

i. Mark the row (√) in which assignment has not been made. In this case, it would be row 3 represented by team 3. It would be called an unassigned row.

TABLE 6.7
Assignments Done Row Wise

	Project 1 (Compact)	Project 2 (Sub-compact)	Project 3 (Intermediate)	Project 4 (Full Size)
Team 1	1	~~0~~	2	0
Team 2	3	(0)	5	4
Team 3	5	~~0~~	3	6
Team 4	0	1	0	2

TABLE 6.8
Assignments Done Column Wise

	Project 1 (Compact)	Project 2 (Sub-compact)	Project 3 (Intermediate)	Project 4 (Full Size)
Team 1	1	~~0~~	2	**(0)**
Team 2	3	**(0)**	5	4
Team 3	5	~~0~~	3	6
Team 4	**(0)**	1	~~0~~	2

ii. Look for any struck-off zero in that row. Row 3 has such zero in cell x_{32}. Most likely, an unassigned row would have a struck off zero because of the row and column reduction procedure.
iii. Mark such a column (√). In this case, that column would be project 2.
iv. In that column find the cell where assignment has been made. In this, it would be cell a_{21}. Mark that row (√). It would be row 2 represented by team 2.
v. Repeat the above steps.
vi. Draw lines through unmarked rows and marked columns. If the number of marked lines is equal to the number of assignments, then the solution has reached an optimal point.

These steps are shown in Table 6.9.

Step 6: Iterate for optimal solution:

Following steps needs to be followed to iterate towards an optimal solution.

i. From the solution shown in Table 6.9 examine the cells that do not lie on the marked lines. These are termed as unoccupied cells
ii. Select the minimum value from these cells. In this case, it would be 3.
iii. Subtract this minimum value from all of the unoccupied cells.
iv. Add this smallest value to the value at intersection cell. In this case intersection cells are x_{21} and x_{42}.

This gives the next feasible solution. Assignments and optimality test would again be applied to this solution as shown in Table 6.10. In row 1, zero in cell x_{14} is assigned. Row 2 has two zeros, so no assignment can be made. Similarly, rows 3 and 4 also have two zeros so no assignment can be made in them as well. Looking column-wise columns 1, 2 and 3 also have more than one zero so no assignment can be made here as well. This results in multiple solutions as any zero out of multiple available can be made as an assignment. Two cases have been elaborated below:

Case 1: In row 2, cell x_{21} is assigned and cell x_{22} and x_{41} are struck off. In row 3, cell x_{32} is assigned and cell x_{33} is stuck off. Finally, assignment is made in cell x_{43} in row 4 as shown in Table 6.11.

Total project completion time would be: $10 + 6 + 2 + 3 = 21$

TABLE 6.9
Optimality Test

	Project 1 (Compact)	Project 2 (Sub-compact)	Project 3 (Intermediate)	Project 4 (Full Size)	
Team 1	1	~~0~~	2	**(0)**	
Team 2	3	**(0)**	5	4	(√) *(iv)*
Team 3	5	~~0~~	3	6	(√) *(i)*
Team 4	**(0)**	1	~~0~~	2	
		(√) *(iii)*			

TABLE 6.10
Iteration 1

	Project 1 (Compact)	Project 2 (Sub-compact)	Project 3 (Intermediate)	Project 4 (Full Size)
Team 1	1	3	2	0
Team 2	0	0	2	1
Team 3	2	0	0	3
Team 4	0	4	0	2

TABLE 6.11
Iteration 2 (Case 1)

	Project 1 (Compact)	Project 2 (Sub-compact)	Project 3 (Intermediate)	Project 4 (Full Size)
Team 1	1	3	2	**(0)**
Team 2	**(0)**	~~0~~	2	1
Team 3	2	**(0)**	~~0~~	3
Team 4	~~0~~	4	**(0)**	2

Case 2: Cell x_{14} is assigned in row1. Then, instead of cell x_{21} cell x_{22} is assigned striking off cell x_{32}. In row 3, cell x_{33} is assigned striking off cell x_{43}. Finally, in row 4, cell x_{41} is assigned as shown in Table 6.12.

As each row and column has been assigned one assignment, so solution is optimal. Solution assigns Team 1 to project 4 = 10 months;

Team 2 to project 2 = 3 months;

Team 3 to project 3 = 5 months;

Team 1 to project 1 = 3 months;

Total project completion time would be: 10 + 3 + 5 + 3 = 21 months.

So, the final solution is the same in both cases.

TABLE 6.12
Iteration 2 (Case 2)

	Project 1 (Compact)	Project 2 (Sub-compact)	Project 3 (Intermediate)	Project 4 (Full Size)
Team 1	1	3	2	**(0)**
Team 2	~~0~~	**(0)**	2	1
Team 3	2	~~0~~	**(0)**	3
Team 4	**(0)**	4	~~0~~	2

Excel Solution

The assignment problem illustrated in Section 6.2 is solved by using solver application in Excel. As discussed, the only difference in the assignment model from transportation is the fixed amount of 1 of supply and demand.

The formulation of model in Excel is shown in Figure 6.2.

In solver, as discussed in the linear programming chapter, the data is specified in all boxes. It is shown in Figure 6.3.

Figure 6.4 shows the final solution.

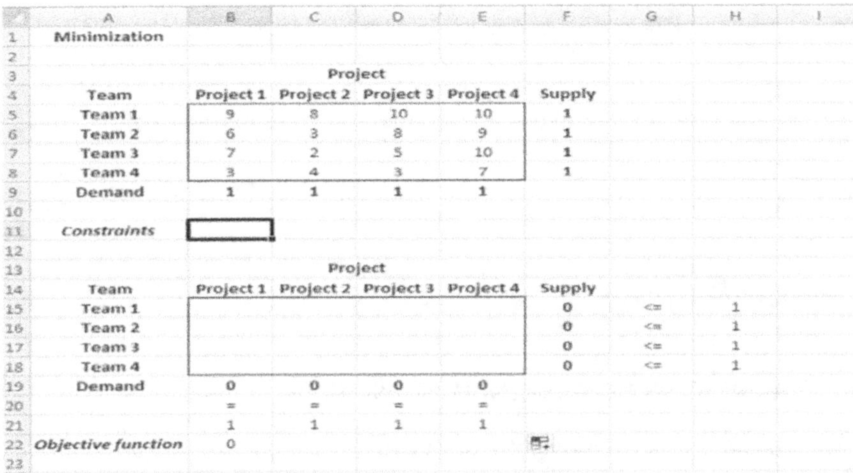

	A	B	C	D	E	F	G	H
1	Minimization							
2								
3				Project				
4	Team	Project 1	Project 2	Project 3	Project 4	Supply		
5	Team 1	9	8	10	10	1		
6	Team 2	6	3	8	9	1		
7	Team 3	7	2	5	10	1		
8	Team 4	3	4	3	7	1		
9	Demand	1	1	1	1			
10								
11	*Constraints*							
12								
13				Project				
14	Team	Project 1	Project 2	Project 3	Project 4	Supply		
15	Team 1					0	<=	1
16	Team 2					0	<=	1
17	Team 3					0	<=	1
18	Team 4					0	<=	1
19	Demand	0	0	0	0			
20		=	=	=	=			
21		1	1	1	1			
22	*Objective function*	0						
23								

FIGURE 6.2 Formulation of model.

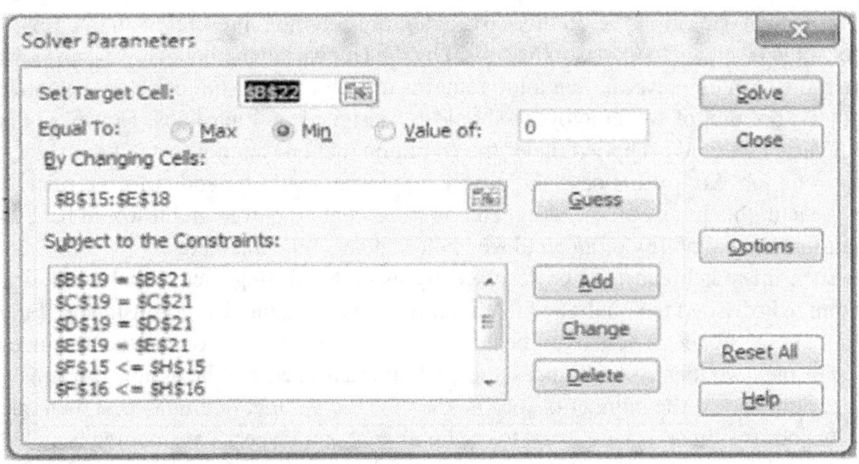

FIGURE 6.3 Solver parameter.

	A	B	C	D	E	F	G	H
1	Minimization							
2								
3			Project					
4	Team	Project 1	Project 2	Project 3	Project 4	Supply		
5	Team 1	9	8	10	10	1		
6	Team 2	6	3	8	9	1		
7	Team 3	7	2	5	10	1		
8	Team 4	3	4	3	7	1		
9	Demand	1	1	1	1			
10								
11	*Constraints*							
12								
13			Project					
14	Team	Project 1	Project 2	Project 3	Project 4	Supply		
15	Team 1	0	0	0	1	1	<=	1
16	Team 2	1	0	0	0	1	<=	1
17	Team 3	0	1	0	0	1	<=	1
18	Team 4	0	0	1	0	1	<=	1
19	Demand	1	1	1	1			
20		=	=	=	=			
21		1	1	1	1			
22	*Objective function*	21						
23								

FIGURE 6.4 Excel solution.

6.6 VARIATIONS OF ASSIGNMENT PROBLEM

6.6.1 Unbalanced Assignment Problem

This type of problem occurs when the number of employees is not equal to the number of jobs. In the transportation model, the same problem occurs when total supply is not equal to total demand. The same resolution procedure of a adding dummy row or column is adopted in the assignment problem. If the number of employees is more than available jobs, then extra employee would sit idle or one job would be assigned to more than one employee. In this case, a dummy column representing an imaginary job would be added to matrix to be solved by the Hungarian method. However, if jobs are more than employees, then a job remains unassigned and might delay the entire project. So, one of the employees should be assigned multiple jobs. However, the Hungarian method is applied under the condition that one employee can be assigned only one job. So, in this case, a dummy employee would be added to the matrix. In the end, this job would be assigned a non-existent employee and it would have to wait for the availability of an employee before it gets implemented.

Such mismatch can also be resolved by using a splitting method where deficit resource is divided to equalize with a number of extra resources. For instance, if the number of assignees is less than the number of jobs by one, then one of the assignees is split into two identical assignees and each new assignee will be assigned one new task. Similarly, if the number of jobs is less than the number of employees, then one of the jobs is split and made equal to the number of employees. This method is called splitting. We will illustrate the resolution of unbalanced problems by using dummy methods in the next section.

6.6.1.1 Case 1: Number of Employees is More than the Number of Jobs

In our illustration of 6.2, suppose MarketOne International LLP organizes five teams instead of four to conduct four projects. This gives manager more choice and instills competition among teams to finish the project at the earliest. The estimated completion time in months pitched by team 5 for projects 1, 2, 3 and 4, respectively, were 5, 6, 4 and 3 months.

Linear programming formulation for changed scenario would be

$$\textbf{Minimize}\quad Z = 9x_{11} + 8x_{12} + 10x_{13} + 10x_{14} + 6x_{21} + 3x_{22} + 8x_{23} + 9x_{24} + 7x_{31} + 2x_{32}$$
$$+ 5x_{33} + 10x_{34} + 3x_{41} + 4x_{42} + 3x_{43} + 7x_{44} + 5x_{51} + 6x_{52} + 4x_{53} + 3x_{54}$$

Subject to:

Assignment of team 1: $x_{11} + x_{12} + x_{13} + x_{14} = 1$
Assignment of team 2: $x_{21} + x_{22} + x_{23} + x_{24} = 1$
Assignment of team 3: $x_{31} + x_{32} + x_{33} + x_{34} = 1$
Assignment of team 4: $x_{41} + x_{42} + x_{43} + x_{44} = 1$
Assignment of team 5: $x_{51} + x_{52} + x_{53} + x_{54} = 1$
Assignment of project 1: $x_{11} + x_{21} + x_{31} + x_{41} + x_{51} \leq 1$
Assignment of project 2: $x_{12} + x_{22} + x_{32} + x_{42} + x_{52} \leq 1$
Assignment of project 3: $x_{13} + x_{23} + x_{33} + x_{43} + x_{53} \leq 1$
Assignment of project 4: $x_{14} + x_{24} + x_{34} + x_{44} + x_{54} \leq 1$

As shown, there is a significant change in constraints of projects. The inequalities are now less than or equal ($\leq$) to rather than equal to (=). This is because, in a changed scenario, the number of teams is more than the number of projects, implying each team would be doing *not more than one* project.

Solution: The matrix form of case 1 would be (Table 6.13):

Step 1: Feasibility check: As number of rows (5) are not equal to number of columns (4) so problem is unbalanced. A dummy column with zero parameters is added, as shown in Table 6.14 to make it balanced.

Step 2: **Row reduction**: This step would give the same result as Table 6.14.

Step 3: **Column reduction**: Reducing the minimum value in each column from all cells of that column would give (Table 6.15):

Step 4: **Make assignments** (Table 6.16):

TABLE 6.13
Estimated Completion Time in Months

	Project 1 (Compact)	Project 2 (Sub-compact)	Project 3 (Intermediate)	Project 4 (Full Size)
Team 1	9	8	10	10
Team 2	6	3	8	9
Team 3	7	2	5	10
Team 4	3	4	3	7
Team 5	5	6	4	3

TABLE 6.14
Estimated Completion Time in Months

	Project 1 (Compact)	Project 2 (Sub-compact)	Project 3 (Intermediate)	Project 4 (Full Size)	Project 5 (Dummy)
Team 1	9	8	10	10	0
Team 2	6	3	8	9	0
Team 3	7	2	5	10	0
Team 4	3	4	3	7	0
Team 5	5	6	4	3	0

TABLE 6.15
Column Reduction

	Project 1 (Compact)	Project 2 (Sub-compact)	Project 3 (Intermediate)	Project 4 (Full Size)	Project 5 (Dummy)
Team 1	6	6	7	7	0
Team 2	3	1	5	6	0
Team 3	4	0	2	7	0
Team 4	0	2	0	4	0
Team 5	2	4	1	0	0

TABLE 6.16
Assignments

	Project 1 (Compact)	Project 2 (Sub-compact)	Project 3 (Intermediate)	Project 4 (Full Size)	Project 5 (Dummy)
Team 1	6	6	7	7	**(0)**
Team 2	3	1	5	6	~~0~~
Team 3	4	**(0)**	2	7	~~0~~
Team 4	**(0)**	2	~~0~~	4	~~0~~
Team 5	2	4	1	**(0)**	~~0~~

Step 5: Optimality (Table 6.17):

As there are only four assignments instead of required five, so solution is not optimal.

Step 6: Iterate for optimal solution (Table 6.18):

Now again make assignments (Table 6.19)

Test optimality (Table 6.20)

TABLE 6.17
Optimality Test

	Project 1 (Compact)	Project 2 (Sub-compact)	Project 3 (Intermediate)	Project 4 (Full Size)	Project 5 (Dummy)	
Team 1	6	6	7	7	(0)	√
Team 2	3	1	5	6	0	√
Team 3	4	(0)	2	7	0	
Team 4	(0)	2	0	4	0	
Team 5	2	4	1	(0)	0	
					√	

TABLE 6.18
Iteration 1

	Project 1 (Compact)	Project 2 (Sub-compact)	Project 3 (Intermediate)	Project 4 (Full Size)	Project 5 (Dummy)
Team 1	5	5	6	6	0
Team 2	2	0	4	5	0
Team 3	4	0	2	7	1
Team 4	0	2	0	4	1
Team 5	2	4	1	0	1

TABLE 6.19
Assignments

	Project 1 (Compact)	Project 2 (Sub-compact)	Project 3 (Intermediate)	Project 4 (Full size)	Project 5 (Dummy)
Team 1	5	5	6	6	(0)
Team 2	2	(0)	4	5	0
Team 3	4	0	2	7	1
Team 4	(0)	2	0	4	1
Team 5	2	4	1	(0)	1

Next solution would be (Table 6.21):

As there are multiple zeros, so more than one solution can be obtained (Table 6.22).

All teams have been assigned a project leading to optimal solution. The minimum possible time along with feasible routes would be (Table 6.23):

Team 1--- Project 5 = 0

Team 2--- Project 1 = 6

TABLE 6.20
Optimality Test

	Project 1 (Compact)	Project 2 (Sub-compact)	Project 3 (Intermediate)	Project 4 (Full Size)	Project 5 (Dummy)	
Team 1	5	5	6	6	**(0)**	
Team 2	2	**(0)**	4	5	~~0~~	√
Team 3	4	~~0~~	2	7	1	√
Team 4	**(0)**	2	~~0~~	4	1	
Team 5	2	4	1	**(0)**	1	
		√			√	

TABLE 6.21
Iteration 2

	Project 1 (Compact)	Project 2 (Sub-compact)	Project 3 (Intermediate)	Project 4 (Full Size)	Project 5 (Dummy)
Team 1	3	5	4	4	0
Team 2	0	0	2	3	0
Team 3	2	0	0	5	1
Team 4	0	4	0	4	3
Team 5	2	6	1	0	3

TABLE 6.22
Assignments (Case 1)

	Project 1 (Compact)	Project 2 (Sub-compact)	Project 3 (Intermediate)	Project 4 (Full Size)	Project 5 (Dummy)
Team 1	3	5	4	4	**(0)**
Team 2	**(0)**	~~0~~	2	3	~~0~~
Team 3	2	**(0)**	~~0~~	5	1
Team 4	~~0~~	4	**(0)**	4	3
Team 5	2	6	1	**(0)**	3

Team 3--- Project 2 = 2
Team 4--- Project 3 = 3
Team 5--- Project 4 = 3
Total time = 14 months
Team 1--- Project 5 = 0
Team 2--- Project 2 = 3
Team 3--- Project 3 = 5

TABLE 6.23
Assignments (Case 2)

	Project 1 (Compact)	Project 2 (Sub-compact)	Project 3 (Intermediate)	Project 4 (Full Size)	Project 5 (Dummy)
Team 1	3	5	4	4	**(0)**
Team 2	~~0~~	**(0)**	2	3	~~0~~
Team 3	2	~~0~~	**(0)**	5	1
Team 4	**(0)**	4	~~0~~	4	3
Team 5	2	6	1	**(0)**	3

Team 4--- Project 1 = 3
Team 5--- Project 4 = 3
Total time = 14 months
Excel Solution
See Figure 6.5

	A	B	C	D	E	F	G	H	I
1	**Assignment problem**								
2		*Project 1 (compact)*	*Project 2 (sub-compact)*	*Project 3 (intermediate)*	*Project 4 (full size)*	*Project 5 (dummy)*	***Supply***		
3	*Team 1*	9	8	10	10	0	1		
4	*Team 2*	6	3	8	9	0	1		
5	*Team 3*	7	2	5	10	0	1		
6	*Team 4*	3	4	3	7	0	1		
7	*Team 5*	5	6	4	3	0	1		
8	***Demand***	1	1	1	1	1			
9									
10	*Objective function*	**Z**	**14**						
11									
12		*Project 1 (compact)*	*Project 2 (sub-compact)*	*Project 3 (intermediate)*	*Project 4 (full size)*	*Project 5 (dummy)*	***Supply***		
13	*Team 1*	0	0	0	0	1	1	=	1
14	*Team 2*	0	1	0	0	0	1	=	1
15	*Team 3*	0	0	1	0	0	1	=	1
16	*Team 4*	1	0	0	0	0	1	=	1
17	*Team 5*	0	0	0	1	0	1	=	1
18	***Demand***	1	1	1	1	1			
19		<=	<=	<=	<=	<=			
20		1	1	1	1	1			

FIGURE 6.5 Excel solution.

6.6.1.2 Case 2: Number of Employees is Less than Number of Jobs

In our illustration of 6.2, suppose MarketOne International LLP organizes four teams to conduct five projects instead of four. The fifth project initiated by the car company pertains to the plan of launching a sports version of the intermediate range of cars. The company wanted to know characteristics of target market, especially rich youth who would like to buy an aspirational product. IQ consultancy can organize only four teams to conduct five projects, implying that a team would conduct at least one or more projects leading to the formulation of the following constraints. Also, the estimated time taken by teams 1, 2, 3 and 4 to conduct the fifth project was 7, 8, 6 and 9 months.

Linear programming formulation for a changed scenario would be

Minimize $Z = 9x_{11} + 8x_{12} + 10x_{13} + 10x_{14} + 7x_{15} + 6x_{21} + 3x_{22} + 8x_{23} + 9x_{24} + 8x_{25}$

$$+ 7x_{31} + 2x_{32} + 5x_{33} + 10x_{34} + 6x_{35} + 3x_{41} + 4x_{42} + 3x_{43} + 7x_{44} + 9x_{45}$$

Subject to:

Assignment of team 1: $x_{11} + x_{12} + x_{13} + x_{14} + x_{15} \leq 1$
Assignment of team 2: $x_{21} + x_{22} + x_{23} + x_{24} + x_{25} \leq 1$
Assignment of team 3: $x_{31} + x_{32} + x_{33} + x_{34} + x_{35} \leq 1$
Assignment of team 4: $x_{41} + x_{42} + x_{43} + x_{44} + x_{45} \leq 1$
Assignment of project 1: $x_{11} + x_{21} + x_{31} + x_{41} = 1$
Assignment of project 2: $x_{12} + x_{22} + x_{32} + x_{42} = 1$
Assignment of project 3: $x_{13} + x_{23} + x_{33} + x_{43} = 1$
Assignment of project 4: $x_{14} + x_{24} + x_{34} + x_{44} = 1$
Assignment of project 5: $x_{15} + x_{25} + x_{35} + x_{45} = 1$

As shown, there is a significant change in the constraints of teams. This is because in the changed scenario, the number of projects is more than the number of teams, implying that one of the teams would be doing more than one project. Thus, the inequalities of teams are now greater than or equal ($\leq$) to rather than equal to (=).

Solution: The matrix form of case 1 would be (Table 6.24):

Step 1: Feasibility check: As the number of columns (5) is not equal to the number of rows (4), so the problem is unbalanced. A dummy row with zero parameters is added, as shown in Table 6.25 to make it balanced.

Step 2 and 3: After row and column reduction, the reduced matrix is (Table 6.26):

TABLE 6.24
Estimated Completion Time in Months

	Project 1 (Compact)	Project 2 (Sub-compact)	Project 3 (Intermediate)	Project 4 (Full Size)	Project 5 (Sporty)
Team 1	9	8	10	10	7
Team 2	6	3	8	9	8
Team 3	7	2	5	10	6
Team 4	3	4	3	7	9

TABLE 6.25
Balanced Model

	Project 1 (Compact)	Project 2 (Sub-compact)	Project 3 (Intermediate)	Project 4 (Full Size)	Project 5 (Sporty)
Team 1	9	8	10	10	7
Team 2	6	3	8	9	8
Team 3	7	2	5	10	6
Team 4	3	4	3	7	9
Team 5 (dummy)	0	0	0	0	0

TABLE 6.26
Reduced Matrix

	Project 1 (Compact)	Project 2 (Sub-compact)	Project 3 (Intermediate)	Project 4 (Full Size)	Project 5 (Sporty)
Team 1	2	1	3	3	0
Team 2	3	0	5	6	5
Team 3	5	0	3	8	4
Team 4	0	1	0	4	6
Team 5 (dummy)	0	0	0	0	0

Step 4: **Making assignments** (Table 6.27):
Step 5: **Optimality test** (Table 6.28):
Step 6: **Iterate for optimal solution** (Tables 6.29 and 6.30):
Solution is optimal. The assignments and monthly times would be (Table 6.31):
Team 1---Project 5 = 7
Team 2---Project 1 = 6
Team 3---Project 2 = 2
Team 4---Project 2 = 3
Team 5---Project 4 = 0
Total time = 18 months

TABLE 6.27
Assignments

	Project 1 (Compact)	Project 2 (Sub-compact)	Project 3 (Intermediate)	Project 4 (Full Size)	Project 5 (Sporty)
Team 1	2	1	3	3	**(0)**
Team 2	3	**(0)**	5	6	5
Team 3	5	~~0~~	3	8	4
Team 4	**(0)**	1	~~0~~	4	6
Team 5 (dummy)	~~0~~	~~0~~	~~0~~	**(0)**	~~0~~

TABLE 6.28
Optimality Test

	Project 1 (Compact)	Project 2 (Sub-compact)	Project 3 (Intermediate)	Project 4 (Full Size)	Project 5 (Sporty)	
Team 1	2	1	3	3	**(0)**	
Team 2	3	**(0)**	5	6	5	√
Team 3	5	~~0~~	3	8	4	√
Team 4	**(0)**	1	~~0~~	4	6	
Team 5 (dummy)	~~0~~	~~0~~	~~0~~	**(0)**	~~0~~	
		√				

TABLE 6.29
Iteration 1

	Project 1 (Compact)	Project 2 (Sub-compact)	Project 3 (Intermediate)	Project 4 (Full Size)	Project 5 (Sporty)
Team 1	2	4	3	3	0
Team 2	0	0	2	3	2
Team 3	2	0	0	5	1
Team 4	0	3	0	4	6
Team 5 (dummy)	0	3	0	0	0

TABLE 6.30
Assignments (Case 1)

	Project 1 (Compact)	Project 2 (Sub-compact)	Project 3 (Intermediate)	Project 4 (Full Size)	Project 5 (Sporty)
Team 1	2	4	3	3	**(0)**
Team 2	**(0)**	~~0~~	2	3	2
Team 3	2	**(0)**	~~0~~	5	1
Team 4	~~0~~	3	**(0)**	4	6
Team 5 (dummy)	~~0~~	3	~~0~~	**(0)**	~~0~~

Solution is optimal. The assignments and monthly times would be:
Team 1---Project 5 = 7
Team 2---Project 2 = 3
Team 3---Project 3 = 5
Team 4---Project 1 = 3
Team 5---Project 4 = 0
Total time = 18 months
Excel Solution
See Figure 6.6.

TABLE 6.31
Assignments (Case 2)

	Project 1 (Compact)	Project 2 (Sub-compact)	Project 3 (Intermediate)	Project 4 (Full Size)	Project 5 (Sporty)
Team 1	2	4	3	3	**(0)**
Team 2	~~0~~	**(0)**	2	3	2
Team 3	2	~~0~~	**(0)**	5	1
Team 4	**(0)**	3	~~0~~	4	6
Team 5 (dummy)	~~0~~	3	~~0~~	**(0)**	0

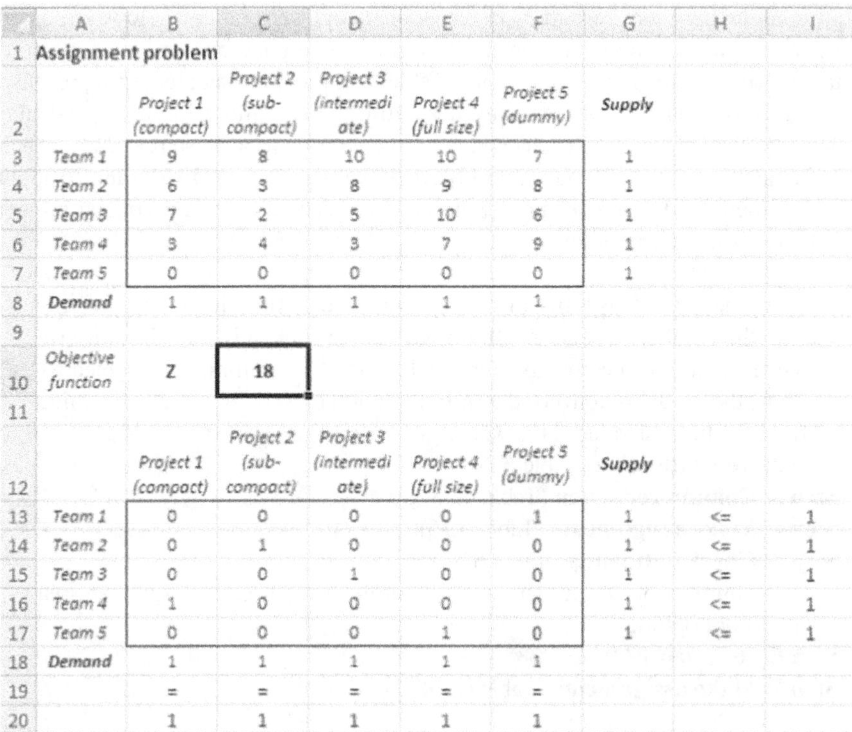

	A	B	C	D	E	F	G	H	I
1	Assignment problem								
2		Project 1 (compact)	Project 2 (sub-compact)	Project 3 (intermediate)	Project 4 (full size)	Project 5 (dummy)	Supply		
3	Team 1	9	8	10	10	7	1		
4	Team 2	6	3	8	9	8	1		
5	Team 3	7	2	5	10	6	1		
6	Team 4	3	4	3	7	9	1		
7	Team 5	0	0	0	0	0	1		
8	Demand	1	1	1	1	1			
9									
10	Objective function	Z	18						
11									
12		Project 1 (compact)	Project 2 (sub-compact)	Project 3 (intermediate)	Project 4 (full size)	Project 5 (dummy)	Supply		
13	Team 1	0	0	0	0	1	1	<=	1
14	Team 2	0	1	0	0	0	1	<=	1
15	Team 3	0	0	1	0	0	1	<=	1
16	Team 4	1	0	0	0	0	1	<=	1
17	Team 5	0	0	0	1	0	1	<=	1
18	Demand	1	1	1	1	1			
19		=	=	=	=	=			
20		1	1	1	1	1			

FIGURE 6.6 Excel solution.

6.6.2 Maximization Problem

Sometimes, the problem is to assign jobs to employees in such a manner to maximize profits or revenues. For instance, while deciding to buy a plot of land in a housing society, an individual would tend to select from multiple plots a location that would have the highest resale value. Thus, if there are 'n' buyers and the same number of available plots, then each buyer has the capacity of buying each plot but would intend to buy, which would have maximum resale value. Therefore, cell values would indicate the resale value of each plot if assigned to a particular individual. Similarly, if a store manager has to decide the location of new store in a mall where he/she has multiple location choices, then the manager would tend to select a location that provides maximum sales or revenues. In such cases, the criteria for assigning is maximization of a parameter like resale value or sales. Such assignment problems can be solved by converting a maximization problem into a minimization problem. This is done by subtracting all the values in the profit matrix from the largest value. This would give values represented as opportunity losses instead of profits. Now such a matrix can be solved by the general Hungarian method.

In illustration of 6.2, MarketOne International LLP, during negotiations, offered the client company to complete the projects of market analysis of four different car models at the earliest by charging extra. Completing a project before deadline would require allocating more resources in terms of manpower and cost. However, steep extra charges associated with such scenarios provide a good opportunity for a consultancy company to earn more revenue. Table 6.32 provides data regarding revenue earned in thousands by the company depending on the allocation of a particular project to a certain team. As each team can conduct each project, so revenue for each combination is shown. The manager of MarketOne International LLP wants to apply the assignment model to make proper allocation of each team to a particular project in order to earn maximum revenue.

Step 1: Check the feasibility of the solution by equating the number of rows with the number of columns. In this case, the problem is feasible. However, the Hungarian method is applicable for the minimization objective. Thus, the model is converted to minimization by subtracting all cell values from the highest value. This would give us cost matrix (Table 6.33).

Step 2: Row reduction (Table 6.34):

Step 3: Column reduction (Table 6.35):

Step 4: Make assignments (Table 6.36):

Step 5: Check optimality (Table 6.37):

As number of assignments are only three as requirement of four so solution is not optimal.

Step 6: Iteration 1 (Table 6.38):

Step 7: Make assignments (Table 6.39):

TABLE 6.32
Estimated Revenue in Thousands

	Project 1 (Compact)	Project 2 (Sub-compact)	Project 3 (Intermediate)	Project 4 (Full Size)
Team 1	51	86	57	65
Team 2	63	61	45	77
Team 3	41	67	55	53
Team 4	56	81	40	43

TABLE 6.33
Estimated Cost in Thousands

	Project 1 (Compact)	Project 2 (Sub-compact)	Project 3 (Intermediate)	Project 4 (Full Size)
Team 1	35	0	29	21
Team 2	23	25	21	9
Team 3	45	19	31	33
Team 4	30	5	46	43

TABLE 6.34
Row Reduction

	Project 1 (Compact)	Project 2 (Sub-compact)	Project 3 (Intermediate)	Project 4 (Full Size)
Team 1	35	0	29	21
Team 2	14	16	12	0
Team 3	26	0	12	14
Team 4	25	0	41	38

TABLE 6.35
Column Reduction

	Project 1 (Compact)	Project 2 (Sub-compact)	Project 3 (Intermediate)	Project 4 (Full Size)
Team 1	21	0	17	21
Team 2	0	16	0	0
Team 3	12	0	0	14
Team 4	1	0	29	38

TABLE 6.36
Assignments

	Project 1 (Compact)	Project 2 (Sub-compact)	Project 3 (Intermediate)	Project 4 (Full Size)
Team 1	21	(**0**)	17	21
Team 2	(**0**)	16	~~0~~	~~0~~
Team 3	12	~~0~~	(**0**)	14
Team 4	1	~~0~~	29	38

TABLE 6.37
Optimality

	Project 1 (Compact)	Project 2 (Sub-compact)	Project 3 (Intermediate)	Project 4 (Full Size)	
Team 1	21	**(0)**	17	21	√
Team 2	**(0)**	16	~~0~~	~~0~~	
Team 3	12	~~0~~	**(0)**	14	
Team 4	1	~~0~~	29	38	√
		√			

TABLE 6.38
Iteration 1

	Project 1 (Compact)	Project 2 (Sub-compact)	Project 3 (Intermediate)	Project 4 (Full Size)
Team 1	20	0	16	20
Team 2	0	17	0	0
Team 3	12	1	0	14
Team 4	0	0	28	37

TABLE 6.39
Assignments

	Project 1 (Compact)	Project 2 (Sub-compact)	Project 3 (Intermediate)	Project 4 (Full Size)
Team 1	20	**(0)**	16	20
Team 2	~~0~~	17	~~0~~	**(0)**
Team 3	12	1	**(0)**	14
Team 4	**(0)**	~~0~~	28	37

This solution is optimal.
Total profits and allocation would be:
Team 1---- Project 2 = 86
Team 2---- Project 4 = 77
Team 3---- Project 3 = 55
Team 4---- Project 1 = 56
Total = $274 in thousands

Excel Solution

See Figure 6.7.

	A	B	C	D	E	F	G	H
1	**Assignment problem**							
2		Project 1 (compact)	Project 2 (sub-compact)	Project 3 (intermediate)	Project 4 (full size)	**Supply**		
3	Team 1	51	86	57	65	1		
4	Team 2	63	61	45	77	1		
5	Team 3	41	67	55	53	1		
6	Team 4	56	81	40	43	1		
7	**Demand**	1	1	1	1			
8								
9	Objective function	**Z**	**274**					
10								
11		Project 1 (compact)	Project 2 (sub-compact)	Project 3 (intermediate)	Project 4 (full size)	**Supply**		
12	Team 1	0	1	0	0	1	<=	1
13	Team 2	0	0	0	1	1	<=	1
14	Team 3	0	0	1	0	1	<=	1
15	Team 4	1	0	0	0	1	<=	1
16	**Demand**	1	1	1	1			
17		=	=	=	=			
18		1	1	1	1			

FIGURE 6.7 Excel solution.

6.6.3 Unacceptable Assignment

In realistic situations, an assignee might not be qualified to do a special kind of job so he/she cannot be assigned that job. For instance, in our illustration of 6.2, suppose each project is broken down into various activities of designing of research (A1), data collection (A2), quantitative analysis & inference (A3) and report preparation & presentation (A4). A team was constituted with four employees (E1, E2, E3 and E4) to perform one of the activities. In our illustration of MarketOne International LLP, activities pertain to market survey of intermediate (largest selling car) model of car and employees are members of team 1. Team member E3 declined to perform activity of data analysis as it involved special statistical skill set. Assigning it to E3 would involve exorbitant training cost, making such assignment infeasible. Moreover, other team members are capable and available to perform the same activity. Such assignment problem variation is classified as unacceptable route problem. In such cases, one or more assignments are infeasible as they would incur substantially high cost. To remove such assignment, an assumed high cost of M is assigned to that route in problems of minimization. Unacceptable routes in problems of maximization are assigned –M so that specific route does not enter the solution.

Table 6.40 indicates time required in weeks to finish an activity by different employees.

TABLE 6.40
Estimated Time in Weeks

	A1	A2	A3	A4
E1	13	45	25	10
E2	30	40	35	32
E3	17	15	M	30
E4	20	14	33	18

Step 1: Check feasibility: As number of rows is equal to number of columns so solution is feasible.
Step 2: Row reduction (Table 6.41):
Step 3: Column reduction (Table 6.42):
Step 4: Make assignments (Table 6.43):
Step 5: Check optimality (Table 6.44):
Step 6: Iteration 1 (Table 6.45):
Step 7: Make assignments (Table 6.46):

So optimal assignments and total time would be:
E1--- A4 = 10
E2--- A3 = 35
E3--- A1 = 17
E4--- A2 = 14
Total = 76 weeks

TABLE 6.41
Row Reduction

	A1	A2	A3	A4
E1	3	35	15	0
E2	0	10	5	2
E3	2	0	M	15
E4	6	0	19	4

TABLE 6.42
Column Reduction

	A1	A2	A3	A4
E1	3	35	10	0
E2	0	10	0	2
E3	2	0	M	15
E4	6	0	14	4

TABLE 6.43
Make Assignments

	A1	A2	A3	A4
E1	3	35	10	**(0)**
E2	**(0)**	10	~~0~~	2
E3	2	**(0)**	M	15
E4	6	~~0~~	14	4

TABLE 6.44
Optimality

	A1	A2	A3	A4	
E1	3	35	10	**(0)**	
E2	**(0)**	10	~~0~~	2	
E3	2	**(0)**	M	15	√
E4	6	~~0~~	14	4	√
		√			

TABLE 6.45
Iteration 1

	A1	A2	A3	A4
E1	3	37	10	0
E2	0	12	0	2
E3	0	0	M	13
E4	4	0	12	2

TABLE 6.46
Assignments

	A1	A2	A3	A4
E1	3	37	10	**(0)**
E2	~~0~~	12	**(0)**	2
E3	**(0)**	~~0~~	M	13
E4	4	**(0)**	12	2

6.7 CREW ASSIGNMENT

6.7.1 Case: Star Airlines

Domestic airlines predominantly follow the practice of flying during day time and resting during the night. During the day, a plane starts from a certain airport early in the day and, according to the schedule, visits a number of stations dropping and carrying passengers at every station. Finally, it retires for the night at the last station, which necessarily might not be the origin station. For instance, a plane starts from Delhi domestic airport (origin) visits Lucknow, Ranchi and then Kolkata (destination). For our illustration purposes, we would define this line of flow as Delhi to Kolkata. The assignment model is very helpful in allocating resources such as (i) type of airplanes and (ii) crew members to various routes that a plane is scheduled to fly during a day. It is important to understand that the application of the assignment model to these two scenarios follows certain assumptions: first, it is assumed that a structured schedule has already been planned. As schedule is a list of flight numbers associated with each route giving departure and arrival station and times. However, a schedule does not allocate type of plane to a particular route. An airline mostly owns different types of planes with differing capacities so allocation of particular airplane to a route is also function of demand and capacity. This process of assigning jobs (different type of planes) to tasks (different routes) is done through the assignment model. Second, each plane in spite of its different capacity has the ability to travel to any of the routes. Third, every plane has to be grounded for certain period during the night for periodic mandatory maintenance. Finally, crew members according to regulations have to be provided minimum and maximum rest time between two consecutive duties.

Considering these mentioned aspects, we have discussed the following two cases: case I discusses the assignment of different types of airplanes to different routes and case II assigns crew members to those routes.

6.7.1.1 Case 1: Assignment of Different Types of Airplanes to Different Routes

Star airlines have five airplanes at its disposal to plan to fly to five destinations. All five flights originate from the same domestic airport in Delhi. One of the considerations in assigning planes to different routes is *minimization of operation cost*. The major component of operation cost is fuel cost. Thus, a nearer destination would consume lesser amount of fuel and in turn incur less cost. Moreover, it would allow aircraft to undertake a number of trips during the day. For simple illustration of the assignment model, we have assumed fuel cost to be directly proportional to the time taken to reach the final destination. Table 6.47 provides the schedule of all planes indicating departure and arrival times. The purpose is to assign different planes to different routes so that each plane takes the minimum possible time to reach its final destination, given that each plane can fly to each destination.

The flying times of each flight on each route are calculated in hours and shown in Table 6.48.

TABLE 6.47
Schedule of Flights

	Flight F1		Flight F2		Flight F3		Flight F4		Flight F5	
	Departure	**Arrival**	**Departure**	**Arrival**	**Departure**	**Arrival**	**Departure**	**Arrival**	**Departure**	**Arrival**
Delhi to Kolkata	04:55	22:35	05:45	20:10	06:50	20:10	07:55	19:20	09:50	21:30
Delhi to Varanasi	08:10	20:25	10:30	22:35	09:50	20:55	08:40	21:35	07:40	21:10
Delhi to Trichy	05:35	20:50	06:30	20:00	05:25	21:10	09:50	22:55	10:20	21:30
Delhi to Mumbai	06:40	21:40	07:30	20:30	04:15	19:30	05:05	22:40	06:30	21:50
Delhi to Chennai	06:25	21:50	07:15	23:25	08:25	20:35	09:55	22:35	06:30	21:50

TABLE 6.48
Flight Times (Hours)

	Delhi to Kolkata	Delhi to Varanasi	Delhi to Trichy	Delhi to Mumbai	Delhi to Chennai
Flight F1	17.6	12.25	15.25	15.0	15.4
Flight F2	14.4	12.1	13.5	13	16.2
Flight F3	13.3	11.1	15.7	17.6	12.2
Flight F4	11.4	12.9	15.3	13.1	12.7
Flight F5	11.7	13.5	11.2	13.0	15.3

Row reduction (Table 6.49):
Column reduction (Table 6.50):
Make assignments (Table 6.51):
Optimality test (Table 6.52):

Iteration 1 (Table 6.53):

Iteration 1 is the most optimal solution and assignments and times would be:
Flight F1--- Delhi to Varanasi: 12.25
Flight F2--- Delhi to Mumbai: 13.00
Flight F3--- Delhi to Chennai: 12.2
Flight F4--- Delhi to Kolkata: 11.4
Flight F5--- Delhi to Trichy: 11.2

Excel Solution

See Figure 6.8.

6.7.1.2 Case 2: Assignment of Crew Members to Routes

This problem is similar to case 1 in that, instead of assigning fleet, crew members have to be assigned to different routes so as to minimize crew rest time. The process of assigning crew members to a particular flight is termed as crew pairing which follows certain requirements. The destination of a flight operated by certain crew is the same as the origin of the same flight for the next destination. Similarly, the origin of a flight operated by a particular crew should be the same as its previous destination. This implies that, at least for our illustration, a crew member is performing two consecutive duties with a rest period between two flights. The resting facilities are provided at both stations. We would illustrate the assignment of crews by

TABLE 6.49
Row Reduction

	Delhi to Kolkata	Delhi to Varanasi	Delhi to Trichy	Delhi to Mumbai	Delhi to Chennai
Flight F1	5.35	0	3.0	2.75	3.15
Flight F2	2.3	0	1.4	0.9	4.1
Flight F3	2.2	0	4.6	6.5	1.1
Flight F4	0	1.5	3.9	1.7	1.3
Flight F5	0.5	2.3	0	0.5	1.8

TABLE 6.50
Column Reduction

	Delhi to Kolkata	Delhi to Varanasi	Delhi to Trichy	Delhi to Mumbai	Delhi to Chennai
Flight F1	5.35	0	3.0	2.7	2.05
Flight F2	2.3	0	1.4	0.4	3.0
Flight F3	2.2	0	4.6	6.0	0
Flight F4	0	1.5	3.9	1.2	0.2
Flight F5	0.5	2.3	0	0	0.5

TABLE 6.51
Assignments

	Delhi to Kolkata	Delhi to Varanasi	Delhi to Trichy	Delhi to Mumbai	Delhi to Chennai
Flight F1	5.35	(**0**)	3.0	2.7	2.05
Flight F2	2.3	~~0~~	1.4	0.4	3.0
Flight F3	2.2	~~0~~	4.6	6.0	(**0**)
Flight F4	(**0**)	1.5	3.9	1.2	0.2
Flight F5	0.5	2.3	(**0**)	~~0~~	0.5

TABLE 6.52
Optimality

	Delhi to Kolkata	Delhi to Varanasi	Delhi to Trichy	Delhi to Mumbai	Delhi to Chennai	
Flight F1	5.35	(**0**)	3.0	2.7	2.05	√
Flight F2	2.3	~~0~~	1.4	0.4	3.0	√
Flight F3	2.2	~~0~~	4.6	6.0	(**0**)	
Flight F4	(**0**)	1.5	3.9	1.2	0.2	
Flight F5	0.5	2.3	(**0**)	~~0~~	0.5	
		√				

TABLE 6.53
Iteration 1

	Delhi to Kolkata	Delhi to Varanasi	Delhi to Trichy	Delhi to Mumbai	Delhi to Chennai
Flight F1	4.95	(**0**)	2.6	2.3	1.65
Flight F2	1.9	0	1.0	(**0**)	2.6
Flight F3	2.2	0.4	4.6	6.0	(**0**)
Flight F4	(**0**)	1.9	3.9	1.2	0.2
Flight F5	0.5	2.7	(**0**)	~~0~~	0.5

	A	B	C	D	E	F	G	H	I
1		Delhi to Kolkata	Delhi to Varanasi	Delhi to Trichy	Delhi to Mumbai	Delhi to Chennai	Supply		
2	Flight F1	17.6	12.25	15.25	15	15.4	1		
3	Flight F2	14.4	12.1	13.5	13	16.2	1		
4	Flight F3	13.3	11.1	15.7	17.6	12.2	1		
5	Flight F4	11.4	12.9	15.3	13.1	12.7	1		
6	Flight F5	11.7	13.5	11.2	13	15.3	1		
7	Demand	1	1	1	1	1			
8									
9	Z	60.05							
10									
11		Delhi to Kolkata	Delhi to Varanasi	Delhi to Trichy	Delhi to Mumbai	Delhi to Chennai			
12	Flight F1	0	1	0	0	0	1	<=	1
13	Flight F2	0	0	0	1	0	1	<=	1
14	Flight F3	0	0	0	0	1	1	<=	1
15	Flight F4	1	0	0	0	0	1	<=	1
16	Flight F5	0	0	1	0	0	1	<=	1
17		1	1	1	1	1			
18		=	=	=	=	=			
19		1	1	1	1	1			

FIGURE 6.8 Excel solution.

considering the to and fro trip of each flight. For instance, a flight originating from Delhi and going to Chennai needs to come back to Delhi by night to complete a round trip. Resting can be done during the initial leg of tour when the flight stops over at Chennai and at Delhi after it has started from Chennai during its final leg of tour. Management adhering to regulations decided to have rest time of at least 5 hours and not more than 11 hours. Taking into consideration, five crew teams denoted as C1, C2, C3, C4 and C5 to be assigned to five flights (F1, F2, F3, F4 and F5) flying on Delhi Chennai route as shown in Table 6.54 estimate optimal assignment with the assumption that each crew team can be assigned to any flight.

TABLE 6.54
Schedule

Flight	Departure from Delhi	Arrival at Chennai	Flight	Departure from Chennai	Arrival at Delhi
Flight F1	06:25	09:15	Flight F1	15:15	18:05
Flight F2	07:15	10:05	Flight F2	16:00	22:55
Flight F3	08:25	11:20	Flight F3	18:05	21:00
Flight F4	08:45	12:40	Flight F4	19:50	01:25
Flight F5	10:20	18:05	Flight F5	18:45	23:45

TABLE 6.55
Flight Times (Hours) with Chennai as Rest Station

	Flight F1	Flight F2	Flight F3	Flight F4	Flight F5
C1	6.0	6.75	8.8	10.6	9.5
C2	5.2	5.9	8.0	9.75	8.7
C3	3.9	4.7	6.75	8.5	7.4
C4	2.6	3.3	5.4	7.2	6.1
C5	21.2	22.1	24	1.75	0.67

Table 6.55 shows waiting/rest time for each crew team with regard to each flight *if rest station is considered to be at Chennai.* The times in the table indicate the difference between arrival and departure from Chennai for each flight. It is important to understand that times depict the resting time of each crew if assigned to particular type of flight. For instance, if C1 is assigned to F1 then rest time is difference between 09:15 and 15:15, which would be 6 hours. Similarly, other times should be read.

If rest station is considered to be at Delhi (Table 6.56):

Considering the constraint of minimum (5 hours) and maximum (11 hours) rest time, least time from both tables is taken and shown in Table 6.57.

TABLE 6.56
Flight Times (Hours) with Delhi as Rest Station

	Flight F1	Flight F2	Flight F3	Flight F4	Flight F5
C1	12.3	13.2	14.3	14.1	21.2
C2	10.3	8.3	9.5	9.8	16.3
C3	9.4	10.25	11.4	11.75	18.25
C4	5.0	5.8	7.0	7.3	13.8
C5	6.7	7.5	8.7	10.2	10.6

TABLE 6.57
Flight Times (Hours)

	Flight F1	Flight F2	Flight F3	Flight F4	Flight F5
C1	6.0	6.75	8.8	10.6	9.5
C2	5.2	5.9	8.0	9.75	8.7
C3	9.4	10.25	6.75	8.5	7.4
C4	5.0	5.8	5.4	7.2	6.1
C5	6.7	7.5	8.7	10.2	10.6

Row reduction (Table 6.58):
Column reduction (Table 6.59):
Make assignments (Table 6.60):
Optimality test (Table 6.61):

Iteration 1 (Table 6.62):
Make assignments (Table 6.63):
Optimality test (Table 6.64):

Iteration 2 (Table 6.65):
Make assignments (Table 6.66):
Optimality test (Table 6.67):

TABLE 6.58
Row Reduction

	Flight F1	Flight F2	Flight F3	Flight F4	Flight F5
C1	0	0.75	2.8	4.6	3.5
C2	0	0.7	2.8	4.55	3.5
C3	2.0	2.85	0.65	1.1	0
C4	0	0.8	0.4	1.4	0.3
C5	0	0.8	2.0	3.5	3.9

TABLE 6.59
Column Reduction

	Flight F1	Flight F2	Flight F3	Flight F4	Flight F5
C1	0	0.05	2.4	3.5	3.5
C2	0	0	2.4	3.45	3.5
C3	2.0	2.15	0.25	0	0
C4	0	0.1	0	0.3	0.3
C5	0	0.1	1.6	2.4	3.9

TABLE 6.60
Assignments

	Flight F1	Flight F2	Flight F3	Flight F4	Flight F5
C1	**(0)**	0.05	2.4	3.5	3.5
C2	~~0~~	**(0)**	2.4	3.45	3.5
C3	2.0	2.15	0.25	**(0)**	~~0~~
C4	~~0~~	0.1	**(0)**	0.3	0.3
C5	~~0~~	0.1	1.6	2.4	3.9

TABLE 6.61
Optimality

	Flight F1	Flight F2	Flight F3	Flight F4	Flight F5	
C1	**(0)**	0.05	2.4	3.5	3.5	√
C2	θ	**(0)**	2.4	3.45	3.5	
C3	2.0	2.15	0.25	**(0)**	θ	
C4	θ	0.1	**(0)**	0.3	0.3	
C5	θ	0.1	1.6	2.4	3.9	√
	√					

TABLE 6.62
Iteration 1

	Flight F1	Flight F2	Flight F3	Flight F4	Flight F5
C1	0	0	2.35	3.45	3.45
C2	0.05	0	2.4	3.45	3.5
C3	2.05	2.15	0.25	0	0
C4	0.05	0.1	0	0.3	0.3
C5	0	0.05	1.55	2.35	3.85

TABLE 6.63
Assignments

	Flight F1	Flight F2	Flight F3	Flight F4	Flight F5
C1	θ	θ	2.35	3.45	3.45
C2	0.05	**(0)**	2.4	3.45	3.5
C3	2.05	2.15	0.25	**(0)**	θ
C4	0.05	0.1	**(0)**	0.3	0.3
C5	**(0)**	0.05	1.55	2.35	3.85

TABLE 6.64
Optimality

	Flight F1	Flight F2	Flight F3	Flight F4	Flight F5	
C1	θ	θ	2.35	3.45	3.45	√
C2	0.05	**(0)**	2.4	3.45	3.5	√
C3	2.05	2.15	0.25	**(0)**	θ	
C4	0.05	0.1	**(0)**	0.3	0.3	
C5	**(0)**	0.05	1.55	2.35	3.85	√
	√	√				

TABLE 6.65
Iteration 2

	Flight F1	Flight F2	Flight F3	Flight F4	Flight F5
C1	0	0	0.8	1.9	1.9
C2	0.05	0	0.85	1.9	1.95
C3	3.6	3.7	0.25	0	0
C4	1.6	1.65	0	0.3	0.3
C5	0	0.05	0	0.8	2.3

TABLE 6.66
Assignments

	Flight F1	Flight F2	Flight F3	Flight F4	Flight F5
C1	~~0~~	~~0~~	0.8	1.9	1.9
C2	0.05	(**0**)	0.85	1.9	1.95
C3	3.6	3.7	0.25	(**0**)	~~0~~
C4	1.6	1.65	(**0**)	0.3	0.3
C5	(**0**)	0.05	~~0~~	0.8	2.3

TABLE 6.67
Optimality

	Flight F1	Flight F2	Flight F3	Flight F4	Flight F5	
C1	~~0~~	~~0~~	0.8	1.9	1.9	√
C2	0.05	(**0**)	0.85	1.9	1.95	√
C3	3.6	3.7	0.25	(**0**)	~~0~~	
C4	1.6	1.65	(**0**)	0.3	0.3	√
C5	(**0**)	0.05	~~0~~	0.8	2.3	√
	√	√	√			

Iteration 3 (Table 6.68):

Make assignments (Tables 6.69 and 6.70):

C1--- Flight F1: 6.0

TABLE 6.68
Iteration 3

	Flight F1	Flight F2	Flight F3	Flight F4	Flight F5
C1	0	0	0.8	1.6	1.6
C2	0.05	0	0.85	1.6	1.65
C3	3.9	4.0	0.55	0	0
C4	1.6	1.65	0	0	0
C5	0	0.05	0	0.5	2.0

TABLE 6.69
Assignments (Solution 1)

	Flight F1	Flight F2	Flight F3	Flight F4	Flight F5
C1	**(0)**	~~0~~	0.8	1.6	1.6
C2	0.05	**(0)**	0.85	1.6	1.65
C3	3.9	4.0	0.55	**(0)**	~~0~~
C4	1.6	1.65	~~0~~	~~0~~	**(0)**
C5	0	0.05	**(0)**	0.5	2.0

TABLE 6.70
Assignments (Solution 2)

	Flight F1	Flight F2	Flight F3	Flight F4	Flight F5
C1	**(0)**	~~0~~	0.8	1.6	1.6
C2	0.05	**(0)**	0.85	1.6	1.65
C3	3.9	4.0	0.55	~~0~~	**(0)**
C4	1.6	1.65	~~0~~	**(0)**	~~0~~
C5	~~0~~	0.05	**(0)**	0.5	2.0

C2--- Flight F2: 5.9
C3--- Flight F4: 9.75
C4--- Flight F5: 6.1
C5--- Flight F3: 8.7
C1--- Flight F1: 6.0
C2--- Flight F2: 5.9
C3--- Flight F5: 7.4
C4--- Flight F4: 7.2

	A	B	C	D	E	F	G	H	I	J
1		Flight F1	Flight F2	Flight F3	Flight F4	Flight F5	Supply			
2	C1	6	6.75	8.8	10.6	9.5	1			
3	C2	5.2	5.9	8	9.75	8.7	1			
4	C3	9.4	10.25	6.75	8.5	7.4	1			
5	C4	5	5.8	5.4	7.2	6.1	1			
6	C5	6.7	7.5	8.7	10.2	10.6	1			
7	Demand	1	1	1	1	1				
8										
9	Objective function									
10	Min Z	35.2								
11										
12		Flight F1	Flight F2	Flight F3	Flight F4	Flight F5				
13	C1	1	0	0	0	0	1	<=	1	
14	C2	0	1	0	0	0	1	<=	1	
15	C3	0	0	0	0	1	1	<=	1	
16	C4	0	0	0	1	0	1	<=	1	
17	C5	0	0	1	0	0	1	<=	1	
18		1	1	1	1	1				
19		=	=	=	=	=				
20		1	1	1	1	1				
21										

FIGURE 6.9 Excel solution.

C5--- Flight F3: 8.7

Excel Solution

See Figure 6.9

6.8 SUMMARY

This chapter discusses the assignment model which are part of network models consisting of nodes and arcs where nodes indicate assignee which needs to be allocated and arcs show the criteria like cost or time on basis of which allocations are to be made. It is a special form of transportation model in which supply and demand are equal to one. The objective of the assignment model is allocation of resources to minimize a performance metric. The model has also been presented as a linear programming model involving constraint functions, criteria and objective function. Variation of model in the form of unbalanced, maximization and unacceptable assignments has been solved by using the Hungarian method and Excel software. Under unbalanced variation, two cases were discussed: the number of employees is more than number of jobs and the number of employees is less than the number of jobs. The chapter also discusses a special case of assignment model in the form of crew assignment. Assignment of types of airplanes to different routes and assignment of crew members to those routes has been discussed.

6.9 CASE 1: ASSIGNING WORKERS TO PROCESSES

Prime auto ltd. manufactures various auto components such as car locks, door handles, grilles, etc. This case illustrates assigning of operators to different processes involved in the assembling process of grilles. This process involves various steps: In step 1 (process P1), grills are picked from raw material bin and primer is applied to the area where emblem of particular car manufacturer (in this case emblem is 'S' mark) needs to be pasted. This step takes on an average of 18 seconds to complete. The second step (process P2) involves pasting of S mark taking on average of 23 seconds. Process P3 in the third step undertakes pressing of pasted mark on the pressing machine. It is kept under press for 2 seconds and then grille is removed from the machine and pasting confirmed. Entire P3 takes on an average 20 seconds. Process P4 involved two steps: first, LH garnish is assembled and second RH garnish is assembled in the grille. This step was completed in 18 seconds. Finally, inspection of entire is grille is done in process P5 which involved confirming all the locks from backside and visually inspecting. This step is done in approximately 21 seconds. Five operators (O1, O2, O3, O4 and O5) were available to conduct the above-mentioned processes. The following table provides production time in seconds for each process with respect to each operator.

		Processing Time of Each Process (Seconds)				
		P1	**P2**	**P3**	**P4**	**P5**
Operators	O1	14	19	20	17	16
	O2	20	27	23	19	24
	O3	7	9	12	13	10
	O4	11	14	12	10	15
	O5	17	23	20	23	22

- Given above data assign operators to different processes so as to minimize total processing time.
- Operator O5 is on leave for next week's schedule. Which process would remain unfinished? Which operator would be most optimal out of remaining four to do unfinished process?
- In (i), it was assumed that all operators are qualified to do all processes. However, the quality manager after analysing past month's data found assignment of processes P5 and P3 respectively to O1 and O3 results in loss of productivity. What would be the optimal assignments if these operators cannot be assigned these processes?
- To fulfil increased demand of grilles from car manufacturer foreman appointed another operator O6, to perform five processes. But O6 could be allocated only P2, P3 or P5 as these processes were considered critical and take significantly higher time than other two processes. The average time taken by operator to perform P2, P3 and P5, respectively, was 20, 19 and 21 seconds. Make optimal assignments.

6.10 CASE 2: ASSIGNING SWINGS TO KIDS

There are five swings in the school and during break time there is huge rush of kids. Each swing can be utilized by only one kid at a time. Due to heavy rush and limited time, school authorities have placed a maximum limit of 5 minutes per kid per swing. The following table gives data regarding maximum time that a kid spends on a particular swing. Every kid would like to take a swing which provides them maximum time.

		Swing Time (Minutes)				
		S1	S2	S3	S4	S5
Kids	K1	5.0	4.9	4.7	4.6	4.9
	K2	4.4	5.0	4.1	4.7	4.2
	K3	4.0	4.3	4.4	4.2	5.0
	K4	4.8	5.0	4.1	4.4	4.0
	K5	4.7	4.1	4.9	5.0	4.4

- Assign a swing per kid so as to maximize swing time.
- According to school rule, each swing would be operated by one helper. Five helpers H1, H2, H3, H4 and H5 were appointed to operate five swings. Each helper intends to take a swing which is utilized for minimum time. The following table gives minimum time in minutes that a swing is utilized with regard to each helper. Make optimal assignment so as to minimize swing time.

		Swing Time (Minutes)				
		S1	S2	S3	S4	S5
Helpers	H1	2.7	4.6	4.7	3.6	3.9
	H2	4.1	4.2	4.0	4.7	5.0
	H3	4.4	2.7	3.6	4.2	4.8
	H4	2.3	4.9	3.4	4.4	4.1
	H5	4.6	4.5	4.4	2.9	2.0

- Helper H3 has to rush for an emergency work leaving behind only four other helpers. Which swing goes unassigned?
- In (ii), kids threw certain tantrums regarding allocation of helpers. To pacify them, H1 and H2 were not assigned to S1; H3 was not assigned to S3; H4 not to S5 and H5 was not allowed to help kid on swing S4. Under these circumstances, how assignments undergo change?

6.11 GLOSSARY

- **Hungarian algorithm**: It is a special purpose method to resolve problems where all demand and supplies equal to one unit termed as assignment problem.

- **Opportunity loss:** The difference between the largest value in each row with each cell leading to row reduction and the largest value in each column with each cell in that column leading to column reduction indicates opportunity loss.
- **Balanced problem:** assignment problem with equal number of rows and columns.
- **Optimality:** An assignment solution is said to be optimal if all jobs are assigned.
- **Unbalanced assignment problem***:* This type of problem occurs when the number of rows is not equal to the number of columns.
- **Maximization problem:** when the problem is to assign jobs to employees in such a manner so as to maximize profits or revenues.
- **Unacceptable assignment:** when an assignee might not be qualified to do a special kind of job so he/she cannot be assigned that job.

6.12 MODEL QUESTIONS

6.12.1 Five racks (R1, R2, R3, R4 and R5) in a grocery store are earmarked for placing five categories of food items (F1, F2, F3, F4 and F5). Each type of food item can be placed in any rack. The collaboration between food item and rack is done based on the distance travelled from entry point to a particular rack. The following table provides such distance data.

		Racks				
		R1	**R2**	**R3**	**R4**	**R5**
Food category	F1	13	20	10	15	21
	F2	11	9	13	8	17
	F3	12	14	15	10	13
	F4	20	22	16	24	25
	F5	14	9	11	10	13

i. Make optimal assignments so as to minimize total distance travelled for placement.

ii. Consider values given as revenue earned by placing a food type in particular rack. Revenue is considered as function of distance travelled as closer a food item placed to the rack close to entry point higher are the chances of its purchase. Which food type should be placed in which rack so that revenues are maximized.

6.12.2 Store accessibility is considered as an important attribute in attracting customers. Four stores (S1, S2, S3 and S4) cater to the demand of consumers in four sectors (Sec-1, Sec-2, Sec-3 and Sec-4). Consumers can access any of the stores. The following table provides distance in kilometres data between each sector and store.

		Stores			
		S1	**S2**	**S3**	**S4**
Sectors	Sec-1	3.6	5.1	4.2	3.4
	Sec-2	3.9	4.8	6.5	5.6
	Sec-3	4.4	3.2	5.2	4.0
	Sec-4	6.7	5.4	4.9	4.1

i. Which would be the most optimal store for customers of each sector so that total distance is minimized?
ii. How do assignments change if management decides to operate stores only within area of 5 km?
iii. Store 3 S3 is under renovation so only three stores S1, S2 and S4 are operational. Due to unbalance between the number of stores and the number of sectors, how do assignments change from (i)?

6.12.3 Five runners (R1, R2, R3, R4 and R5) of a relay race team compete for four positions (P1, P2, P3 and P4). Each runner can be placed at any of the position. The following data gives minimum time in seconds taken to complete a runner's leg of race. Make optimal assignment which minimizes total time.

		Positions			
		P1	**P2**	**P3**	**P4**
Runner	R1	6.1	6.7	6.5	6.4
	R2	6.3	6.2	6.1	6.0
	R3	5.9	5.8	6.4	6.2
	R4	6.6	6.5	5.7	6.3
	R5	6.7	6.2	6.0	5.9

6.12.4 Lex Associates manages various dealerships of a particular car company. The company wants to assign five potential dealers to five geographic areas of a state. A particular area would be allotted to a particular dealer depending on sales accrued on a monthly basis. Each dealer can perform in each area. To decide allocation depending on background of each dealer, the company asked them to provide potential sales estimate for each area. Sales data is given in the following table.

		Geographic Areas				
		A1	**A2**	**A3**	**A4**	**A5**
Dealers	D1	15	17	14	11	19
	D2	20	16	18	12	13
	D3	17	13	19	19	14
	D4	12	18	20	22	17
	D5	19	20	16	18	21

i. Find optimal assignment so as to maximize sales.
ii. To instill competition, the company decided to increase number of potential dealers from five to six for same sales area. The sixth dealer provides sales estimate of 23, 21, 17, 19 and 20 cars per month for five areas respectively. How assignments change and which dealer would be left out?
iii. Distance being one of important criteria in sales estimation Lex Associates decided to restrict certain dealers to bid for certain areas due to their inaccessibility. This changed sales estimate from (i) in the following way. Make optimal assignments.

		Geographic Areas				
		A1	A2	A3	A4	A5
	D1	18	17	21	20	Can't bid
	D2	Can't bid	24	20	25	19
Dealers	D3	19	18	19	Can't bid	20
	D4	22	21	Can't bid	24	20
	D5	21	Can't bid	22	23	19

6.12.5 A bus service operating five buses between Delhi and Chandigarh airport and back is given below. Five crew teams have to be assigned to each of the bus with the condition that a crew should rest for minimum of 4 hours and maximum of 10 hours. With rest facilities available at both stations, estimate optimal assignment of crew teams to particular routes in order to minimize rest time.

	Delhi to Chandigarh Airport			Chandigarh to Delhi Airport	
Route	**Departure**	**Arrival**	**Route**	**Departure**	**Arrival**
1.	07:00	11:00	a	18:00	22:00
2.	07:40	11:40	b	19:10	23:10
3.	08:00	12:00	c	19:40	23:40
4.	09:30	13:30	d	21:00	01:00
5.	10:15	14:15	e	23:30	03:30

6.12.6 Following table shows estimated job completion time in days of three jobs by three employees. The objective is to assign employees to given jobs in such a way that total job completion time is minimized. Estimate by using the Hungarian method.

Employees	J1	J2	J3
	Jobs		
E1	10	15	9
E2	9	18	5
E3	6	14	3

6.12.7 A firm needs to assign three project leaders to three projects. Based on different backgrounds and experiences of the leaders, various leader–client assignments differ in terms of projected completion times. The possible assignments and estimated completion times in days are given. Estimate minimum total completion times by assigning projects to respective clients.

	Client		
Project leader	**C1**	**C2**	**C3**
A	10	16	32
B	14	22	40
C	22	24	34

6.12.8 Project completion times where four project leaders have to be assigned three available projects are given. Find the best possible assignments. Are there multiple solutions? Find, if yes.

	Client		
Project leader	**C1**	**C2**	**C3**
A	10	15	9
B	9	18	5
C	6	14	3
D	8	16	6

6.12.9 A distributor distributes a product in four territories (P, Q, R, S) each of which has to be assigned a sales representative (A, B, C, D). The selected criteria for assignment were revenue generated by each sales representative in each territory. Estimated revenues from past experience in have been shown. The objective is to assign a territory to a particular sales representative so that maximum revenues can be generated.

Sales Representative	Sales Territory			
	P	Q	R	S
A	44	80	52	60
B	60	56	40	72
C	36	60	48	48
D	52	76	36	40

6.12.10 A company is faced with the problem of assigning four machines to six different jobs. The profits are estimated as follows:

Job	Machine			
	A	B	C	D
1	3	6	2	6
2	7	1	4	4
3	3	8	5	8
4	6	4	3	7
5	5	2	4	3
6	5	7	6	4

Solve the problem to maximize total profits.

6.12.11 Find the proper assignment of the assignment problem whose cost matrix is given as under.

	1	2	3	4	5
A	10	6	4	8	3
B	2	11	7	7	6
C	5	10	11	4	8
D	6	5	3	2	5
E	11	7	10	11	7

6.12.12 Find the assignment of salesman to various districts which will yield maximum profit.

	1	2	3	4
A	16	10	14	11
B	14	11	15	15
C	15	15	13	12
D	13	12	14	15

6.12.13 Solve the following assignment problem where five operators are to be assigned five jobs so as to minimize operating time given in minutes.

	J1	J2	J3	J4	J5
O1	10	15	11	8	5
O2	15	14	10	9	6
O3	7	8	14	13	11
O4	10	12	9	11	10
O5	6	9	14	12	12

6.12.14 A foreman wants to assign six workers to six machines in order to maximize productivity measured in terms of number of units produced in given time. The following table gives the data regarding number of units produced if each worker works on each type of machine. Find an optimal assignment which would maximize productivity.

	M1	M2	M3	M4	M5	M6
W1	8	5	7	4	8	9
W2	4	6	9	7	7	10
W3	9	7	6	9	11	6
W4	12	10	8	11	6	7
W5	6	9	10	8	9	12
W6	10	9	12	10	8	8

6.12.15 A toy set is complete by packing five small toys. Each toy is packed by a worker on the packing line. So, five workers W1, W2, W3, W4 and W5 are required to complete the set. Each worker can pack any of the small toys (T1, T2, T3, T4 and T5). Cost of packing and revenue earned from an entire lot of toy sets with regard to each worker is given. Make an optimal assignment so as to maximize profit.

	Revenue ($)					Cost ($)				
	T1	T2	T3	T4	T5	T1	T2	T3	T4	T5
W1	50	70	80	60	75	30	50	30	35	45
W2	80	50	60	40	65	40	25	30	20	35
W3	85	55	70	65	75	40	30	35	30	40
W4	50	65	80	80	70	25	40	40	35	35
W5	55	67	82	70	81	25	30	41	38	42

6.12.16 B & R limited manufactures and sells hygiene products such as soaps and hand wash. Company with the intention to expand its portfolio and improve its competitive position in the existing territory acquired a famous oral care product range. This led to increase in their product portfolio from three to four product lines (P1, P2, P3 and P4). Management faced with a problem of reallocating their existing sales personnel of three sales team (S1, S2 and S3). Sales revenue of three existing product lines and with four products with regard to different sales teams. Make and differentiate optimal assignment in both cases.

	Sales Revenue (Thousands)			Sales Revenue (Thousands)			
	P1	**P2**	**P3**	**P1**	**P2**	**P3**	**P4**
S1	20	23	25	17	20	22	35
S2	26	32	27	24	30	22	39
S3	34	20	30	30	15	27	26

6.12.17 Due to high rental values, a mall owner faced with a problem of low occupancy of showrooms. The manager reduced the rent of six vacant showrooms (S1, S2, S3, S4, S5 and S6). However, only four retailers (R1, R2, R3 and R4) showed interest. Loss in revenue due to reduction in rent if it is assumed that any of the showrooms can be allocated to any of the retailers is given. Make an optimal assignment of showrooms to retailers so as to minimize loss in revenue.

	R1	**R2**	**R3**	**R4**
S1	5	7	3	8
S2	6	2	5	4
S3	4	9	6	9
S4	7	5	4	8
S5	6	4	6	5
S6	7	9	5	6

6.12.18 Four teachers (T1, T2, T3 and T4) specializing in the subject of Business Statistics needs to be assigned to four classes (C1, C2, C3 and C4). As each teacher can teach the same subject, so they can be allocated to any of the class with only one exception. Teacher T3 cannot be given class C4 as it involved Ph.D students and only Ph.D holder teachers can teach to C4. T3 was not a Ph.D. Head of department assigns classes based on feedback score obtained from students of previous batches regarding each teacher. Score varied from minimum value of '1' to maximum of '10'. The following table provides average scores for each teacher and each class. Make an optimal assignment which assigns by maximizing the feedback scores.

	C1	**C2**	**C3**	**C4**
T1	6.5	5.8	7.2	6.0
T2	6.4	7.4	8.1	8.2
T3	7.1	6.0	8.3	M
T4	8.5	6.9	6.5	7.1

7 Network Model III

Travelling Salesman, Vehicle Routing and Shortest Path Problem

7.1 INTRODUCTION

This chapter is devoted to discussing three special cases of network flow models. The models have been considered as complex due to their practicality and multiplicity of constraints. In the travelling salesman problem, a network is constructed where instead of moving only from source to destination, the entity which could be a salesman or a vehicle has to come back to origin. This aspect was missing in previously discussed transportation and assignment problems. When capacity constraints of vehicle are incorporated, the problem becomes more complex as vehicles cannot visit all cities fulfilling their demand and then come back to origin. So, a vehicle routing problem (VRP) involves the number of vehicle and routes. However, in these cases, sources and destinations are only one, whereas, in previous flow models, they were many. Also how to find the shortest path in case of the availability of multiple alternative paths for a vehicle to traverse was discussed in this chapter.

7.2 TRAVELLING SALESMAN PROBLEM

The travelling salesman problem deals with the problem of estimating total shortest distance that a salesman would travel in visiting each of specified group of cities only once and then returning to origin city. In general, in a square matrix A of order 'n' where n represents the number of destinations/cities, the salesman would start off from home city and visit all other n−1 cities before coming back to his home city. The entire tour must be planned in a manner which minimizes total travel cost or time. The salesman in its first selection would select destination which is closest or takes the least time to reach and then the next selection would be closest to the first one. This rule would be followed in all selections.

The problem is similar to assignment problem discussed in previous sections. Both problems are combinatorial in nature where personnel qualified to do every type of job are assigned to exactly one job. In the assignment method, there are n persons and n jobs. Each person is allocated a job. In salesman problem, there is still allocation, but only one person who has to be assigned to different jobs or cities. Most importantly, assignment is a non-cyclic method of allocating jobs in which

DOI: 10.1201/9781003212966-7

each person can be assigned any of the jobs, whereas the salesman problem is cyclic in which the salesman must make a cyclic tour of all cities and return to the city of origin from where the tour began.

Thus, mathematically, the solution of the salesman problem is done by applying the same transformation methods as done in the assignment problem. These methods are as follows:

- A salesman problem is a square matrix.
- The minimum value in each row or column is subtracted from all other elements in the respective row or column, making at least one element to be zero and all other elements positive.
- An important distinction in the salesman problem is that all diagonal elements give value of infinity. This is because, according to the concept of the travelling salesman problem, a salesman starting from a particular place would not visit it again, so to remove such an element from the solution, the element is given an arbitrarily very high value which is denoted by infinity (∞).
- A salesman problem is termed as optimal if every element above diagonal elements and element in the last row and the first column is zero or an assignable element. If that is obtained in any iteration of the problem, then cyclic allocation can be achieved, making the solution optimal.

For assignment model:

$$\sum_{i=1}^{n} X_{ij}=1$$

$x_{ij}=1$ or 0 for all $i=1, 2\ldots, n$

where x_{ij} denotes whether an assignment is been made or not. The assignment of ith personnel can be done on j^{th} job. But for the travelling salesman problem, if there are five cities (1, 2, 3, 4 and 5) that salesman has to visit starting from 1 before returning to 1, then assignment activity would be denoted by

$$x_{12}=x_{23}=x_{34}=x_{45}=x_{51}=1$$

This is only one possible tour, but there can be different possible tours, but the method is applied in a fashion that only a tour is selected which fulfils the criteria of minimizing time or distance. Another example of a possible tour could be:

$$x_{13}=x_{32}=x_{24}=x_{45}=x_{51}=1$$

It is important to note that in both examples, the tour was in a cyclic manner and ultimately returns to the destination from where it started.

Applications:

The following are some very common applications of the travelling salesman problem:

- **School bus routing problem**: A school bus every morning starts from school, picks children from pre-decided stations and then comes back to school. For instance, if the origin is marked as O and five picking stations as S1, S2, S3, S4 and S5, then the bus would start from O, follows a route covering all stations once which is necessary and reaches back at O. There can be a variety of possible routes, but the most optimal route following certain criteria of either minimizing time or distance is selected.
- **Job-machine problem**: Suppose five parts forming a set are to be machined with the same tool and on the same machine. Thus, the machine follows a repetitive process. There is no set precedence for the selection of parts. An operator can initiate the process with any part and can follow with any part. The criteria of selection would be minimum processing time so that total cycle time for the entire set is minimum. The same sequence would again be followed by an incoming set of five parts.
- **Distribution problem**: This is illustrated by considering a pizza owners' need to deliver a certain number of orders to different places. Delivery of one order and coming back to the restaurant (origin station) is a costly proposition. Thus, the manager intends to club a number of orders to different residential places in the same area, if possible. The distributor would first select the residence closest to the restaurant and then the second residence which is closest to the first. Similarly, all residences would be delivered. The last residence would be farthest from the restaurant. Finally, the distributor would return back to the origin station after delivering the last order. Thus, travelling salesman is a form of distribution problem.

The above examples imply the application of the travelling salesman problem to all those allocation problems, which follows the characteristics of a finite number of cities being visited only once in a cyclic manner and then coming back to the city from where the tour was started. The following illustration explains the method of solving such problems by using a Hungarian algorithm.

Case: Musashi Auto Parts Michigan, Inc.

To understand the problem of travelling salesman, the case of Musashi Auto Parts from Transportation problem is modified here. Car components from Southfield (city 1) plant of Musashi Auto Parts are delivered to three manufacturing plants of Honda at Marysville (city 2), East Liberty (city 3) and Greensburg (city 4). It is assumed for this example that only one truck is loaded with full capacity and has to deliver auto components to all three plants. Distances between every city are shown in Table 7.1. These distances imply, for instance, that from Southfield to Marysville distance truck needs to travel is 171 miles; between Southfield and East Liberty is 165 miles and so

TABLE 7.1
Distance in Miles

	Southfield (1)	Marysville (2)	East Liberty (3)	Greensburg (4)
Southfield (1)	---	37	53	156
Marysville (2)	37	---	17	161
East Liberty (3)	53	17	---	160
Greensburg (4)	156	161	160	---

on. The truck starting from Southfield has to deliver all three manufacturing plants before coming back to origin. Importantly, diagonal elements are shown with a dash (---). According to the problem, a salesperson, in this case, a truck has to leave a city and then go to another city. The movement cannot be within one city. Therefore, any movement within a city should not consume any value. Whether it should be zero or infinity will be discussed later.

The purpose is to find a route that would minimize travelling distance while visiting and delivering components to all three plants.

The mathematical formulation involves certain notations:

X: would represent cities

i: indicates row values

j: indicates column values

So X_{ij} would imply travelling from city i to city j

D_{ij}: distance between cities i and j

The objective function of minimizing distance is represented as $Z = \sum\sum D_{ij}X_{ij}$

Fulfilment of objective function is subject to certain constraints:

1. $X_{ij} = 0$ or 1, meaning that either the city has been visited (1) or not (0).
2. The sum of visits from one particular city to all other remaining cities *row wise* would be 1.

$$\sum_{j=1}^{n} X_{ij} = 1 \text{ for all i}$$

This constraint implies that if a salesperson is in city 'i', then he or she can only go to one other city out of the remaining cities. For instance, in the above example, for i = 1 and for j = 1,2,3,4 the equation would be:

$$x_{11} + x_{12} + x_{13} + x_{14} = 1; x_{21} + x_{22} + x_{23} + x_{24} = 1;$$

$$x_{31} + x_{32} + x_{33} + x_{34} = 1; x_{41} + x_{42} + x_{43} + x_{44} = 1;$$

It is important to reiterate here that the sum indicates whether a particular city has been visited or not and not the values of time or distance. Also, the first equation shows that a salesperson has moved from city 1 to 1 (x_{11}) or city 1 to 2 (x_{12}) or city 1 to 3 (x_{13}) or city 1 to 4 (x_{14}). Only one of these movements is possible and not all in combination as after moving from 1 to any other city (2,3 or 4) he or she has to go some other city. So, the sum of the equation would be of one time 1 and three time zeroes implying equal to one.

3. Similarly, the sum of visits from one particular city to all other remaining cities *column-wise* would be one.

$$\sum_{i=1}^{n} X_{ij} = 1 \text{ for all j}$$

This implies that if a person is in city 'j', then he/she would have come to this city from one of the remaining cities. Thus, if person is in city 2, then he or she would have come from some other cities like 1, 3 or 4. The above equation when expanded would illustrate this. For instance, for j = 1, and for i = 1, 2, 3 and4, equation would be:

$$x_{11} + x_{21} + x_{31} + x_{41} = 1; x_{12} + x_{22} + x_{32} + x_{42} = 1;$$

$$x_{13} + x_{23} + x_{33} + x_{43} = 1; x_{14} + x_{24} + x_{34} + x_{44} = 1;$$

4. According to the travelling salesman problem, every city has to be surely visited once, implying that no city should be repeated and also no city should be left out. For instance, a route such as 1---2---1 or 1---2---3---1 or 1---1 are called sub-tours as they are either repeating or has left certain cities in the route. Elimination of these sub-tours is done by creating constraints that are discussed here. The above-mentioned three sub-tours are shown in Figure 7.1.

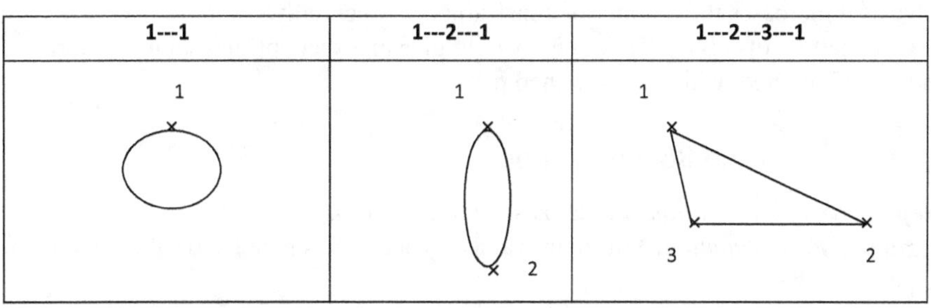

FIGURE 7.1 Sub-tours.

TABLE 7.2
Distance in Miles

	Southfield (1)	Marysville (2)	East Liberty (3)	Greensburg (4)
Southfield (1)	∞	37	53	156
Marysville (2)	37	∞	17	161
East Liberty (3)	56	17	∞	160
Greensburg (4)	156	161	160	∞

Following understanding is required to eliminate such sub-tours.

Sub-tour 1---1 implies that truck started from Southfield and came back to the same city without visiting any other city. According to $x_{ij}=0$ or 1, in such case, value of x_{ii} should be equal to 0. But tour length, i.e. movement indicated by x_{11} would be equal to 1, thus violating the constraint and making solution infeasible. Also, the condition of not visiting the same city twice is violated. Thus, $\sum x_{11}$ is put as equal to zero. As condition is not met, so sub-tour would not come into the solution.

Sub-tour 1---2---1 comprises two movements from 1 to 2 and then 2 to 1. So, total tour length would be 2. To remove such sub-tour, the sum of tour lengths is put as less than or equal to 1, i.e. $x_{12}+x_{21}\leq 1$

$1+1=2$ which is not less than or equal to 1. Thus, this sub-tour is also eliminated.

Finally, for sub-tour 1---2---3---1, tour length would be 3 as it comprises three tours (Figure 7.1) from 1 to 2; 2 to 3 and 3 to 1. To remove such sub-tour, the sum of tour lengths is put as less than or equal to 2, i.e. $x_{12}+x_{23}+x_{31}\leq 2$

$1+1+1=3$, which is not less than or equal to 2.

Thus, this sub-tour is also eliminated.

So, to remove such sub-tours, put concerned x_{ij} as zero. However, in matrix as shown in Table 7.1, a TSP problem requires minimization of cost, distance or time. So, instead of putting x_{ii} as zero, objective function criteria are put as infinity, implying allocating such huge cost that sub-tour does not enter the solution. This is shown in Table 7.2. The same concept is used when a TSP is solved in the next section.

So, to generalize, constraints could be created by comparing left-hand side with value of n−1 in the right-hand side. So, nine such constraints need to be created in the case of ten city problems. This aspect in TSP makes it a very difficult problem to solve as it increases the number of constraints exponentially.

One method of solving TSP which would give an exact optimal solution is branch and bound method, which is explained here.

7.2.1 Branch and Bound Method

Step 1: *Model formulation:* Model is shown in Table 7.3.

Step 2: *Row reduction*: Minimum value of each row is noted and subtracted from respective rows.

Minimum value represents reduction in time for each row. So, the reduced distance row-wise would be $37+17+17+156=227$

TABLE 7.3
Row Reduction

	Southfield (1)	Marysville (2)	East Liberty (3)	Greensburg (4)	Minimum Value
Southfield (1)	∞	37	53	156	37
Marysville (2)	37	∞	17	161	17
East Liberty (3)	53	17	∞	160	17
Greensburg (4)	156	161	160	∞	156
Row reduction					
Southfield (1)	∞	0	16	119	
Marysville (2)	20	∞	0	144	
East Liberty (3)	36	0	∞	143	
Greensburg (4)	0	5	4	∞	

TABLE 7.4
Column Reduction

	Southfield (1)	Marysville (2)	East Liberty (3)	Greensburg (4)
Southfield (1)	∞	0	16	119
Marysville (2)	20	∞	0	144
East Liberty (3)	36	0	∞	143
Greensburg (4)	0	5	4	∞
Minimum value	0	0	0	119
Column reduction				
Southfield (1)	∞	0	16	0
Marysville (2)	20	∞	0	25
East Liberty (3)	36	0	∞	24
Greensburg (4)	0	5	4	∞

Step 3: *Column reduction*: Minimum value is calculated from each column (Table 7.4).

So, the reduced distance column-wise would be 119

Total reduced distance = 227 + 119 = 346 miles

Step 4: *Branch network*: As truck would start from Southfield and can visit any of the three cities Marysville (2), East Liberty (3) and Greensburg (4) so following branch network is created.

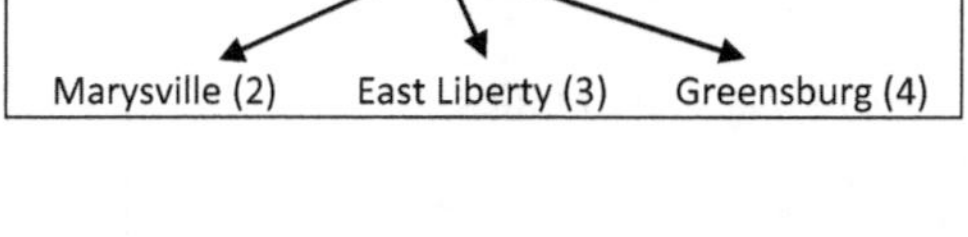

TABLE 7.5
Southfield-Marysville Route

	Southfield (1)	Marysville (2)	East Liberty (3)	Greensburg (4)	Minimum Value
Southfield (1)	∞	∞	∞	∞	---
Marysville (2)	∞	∞	0	25	0
East Liberty (3)	36	∞	∞	24	24
Greensburg (4)	0	∞	4	∞	0
Row reduction					
Southfield (1)	∞	∞	∞	∞	---
Marysville (2)	∞	∞	0	25	0
East Liberty (3)	12	∞	∞	0	0
Greensburg (4)	0	∞	4	∞	0
Minimum value	0	-	0	0	0

Step 5: *Calculate travel distance for each route:*

For Southfield-Marysville route: All values in Southfield-Marysville route are replaced by infinity value as discussed in explanation of sub-tour constraints. Also as route is from city 1 to city 2 so truck cannot go back to city 2 to city 1 so value in that cell is also put to infinity. This is shown in Table 7.5.

So, total reduced distance = distance between Southfield to Marysville in Table 7.4 + reduced distance + distance reduced after branching in Table 7.5.

$$= 0 + 346 + (24 + 0)$$

$$= 370 \text{ miles}$$

For Southfield-East Liberty route:

So, total reduced distance = distance between Southfield to East Liberty in Table 7.3 + reduced distance + distance reduced after branching in Table 7.6

$$= 16 + 346 + (20 + 5)$$

$$= 387 \text{ miles}$$

For Southfield-Greensburg route:

So, total reduced distance = distance between Southfield to Greensburg in Table 7.3 + reduced distance + distance reduced after branching in Table 7.7

$$= 0 + 346 + (4 + 20)$$

$$= 370 \text{ miles}$$

TABLE 7.6
Southfield-East Liberty Route

	Southfield (1)	Marysville (2)	East Liberty (3)	Greensburg (4)	Minimum Value
Southfield (1)	∞	∞	∞	∞	---
Marysville (2)	20	∞	∞	25	20
East Liberty (3)	∞	0	∞	24	0
Greensburg (4)	0	5	∞	∞	0
Row reduction					
Southfield (1)	∞	∞	∞	∞	
Marysville (2)	0	∞	∞	5	
East Liberty (3)	∞	0	∞	24	
Greensburg (4)	0	5	∞	∞	
Minimum value	0	0	0	5	
Column reduction					
Southfield (1)	∞	∞	∞	∞	
Marysville (2)	0	∞	∞	0	
East Liberty (3)	∞	0	∞	19	
Greensburg (4)	0	5	∞	∞	

TABLE 7.7
Southfield-Greensburg Route

	Southfield (1)	Marysville (2)	East Liberty (3)	Greensburg (4)	Minimum Value
Southfield (1)	∞	∞	∞	∞	---
Marysville (2)	20	∞	0	∞	0
East Liberty (3)	36	0	∞	∞	0
Greensburg (4)	∞	5	4	∞	4
Row reduction					
Southfield (1)	∞	∞	∞	∞	
Marysville (2)	20	∞	0	∞	
East Liberty (3)	36	0	∞	∞	
Greensburg (4)	∞	1	0	∞	
Minimum value	20	0	0	---	
Column reduction					
Southfield (1)	∞	∞	∞	∞	
Marysville (2)	0	∞	0	∞	
East Liberty (3)	16	0	∞	∞	
Greensburg (4)	∞	1	0	∞	

Step 6: *Branch network:* Branch network is extended further.

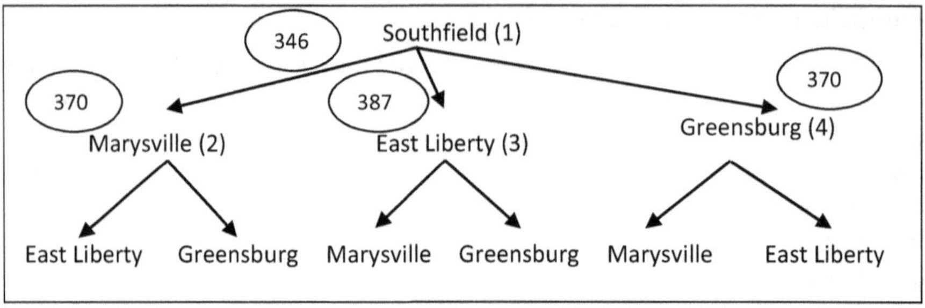

From Southfield point of origin of truck, it could go to any of three remaining cities. However, due to the least distance route which is in this case is Southfield-Marysville and Southfield-Greensburg route with travel distance of 370 miles so routes from city 1 to 2 and city 1 to 4 are feasible. In subsequent steps, the solution for both routes is examined.

After visiting city 2, truck could move to either East Liberty or Greensburg. Also, from city 4, i.e. Greensburg truck could move to Marysville or East Liberty. This is decided by finding travel distance of each route.

Step 7: *Calculate travel distance for each route:*

For Marysville-East Liberty route:

No further reduction is required as minimum values both row and column wise are zeroes.

So, total reduced distance = distance between Marysville-East Liberty in Table 7.4 + reduced distance at Marysville + distance reduced after branching in Table 7.8

$$= 0 + 370 + (0 + 0)$$

$$= 370 \text{ miles}$$

For Marysville- Greensburg route:

TABLE 7.8
Marysville-East Liberty Route

	Southfield (1)	Marysville (2)	East Liberty (3)	Greensburg (4)	Minimum Value
Southfield (1)	∞	∞	∞	∞	---
Marysville (2)	∞	∞	∞	∞	---
East Liberty (3)	12	∞	∞	0	0
Greensburg (4)	0	∞	∞	∞	0
Minimum value	0	---	---	0	

TABLE 7.9
Marysville- Greensburg Route

	Southfield (1)	Marysville (2)	East Liberty (3)	Greensburg (4)	Minimum Value
Southfield (1)	∞	∞	∞	∞	---
Marysville (2)	∞	∞	∞	∞	---
East Liberty (3)	12	∞	∞	∞	12
Greensburg (4)	0	∞	4	∞	0
Row reduction					
Southfield (1)	∞	∞	∞	∞	
Marysville (2)	∞	∞	∞	∞	
East Liberty (3)	0	∞	∞	∞	
Greensburg (4)	0	∞	4	∞	
Minimum value	*0*	---	*4*	---	
Column reduction					
Southfield (1)	∞	∞	∞	∞	
Marysville (2)	∞	∞	∞	∞	
East Liberty (3)	0	∞	∞	∞	
Greensburg (4)	0	∞	0	∞	

So, total reduced distance = distance between Marysville-Greensburg in Table 7.4 + reduced distance at Marysville + distance reduced after branching in Table 7.9.

$$= 25 + 370 + 12 + 4$$

$$= 411 \text{ miles}$$

For Greensburg-Marysville route:

So, total reduced distance = distance between Greensburg- Marysville in Table 7.7 + reduced distance at Greensburg + distance reduced after branching in Table 7.10.

$$= 1 + 370 + 16$$

$$= 387 \text{ miles}$$

For Greensburg-East Liberty route:

So, total reduced distance = distance between Greensburg- East Liberty in Table 7.7 + reduced distance at Greensburg + distance reduced after branching in Table 7.11.

$$= 0 + 370 + 0$$

$$= 370 \text{ miles}$$

TABLE 7.10
Greensburg-Marysville Route

	Southfield (1)	Marysville (2)	East Liberty (3)	Greensburg (4)	Minimum Value
Southfield (1)	∞	∞	∞	∞	---
Marysville (2)	0	∞	0	∞	0
East Liberty (3)	16	∞	∞	∞	16
Greensburg (4)	∞	∞	∞	∞	---
Row reduction					
Southfield (1)	∞	∞	∞	∞	
Marysville (2)	0	∞	0	∞	
East Liberty (3)	0	∞	∞	∞	
Greensburg (4)	∞	∞	∞	∞	

TABLE 7.11
Greensburg-East Liberty route

	Southfield (1)	Marysville (2)	East Liberty (3)	Greensburg (4)	Minimum value
Southfield (1)	∞	∞	∞	∞	---
Marysville (2)	0	∞	0	∞	0
East Liberty (3)	16	0	∞	∞	0
Greensburg (4)	∞	∞	∞	∞	---
Minimum value	0	0	0	---	

Step 8*: Branch network:* Branch network is extended further.

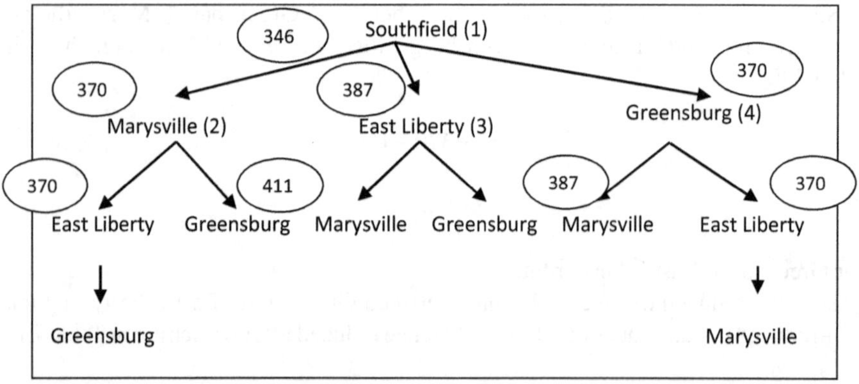

Step 9: Calculate travel distance for each route:
For East Liberty- Greensburg route:

TABLE 7.12
East Liberty- Greensburg route

	Southfield (1)	Marysville (2)	East Liberty (3)	Greensburg (4)	Minimum Value
Southfield (1)	∞	∞	∞	∞	---
Marysville (2)	∞	∞	∞	∞	---
East Liberty (3)	∞	∞	∞	∞	---
Greensburg (4)	0	∞	∞	∞	---
Minimum value	0	---	---	---	

TABLE 7.13
East Liberty- Marysville route

	Southfield (1)	Marysville (2)	East Liberty (3)	Greensburg (4)	Minimum value
Southfield (1)	∞	∞	∞	∞	---
Marysville (2)	0	∞	∞	∞	0
East Liberty (3)	∞	∞	∞	∞	---
Greensburg (4)	∞	∞	∞	∞	---
Minimum value	0	---	---	---	

So, total reduced distance = distance between East Liberty-Greensburg in Table 7.7 + reduced distance at East Liberty + distance reduced after branching in Table 7.12.

$$= \text{undefined} + 370 + (0)$$

$$= 370 \text{ miles}$$

For East Liberty- Marysville route:

So, total reduced distance = distance between East Liberty-Marysville in Table 7.7 + reduced distance at East Liberty + distance reduced after branching in Table 7.13.

$$= 0 + 370 + (0)$$

$$= 370 \text{ miles}$$

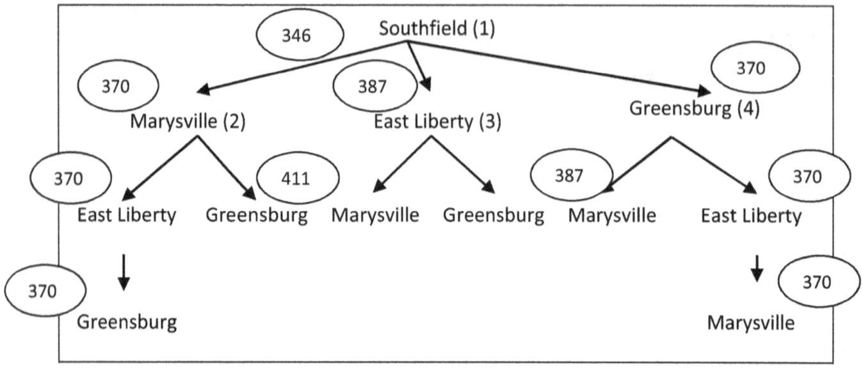

It is important to understand here that these times calculated and represented in circles represent the lower bound, implying that a minimum of 370 miles would be taken by the truck to make a to-and-fro trip consisting of four cities. To find the upper bound, we add the actual distances for travelling between cities for the identified feasible route.

Thus, the objective function, i.e. of minimum travel distance Z could be starting from Southfield the truck goes to Marysville (37 miles) then to East Liberty (17 miles) and finally to Greensburg (160 miles) before coming back to Southfield (156 miles) traversing a total distance of 370 miles. Second feasible tour could be: from Southfield to Greensburg (156 miles) + Greensburg to East Liberty (160 miles) + East Liberty to Marysville (17 miles) and + Marysville to Southfield (37 miles) traversing total distance of 370 miles. So, both routes are feasible and equivalent.

Excel Solution

Step 1: As shown in the following figure, the input data of the above illustration in Excel spreadsheet. The array from A1 to E5 shows this data (Figure 7.2).

Step 2: Constraints are represented by nodes from 1 to 4 and then again back to 1, implying that each station has to be visited only once. This is calculated by applying the following formula shown in B8 and then applied to all nodes till E8.

	A	B	C	D	E	F
1		*Southfield (1)*	*Marysville (2)*	*East Liberty (3)*	*Greensburg (4)*	
2	*Southfield (1)*	---	37	53	156	
3	*Marysville (2)*	37	---	17	161	
4	*East Liberty (3)*	53	17	---	160	
5	*Greensburg (4)*	156	161	160	---	
6						

FIGURE 7.2 Input data table.

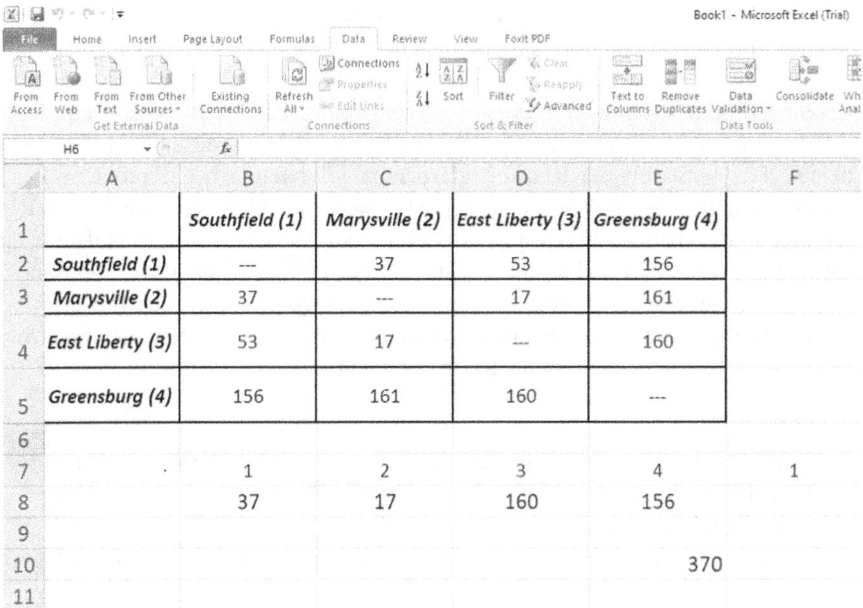

	A	B	C	D	E	F
1		Southfield (1)	Marysville (2)	East Liberty (3)	Greensburg (4)	
2	Southfield (1)	---	37	53	156	
3	Marysville (2)	37	---	17	161	
4	East Liberty (3)	53	17	---	160	
5	Greensburg (4)	156	161	160	---	
6						
7		1	2	3	4	1
8		37	17	160	156	
9						
10					370	
11						

FIGURE 7.3 Excel solution.

=INDEX($B$2:$E$5,B7,C7)
$B$2:$E$5 implies data array
B7 indicates node to be compared
C7 indicates node with which it has to be compared.

Step 3: Sum of these values indicates objective function of minimizing the travel cost shown in cell E10 referred as =SUM(B8:E8).

Step 4: Solver solution: In solver application

Objective function of minimization is set as cell H7.
By changing variable cells: $G$2:$G$5.
Constraints: $G$2:$G$5=All different
Solving Method: Evolutionary
Calculation will provide the answer matching with solution provided by branch and bound method of travelling from 1---2---3---4=370 miles (Figure 7.3)

7.3 VEHICLE ROUTING PROBLEM

In TSP, the salesperson or truck, as was in this case, who travels to different cities is one. For instance, in the illustrated case, one truck of full capacity travels to three different cities before coming back to origin. However, a company, in most likelihood, has a number of salespersons. Similarly, auto components are delivered through a number of trucks with a certain capacity to fulfil certain demand of different plants.

Thus, in VRP, primarily two decisions needed to be made; one routing problem where the purpose is to identify a route that goes to all stations by covering the minimum distance; second is allocation problem where the purpose is to allocate trucks to different routes depending on their capacity and demand of destinations. Thus, TSP is purely a routing problem, whereas VRP is a combination of routing and allocation problem. VRP with only one truck or salesperson will become a TSP problem.

Continuing with the illustration used in TSP, we will explain the solution of VRP. According to the chapter of the transportation model, the demand for manufacturing plant in Marysville, Greensburg and East Liberty are 65, 45 and 10 (in thousands) units. Suppose at Southfield Musashi Auto Parts employs two trucks for transporting units, each with a maximum capacity of 70 (in thousands) units. Before solving, a few points that should be understood is that in a VRP problem:

- There is single origin
- From single origin, vehicles are more than one with known and fixed capacity
- Destinations are many with different demands
- Objective is to identify tours for each vehicle so that it covers a fixed number of cities by traversing minimum distance.

The VRP solution is illustrated by using two methods:

Heuristic method: In this method, routing is first decided by considering problem as a travelling salesman problem. Then, according to the demand of destinations and capacity of vehicles, allocation of cities or cluster of cities a truck can visit before coming back to origin are decided. This approach is termed as *route first and clusters second*. For instance, in Figure 7.4, origin is Southfield and according to the TSP method, in previous section, route identified which would entail minimum travelling distance was Southfield to Marysville; Marysville to Greensburg; Greensburg to East Liberty; East Liberty to Southfield. The figure also shows the demand of each Honda plant.

So, the first aspect of a VRP, i.e. identifying appropriate route has been solved. Now allocation problem is solved by allocating trucks to various routes. Truck1 with capacity of 70 units can be allocated to Marysville whose demand was less than 70. However, it cannot go further to any other city as it does not have the capacity to fulfil any of plants' demand. Truck2 could be allocated to deliver to cities of Greensburg and East Liberty before coming back. Importantly, a VRP can be visualized as a combination of TSP problems. In this case, there are two routes: first,

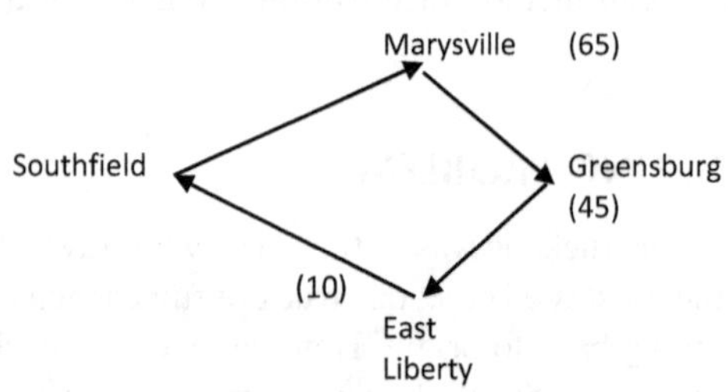

FIGURE 7.4 Heuristic method.

Southfield---Marysville---Southfield and second is Southfield---Greensburg---East Liberty---Southfield.

Therefore, solutions of both routes are feasible as they fulfil the conditions of routing and supply and demand constraints. However, it is an optimal solution or a better solution in terms of reduction in travel distance or further minimization of travelling cost can be found is difficult to judge from this method. For instance, Truck2 can first go to East Liberty and then to Greenburg before coming back. Such issues become numerous in case of the increase in the number of cities and thus routes. For that, it is better to use the following method of Clark–Wright savings algorithm.

Linear programming formulation of the VRP model:

The VRP model can be formulated by using integer linear programming model where decision variable is coded as a binary variable, implying either a route has been visited (1) or not (0). The visited route would contribute to deciding a route with any of the metrics of either distance or cost. In this illustration, the purpose is to find the route from alternative routes available, which minimizes distance taking into consideration demand and capacity of vehicles. A vehicle cannot be allocated to minimum distance or maximum saving in distance path if it does not have capacity to fulfil the demand. The following notations are used to illustrate VRP in a linear programming form.

$x_i = 1$ if route 'i' is visited and
$= 0$ otherwise
$d_i =$ distance between demand nodes
$j =$ indicates nodes
$D_j =$ total number of nodes visited
$k =$ number of vehicles
$a_{ij} = 1$ if node 'j' is in route 'i' and
$= 0$ otherwise
$Q_k =$ capacity of vehicle k
Objective function:

$$\text{Minimize } Z = \sum_{i} d_i x_i$$

Subject to following constraints:

Constraint 1: Minimum number of routes to be visited falling on the shortest path

$$\sum a_{ji} x_i \geq D_j$$

For instance, for node $j = 2$, and if route $i = 3$ is visited involving node 2, then equation would be $a_{23} = 1$; and $x_3 = 1$ implying $a_{23}x_3 = 1*1 = 1 = D$

In case of a path 4---5---6 falling on the shortest path for a vehicle, there are nodes 4, 5 and 6 and routes 4---5, 5---6 and 6---4 so equation would be

$a_{45}x_5 + a_{56}x_6 + a_{64}x_4 = 1*1 + 1*1 + 1*1 = 3$, implying total number of nodes visited in the path

Constraint 2: A route 'i' can be visited by vehicle 'k' or not.

$$\sum_{i} x_{ik} \geq 1$$

For $k = 1$, vehicle 1 visits a route $i = 1$, then $x_{11} = 1$; otherwise, it is zero. This happens for all the routes

Constraint 3: A vehicle visits at least one route.

$$\sum_{k} x_{ik} \geq 1$$

For $i = 1$, route 1 is visited by either of number of vehicles once.

Constraint 4: Vehicle visits a node if its capacity is greater or equal to demand of that node.

$$\sum_{j} Q_k x_{ik} - \sum_{j} D_j x_{ik} \geq 0$$

7.3.1 Clark–Wright Savings Algorithm

This method is based on calculating savings that could be availed in visiting two cities through two different routes. The first tour involves starting from origin, going to a city, coming back to origin and then starting again to visit other city. This is called as *current tour.* In the second route vehicle, after visiting a city from origin does not come back and directly goes to another city before coming back. This is called as *joined tour.*

For instance, route of Truck2 is Southfield to East Liberty to Greenburg and back to Southfield. So, two routes that it can use are shown in Figures 7.5 and 7.6.

In the case of current tour, total distance travelled = 2*156 + 2*53 = 418 miles.

In the case of joined tour, total distance travelled = 156 + 160 + 53 = 369 miles.

So, joined tour has less distance to travel, so savings in distance would be 418 − 369 = 49 miles.

This shows that it would always be better to go through a round tour rather than visiting individually. Thus, in this method, the first step is to identify cluster of cities

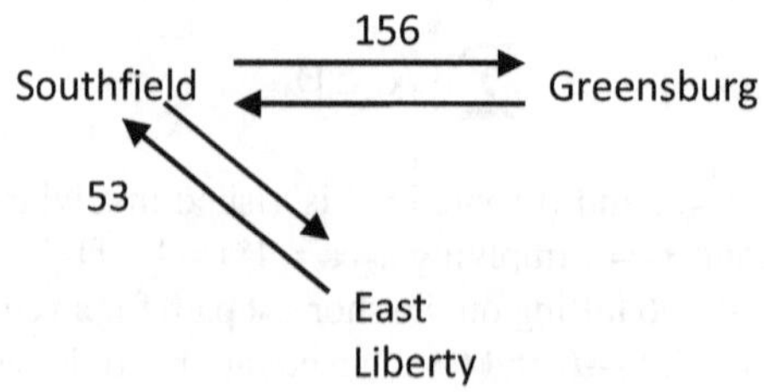

FIGURE 7.5 Current tour.

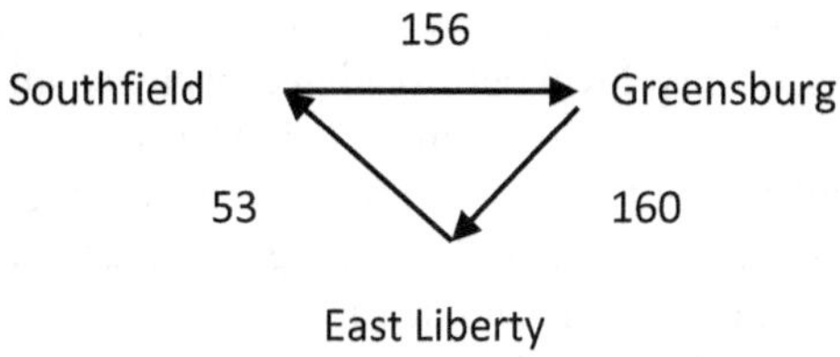

FIGURE 7.6 Joined tour.

and then routing; for this reason, this method is called *cluster first route second*. Now if origin city (Southfield) is denoted by city '0'; Greensburg as city 1 and East Liberty by city 2, then saving equations can be created as:

Current tour distance $= 2c_{01} + 2c_{02}$

Joined tour distance $= c_{01} + c_{12} + c_{20}$

If $2c_{01} + 2c_{02} > c_{01} + c_{12} + c_{20}$

$$c_{01} + c_{20} - c_{12} > 0$$

then joined tour should be used. Solution is approached by considering the following points:

- As in this method clustering/allocation is done first and routing later, so saving values is calculated for every pair of nodes.
- These saving values are then arranged in descending order. Highest saving pairs are first allocated to trucks and taking into consideration capacity and demand comparison further allocations are made.
- It is important to make sure that while clustering vehicle capacity is not violated.

Illustration: To illustrate Clark–Wright savings algorithm, we expand the example of Musashi Auto Parts. In this case, auto parts are delivered from the same plant of Southfield, Michigan (O) to seven plants of Honda: Marysville, Ohio (i); East Liberty, Ohio (ii); Greensburg Indiana (iii); Timmonsville, South Carolina (iv); Swepsonville, North Carolina (v); Greensboro, North Carolina (vi) and Lincoln, Alabama (vii). Demand of each of plants in thousand units is estimated as 60, 10, 45, 30, 25, 15 and 20. Company has employed three trucks with a capacity of 70 thousand units to deliver similar parts to all seven destinations. The distances between origin, i.e. Southfield to all seven stations and between these stations are shown in Table 7.14. By using Clark–Wright method, the most optimal and feasible route for each of the trucks needs to be estimated.

The last column indicates distance between origin, i.e. from Southfield to other cities, whereas other columns show distance between different cities.

Step 1: Find savings by using the formula explained above. For instance, savings in distance by joining cities 1 and 2 would be:

$$1-2: c_{01} + c_{20} - c_{12} = 37 + 53 - 17 = 73 > 0$$

A positive value indicates that a truck going from origin to city 1 instead of coming back to origin should go to city 2 without violating capacity conditions and then come back to origin. By doing so, saving of 73 miles could be achieved. Similarly, savings for all pairs are calculated and shown in Table 7.14 below the diagonal matrix. A negative value as was found between Greensburg and East Liberty implies otherwise. Thus, joining these two cities is not feasible.

Step 2: These saving values are then arranged in descending order (Table 7.15) showing that joining city 6 with city 5 would give maximum savings and so on.

Step 3: Routes are allocated depending on the capacity. The available capacity of each truck is 70,000 units. Relating demand with capacity and taking into consideration savings of distance by joining certain routes, allocation of trucks to particular routes is done (Table 7.16).

TABLE 7.14
Distances and Savings

Cities	1	2	3	4	5	6	7	O
1	---	17	161	561	459	435	592	37
2	73	---	160	576	474	450	610	53
3	32	−272	---	640	540	516	500	156
4	7	8	47	---	172	148	388	531
5	6	7	44	787	---	27	462	428
6	6	7	44	787	805	---	437	404
7	29	27	240	727	550	551	---	584

TABLE 7.15
Descending Order of Savings

Route	6–5	6–4	5–4	7–4	7–6	7–5	7–3	2–1	4–3
Savings	805	787	787	727	551	550	240	73	47
Route	5–3	6–3	3–1	7–1	7–2	4–2	4–1	5–2	6–2
Savings	44	44	32	29	27	8	7	7	7
Route	5–1	6–1	3–2						
Savings	6	6	−272						

TABLE 7.16
Route Allocation

Route		Demand	Savings (Miles)	Capacity Used	Capacity Available
Truck1	6–5	15	805	40	30
		25			
	6–4	30	787	30	0
Truck2	7–3	20	240	65	5
		45			
Truck3	2–1	10	73	70	0
		60			

So, Truck1 starting from origin goes to Greensboro first and delivers 15 units. After that, as capacity is still unused, so it is allocated station Swepsonville where it delivers 25 units. This tour results in savings of 805 miles of distance. Now because of the remaining capacity, Truck1 can visit another city that is connected with Swepsonville. From Table 7.16, it moves to Timmonsville that further results in savings of 787 miles. Here, Truck1 delivers remaining 30 units. As it has exhausted its capacity, so Truck1 would come back to origin. Also, as all these three cities have been visited by Truck1, so Truck2 cannot go to any of these cities again. Thus, any other city combined with these three visits is not feasible so not included in the solution. For instance, combining Lincoln (7) with Timmonsville (4) gives the next best savings of 727 miles but $7-4$ combination is not feasible as city 4 has already been visited and its requirement is fulfilled.

$$\text{Route for truck1 would be} = O-6-5-4-O$$

$$\text{Distance travelled} = 404 + 27 + 172 + 531 = 1{,}134 \text{ miles}$$

Thus, the next combination of cities would be $7-3$ i.e. Lincoln with Greensburg. It would entail a saving of 240 miles. Truck2 would begin its journey from origin and will visit Lincoln and deliver 20,000 units before visiting Greensburg and deliver 45,000 units. In this case, capacity of 5 units remains unutilized.

$$\text{Route for truck 2 would be} = O-7-3-O$$

$$\text{Distance travelled} = 584 + 500 + 156 = 1{,}240 \text{ miles}$$

Finally, Truck3 would be allocated route of East Liberty and Marysville $(2-1)$ resulting in savings of 73 miles. So,

$$\text{Route for truck3 would be} = O-2-1-O$$

$$\text{Distance travelled} = 53 + 17 + 37 = 107 \text{ miles}$$

Total savings:

If no routes have been joined, then total distance to be travelled would be twice the sum of all distances from origin to different cities as shown in the last column of Table 7.14.

So, the total distance travelled without joining

$$= 2 * (37+53+156+531+428+504+584)$$

$$= 4{,}386 \text{ miles.}$$

So, the total distance travelled with joining $\begin{aligned} &= 1{,}134+1{,}240+107 \\ &= 2{,}481 \text{ miles} \end{aligned}$

$$\text{Savings} \begin{array}{l} = 4{,}386 - 2{,}481 \\ = 1{,}905 \text{ miles} \end{array}$$

This savings is the same as the sum of savings as shown in Table 7.16.

The Clark–Wright savings algorithm provides more accurate solution of VRP. However, it can become very complex with an increase in the number of cities and vehicles with different capacities and demands.

7.4 SHORTEST PATH PROBLEM: DIJKSTRA'S ALGORITHM

Shortest path problem (SPP) is another network model including nodes and arcs where nods represent destinations and arcs represent distance between two destinations. The purpose of SPP is to find the shortest route out of available multiple routes. In a VRP problem, the vehicle begins its journey from a station, traverses a path moving to different stations and then comes back to origin. However, a SPP solution limits its calculations to finding the shortest path a vehicle traverses between two nodes without focusing on its backward path back to origin.

Linear programming formulation:

The purpose of the SPP is to find the shortest route out of possible multiple routes that covers the smallest distance from origin to destination passing through multiple stations on its way. Thus, the objective function would be of minimization of total distance travelled from origin to destination. Decision variable would be represented by binary value of either '1' or '0' meaning whether or not a particular city has been visited, respectively. Thus:

$$\text{Minimize} \sum_{i} \sum_{j} C_{ij} X_{ij}$$

where c_{ij} represents the distance between two nodes from i to j and x_{ij} indicated whether a city has been visited or not taking a value of 1 if it is in the shortest path and 0 if not.

Objective of minimization is subject to constraints of net flow of each city being visited. From the source there is only outward movement, so net flow would be positive 1; for destination node, there is only inward movement from one city, making a net flow of –1; for intermediary nodes, there would be incoming and outgoing making a net flow of zero. This is shown in the following equations for each of the node following a rule of

(Flow out of node) – (Flow into a node) = Net supply at a node.

Thus for:

$$\text{Southfield (S)}: x_{sm} + x_{sel} + x_{sgin} = 1$$

$$\text{Marysville (M)}: x_{mgnc} + x_{mel} - x_{ms} = 0$$

$$\text{East Liberty (EL)}: x_{elgin} + x_{elgnc} + x_{elsnc} - x_{elm} - x_{els} = 0$$

$$\text{Greensburg (Gin)}: x_{ginsnc} + x_{gint} - x_{gins} - x_{ginel} = 0$$

$$\text{Greensboro (Gnc)}: x_{gncl} + x_{gncsnc} - x_{gncm} - x_{gncel} = 0$$

$$\text{Swepsonville (Snc)}: x_{sncl} + x_{snct} - x_{sncgnc} - x_{sncel} - x_{sncgin} = 0$$

$$\text{Timmonsville (T)}: x_{tl} - x_{gint} - x_{snct} = 0$$

$$\text{Lincoln (L)}: - x_{gncl} - x_{sncl} - x_{tl} = -1$$

Dijkstra's algorithm provides a quantitative and precise method of achieving the aim of finding the shortest path. The algorithm is important in providing bases for further advanced analysis. An example is illustrated to understand the application of Dijkstra's algorithm. By using distance between various auto plants of Honda as given in Table 7.14, the following graph (Figure 7.7) shows connectivity between various stations and routes a vehicle can adopt by initiating its journey from Southfield (S), Michigan.

According to the algorithm, the node or station from which journey would begin becomes the active node. In this case, Southfield (S) is the first station. From S, the vehicle can move to M by traversing a distance of 36; to EL by travelling a distance of 53 and to Gin distance would be 156. The shortest or minimum distance travelled would be to M, so it becomes the next active node. Now from M, the vehicle can move to Gnc and EL. Distances would be:

M---Gnc = 36 + 435 = 471 and

M---EL = 36 + 15 = 51

Importantly, earlier distance from S---EL was 53, but by travelling via M distance is only 51. Thus value at EL as active node would be 53. Now out of EL, Gin and Gnc minimum value is at EL = 51 so out of three nodes EL becomes the active node. From EL, vehicle can move to Gin, Snc or Gnc. Distance would be:

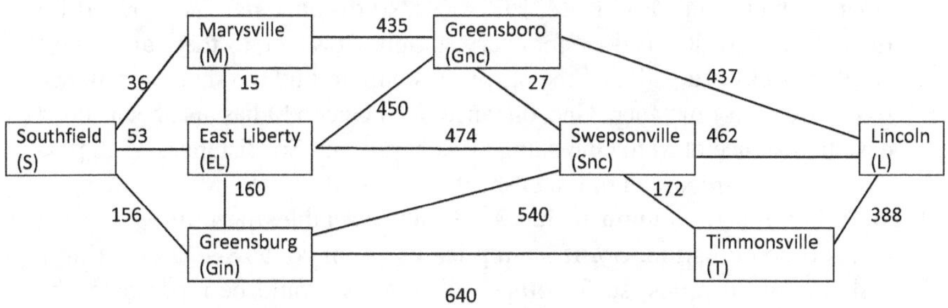

FIGURE 7.7 Network diagram

EL --- Gin = 51 + 160 = 211; as it is larger than earlier value of 156 so value at Gin remains 156

EL---Snc = 51 + 474 = 525;

EL---Gnc = 51 + 450 = 501; as it is larger than 471 so value at Gnc remains as 471.

Out of above three, distance at Gin of 156 becomes the active node.

From Gin node, two paths diverge; one to Snc and other to T. Distances between:

Gin---S = 156 + 540 = 696

Gin---T = 156 + 640 = 796

Now out of Gnc, Snc and T minimum distance is at Gnc with a value of 471 so it becomes active node.

From Gnc vehicle can move to S and L. Distances between them are:

Gnc---Snc = 498; earlier distance value at Snc was 525 which now would be changed to 498.

Gnc---L = 471 + 437 = 908

Now out of Snc, T and L minimum distance value is at Snc (498) making it next active node.

From Snc vehicle can move to L or T. Distances would be:

Snc---L = 960

Snc---T = 670; as this is less than earlier value of 796 so distance value at T would be 670.

Out of T (670) and L (960) minimum value is at T making it active node. From T, vehicle can only move to L by traversing a distance of 388. So distance would be:

T---L = 670 + 388 = 1,058; this value is more than distance travelled from Snc to L = 960. So last destination would be 960 miles at Lincoln.

Now shortest path is found by moving backwards from Lincoln to start node of Southfield.

L --- Gnc --- M --- S = 437 + 435 + 36 = 908 miles

Excel Solution

Using this information, the above illustration is also solved by using Excel.

Step 1: In Excel spreadsheet, the first column indicates all stations from which vehicle will start its journey. The second column includes stations where it ends its journey. The third column includes the distance data between two stations. Importantly, from-to distances are included for all directions as there is no direct path. For example, starting station is Southfield (S) from where vehicle can travel to M, EL or Gin so distance data is included for them. Similarly, it was done for every station. However, as there are no forward or backward path and the purpose is only to find the shortest distance so in case of, for instance, Gnc, distance from Gnc to M has also been added despite distance of M to Gnc being already present. So, distances of all possible paths are recorded in Excel sheet.

Step 2: The fourth column includes decision variables pertaining to each from-to combination. So, if x_1 implies usage of route S---M covering a distance of 36 miles, so for other routes, they would be represented by x_2, $x_3 \ldots x_{27}$.

	A	B	C	D	E	F	G	H	I	J
1	From	To	Distance	Decision variables		Nodes	Constraints			
2	S	M	36			S	0	=	1	
3	S	EL	53			M	0	=	0	
4	S	Gin	156			EL	0	=	0	
5	M	EL	15			Gin	0	=	0	
6	M	Gnc	435			Gnc	0	=	0	
7	EL	Gnc	450			Snc	0	=	0	
8	EL	Gin	160			T	0	=	0	
9	EL	Snc	474			L	0	=	-1	
10	EL	M	15							
11	Gin	Snc	540		*Objectiive function (Z)*					
12	Gin	T	640		*Minimize*	0				
13	Gin	EL	160							
14	Gnc	Snc	27							
15	Gnc	L	437							
16	Gnc	M	435							
17	Gnc	EL	450							
18	Snc	L	462							
19	Snc	T	172							
20	Snc	Gnc	27							
21	Snc	EL	474							
22	Snc	Gin	540							
23	T	L	388							
24	T	Snc	172							
25	T	Gin	640							
26	L	Gnc	437							
27	L	Snc	462							
28	L	T	388							
29										

FIGURE 7.8 Excel sheet.

Step 3: Colum G indicates constraints. In SPP, the constraint is that a particular station would be visited only once. For that, a formula was created which was:

=sumif(a2:a28,f2,d2:d28)-sumif(b2:b28,f2,d2:d28)

The same formula is applied for all other nodes.

Step 4: Column I implies right-hand side of constraint equation. For node S, it is equal to 1 which means vehicle has started from S, i.e. outward journey begins from S. For node L it is -1 indicating only incoming vehicle and no outgoing. For all other nodes it is 0 representing both incoming and outgoing vehicle thus nullifying each others' effect making value to be zero.

Step 5: Objective function is of minimization, i.e finding a route which entails minimum distance from Southfield to Lincoln. In cell F12 formula was created which was:

=SUMPRODUCT(C2:C28,D2:D28)

Step 6: As discussed in previous chapter solver application was applied to obtain the solution shown in Figure 7.8 (Figure 7.9).

7.5 SUMMARY

Out of various variants, this chapter has discussed three special, practical and complex network flow models. Travelling salesman problem is considered to be very vital in both theoretical and industrial applications though solving of it becomes increasingly complex with increase in number of nodes. Problem deals with a salesman or

From	To	Distance	Decision variables		Nodes	Constraints		
S	M	36	1		S	1	=	1
S	EL	53	0		M	0	=	0
S	Gin	156	0		EL	0	=	0
M	EL	15	0		Gin	0	=	0
M	Gnc	435	1		Gnc	0	=	0
EL	Gnc	450	0		Snc	0	=	0
EL	Gin	160	0		T	0	=	0
EL	Snc	474	0		L	-1	=	-1
EL	M	15	0					
Gin	Snc	540	0	*Objectiive function (Z)*				
Gin	T	640	0	*Minimize*	908			
Gin	EL	160	0					
Gnc	Snc	27	0					
Gnc	L	437	1					
Gnc	M	435	0					
Gnc	EL	450	0					
Snc	L	462	0					
Snc	T	172	0					
Snc	Gnc	27	0					
Snc	EL	474	0					
Snc	Gin	540	0					
T	L	388	0					
T	Snc	172	0					
T	Gin	640	0					
L	Gnc	437	0					
L	Snc	462	0					
L	T	388	0					

FIGURE 7.9 Solver solution.

a vehicle starting from origin, visiting every other city (node) only once and then coming back to source. Branch and bound method has been discussed for solution of these problems due to its robustness in providing an appropriate solution. This problem occurs quite frequently in case of logistics. However, in those cases, instead of one entity, there are a number of vehicles with known and limited capacity. One vehicle could not visit all the cities or meet demand of all centres due to limited capacity constraint and so it starts from origin, visits few of cities and then come back. Such a problem is termed as VRP that forms a combination of number of travelling salesman problems. The objective in this case is to allocate a vehicle to which cities so that savings in terms of distance travelled or time taken is maximum. Clark–Wright savings algorithm has been discussed to solve VRPs. Finally, the SPP was discussed using Dijkstra's algorithm to find the shortest path out of multiple paths from a source to reach to a destination without the need to come back to origin.

7.6 GLOSSARY

- **Travelling salesman:** problem estimating total shortest distance that an entity would travel from source to destination and then back to source after visiting each destination only once.
- **Node**: each source or destination is represented by a node.
- **Arc**: flow of activity between two nodes which could be distance, cost or time is represented by arc.
- **Sub-tour:** A tour that comes back to source without visiting all destinations.

- **Branch and bound method:** provides an appropriate method of solving travelling salesman problem by following network of each destination to further nodes and calculating minimum value of flow of activity.
- **Vehicle routing problem:** is a version of TSP where multiple vehicles with fixed capacity are involved in fulfilling demand of various destinations. A VRP could be interpreted as a combination of multiple TSPs.
- **Clark–Wright savings algorithm:** is based on calculating savings that could be availed in visiting two cities through two different routes.
- **Shortest path problem:** a SPP solution finds the shortest path out of available multiple routes a vehicle traverses between two nodes without focusing on its backward path back to origin.
- **Dijkstra's algorithm:** provides a quantitative and precise method of achieving the aim of finding the shortest path.

7.7 CASE STUDY: JTEKT CORPORATION

The most important reasons of the redesign of facility layouts are the continuously fluctuating customer demands and changing market environment. Changes in the product portfolio, production volume, as well as changes in the manufacturing process and technology can result in bad utilization of space, huge work in progress at the plant, high material handling distances, etc. Thus, in order to minimize the material movement, improving bottleneck capacities and being flexible to fulfil the new customer demand or new customer order, it is necessary to standardize the layout of the plant. Hence, JTEKT Corporation, formerly known as Sona Koyo, in India decided that these drawbacks could be overcome by establishing and implementing systematic layout planning or standardizing the layout of plant facility.

JTEKT India Limited (formerly known as Sona Koyo Steering Systems) is a part of JTEKT Corporation Japan and operates as part of JTEKT Group India. It is engaged in the business of manufacturing/production, supply and sale of steering systems and other auto ancillaries to almost all Indian passenger car and utility vehicle manufacturers. The shop floor involving production of ancillaries such as emblems, handles and grills due to the above-mentioned reason was facing bottlenecks of material movement and high work in process inventory leading to poor productivity results. The management in order to improve productivity decided to install an automated conveyor system of movement of material starting from raw material section (i) to every other work centre in the shop floor. Importantly, the process of manufacturing of these parts follow similar process of moulding (ii), plating (iii), painting (iv), assembly (v), inspection (vi), packing (vii) and finished goods store (viii). The movement and relationship between all work centres in existing layout is shown in Figure 7.10. The conveyor belt would be designed in a manner that it would start its journey from raw material store, would visit all work centres only once to deliver required components and come back to raw material store for another round. Without discussing other aspects of the design, this case is intended to find a route for conveyor system that would minimize travel distance after visiting all stations. The distance data between all eight centres is shown in Table 7.17.

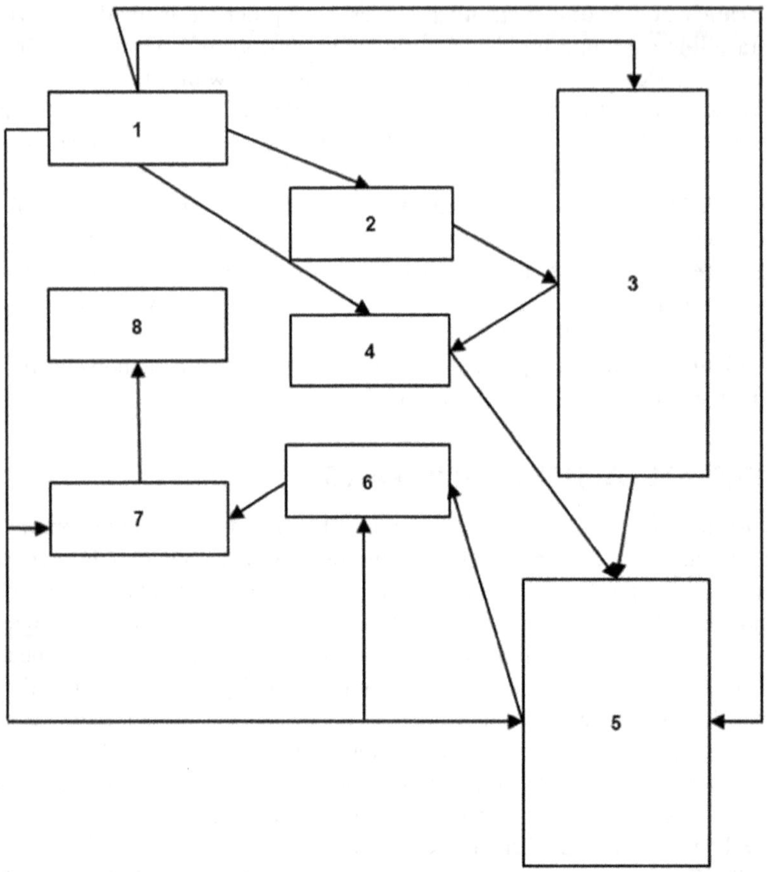

FIGURE 7.10 Existing layout and relationship among departments.

TABLE 7.17
Distance (in Feet) in Existing Layout

Stations	2	3	4	5	6	7	8
1	50	300	400	300	600	400	200
2		60	40	370	250	460	510
3			500	300	450	530	620
4				50	40	150	320
5					70	330	550
6						400	480
7							420

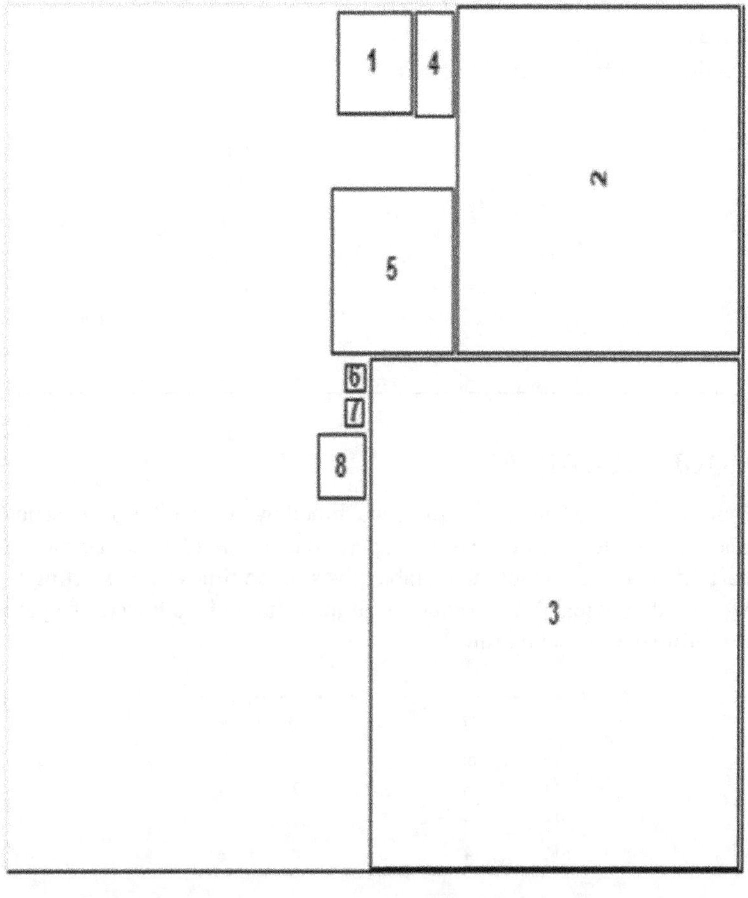

FIGURE 7.11 Alternative layout with circuitous distribution system.

Management was provided an alternative layout design by the shop floor supervisor, which would be more effective in reducing wasteful material movement. The layout is shown in Figure 7.11. It was an improved version of existing layout as indicated by data distance between all work centres, which were much less. In alternative layout, the route of automated distribution was intended to be circuitous rather than jumbled. In a new proposed layout, no work centre was stationed between other workstations as was in case of painting section in existing layout. Moreover, sub-tours of distribution system were not allowed implying once automated conveyor system initiates from raw material store it has to cover all workstations before coming back to origin. The changes in distances between workstations of alternative layout are shown in Table 7.18. Decision makers intend to compare appropriateness of distribution system in two layouts that would minimize the travel distance.

TABLE 7.18
Distance (in Feet) in Alternative Layout

2	3	4	5	6	7	8
50	180	250	100	400	200	120
	10	30	200	170	320	350
		200	150	280	370	410
			60	20	80	190
				20	180	390
					10	50
						10

7.8 MODEL QUESTIONS

7.8.1 Four jobs J1, J2, J3 and J4 are to be machined by an operator on a particular machine. All jobs require similar operation, so any of it can be picked to start the process. The following table gives set up times when shifting from one job to another. Which sequence of jobs should be adopted by operator to minimize total set up time?

	J1	J2	J3	J4
J1	∞	6	8	4
J2	6	∞	7	4
J3	9	7	∞	8
J4	4	5	8	∞

7.8.2 A fast food restaurant needs to serve five areas (A, B, C, D and E) in the city. Restaurant follows the policy of delivering order within 30 minutes otherwise meal has to be given free. If it is assumed that all the orders are received at the same time, so delivery man starting from A has to deliver all other orders within 30 minutes, implying total time to be less than the limit. The manager wants to estimate optimal route which would minimize travel time with the condition that delivery man has to come back to its location of origin. The times of various route permutations is given. It could be read that tour starting from A takes 4 minutes to go to B, 7 minutes to C and so on.

	A	B	C	D	E
A	∞	4	7	8	4
B	8	∞	5	9	3
C	9	7	∞	5	8
D	11	5	8	∞	6
E	1	6	4	9	∞

7.8.3 Solve following travelling salesman problem so as to minimize cost of travelling four destinations per cycle.

		To			
		A	B	C	D
From	A	∞	15	20	24
	B	14	∞	10	12
	C	9	11	∞	17
	D	8	16	13	∞

7.8.4 Find optimal sequence of five parts forming one component so as to minimize production time.

	P1	P2	P3	P4	P5
P1	∞	7	9	11	4
P2	10	∞	5	9	3
P3	12	11	∞	7	10
P4	15	9	11	∞	12
P5	8	10	12	14	∞

7.8.5 Certain fixed number of homogeneous units needs to be delivered from city 1 to city 7. The number of alternative routes is possible to reach city 7 by passing through five more cities denoted as city 2 to city 6. Flow or distance between cities in miles of all possible routes is given. Find the shortest path by applying Dijkstra's algorithm after drawing the network flow diagram.

Cities	1---2	1---3	2---3	2---4	2---5	3---4	3---5	3---7	4---6	5---6	6---7
Distance (miles)	5	6	6	9	4	8	2	25	15	12	12

7.8.6 Newspapers for a particular day are gathered at one location in a city. This location is termed as origin (O). The vendor has to collect papers and distribute to various sub centres in the city. There are six sub-centres marked as 1 to 6. Distances from origin to all sub-centres and between them are given. Also demand of each sub-centre in thousand units from past data has been estimated as 7, 3, 10, 3, 8 and 4 units for sub-centre 1 from 6 respectively. To deliver newspapers vendor has three mini-trucks each with capacity of 12 ('000) units. The vendor needs to find a route for each vehicle, which minimizes total distance travelled. Also, calculate the idle capacity of each vehicle.

	1	2	3	4	5	6	Origin
1	---	6	9	4	24	21	5
2		---	8	2	23	25	6
3			---	13	16	27	14
4				---	12	17	9
5					---	12	29
6						---	41

7.8.7 In emergency, an ambulance from location 1 needs to reach to location 5. In between, there are three high traffic density locations, which could delay its journey. These are marked as locations 2, 3 and 4. This results in alternative routes which the ambulance can utilize to finish its journey at the earliest. The distances between each possible location are given. Estimate the shortest possible route of all available routes.

Location	1---2	1---4	1---3	2---3	2---4	2---5	3---4	3---5	4---5
Distance (miles)	5	4	7	6	4	5	3	5	4

What would be the shortest path if origin is location 2 and destination remains the same?

7.8.8 For the following data, draw the network and find the shortest path for a truck starting from wholesaler (marked as 1) to a retailer (marked as 6). There are multiple routes available to complete the journey. Trucker needs to select a route which results in minimum time. The flow in this case indicates time in minutes between two locations.

Routes	1–2	1–3	2–4	2–7	3–5	4–7	5–6	6–7
Time (in minutes	22	18	14	27	17	13	7	20

7.8.9 School van of Learning Ladder kindergarten leaves the school (marked as 1) to pick kids from seven destinations (marked from 2 to 8) before coming back to school. Find the appropriate route which results in minimum distance travelled. The following table provides distance data in miles between each station.

	1	2	3	4	5	6	7	8
1	∞	3	4	3	7	9	12	14
2		∞	4	2	4	6	9	9
3			∞	2	4	5	8	9
4				∞	3	7	10	8
5					∞	2	7	5
6						∞	3	2
7							∞	2

7.8.10 From following data giving times in minutes what route a cab driver should take to travel from starting destination of location 1 to reach location 10? What would be the shortest route and minimum time if starting location is ‘2’ instead of ‘1’? Also draw the network diagram.

Routes	1–2	1–3	1–4	1–5	2–3	2–7	3–4	3–6	4–6	4–5
Time (minutes)	15	17	12	7	17	8	7	5	5	6
Routes	5–6	5–9	6–7	6–8	7–8	7–10	8–9	8–10	9–10	
Time (minutes)	11	13	6	4	4	6	7	9	7	

7.8.11 Demand of 10 stores (marked from 2 to 11) and distances in miles between each store and from origin (marked as 1) are given. Estimate number of vehicles each with a capacity of 200 units required to fulfil demand of each store by covering minimum distance. Also find the most optimum route and distance travelled by each vehicle.

Demand	Stores	2	3	4	5	6	7	8	9	10	11	Origin (1)
40	*2*	---	4	6	9	8	9	9	13	12	15	5
55	*3*		---	4	7	6	6	9	10	11	13	5
35	*4*			---	5	5	8	11	14	14	18	4
90	*5*				---	6	8	11	14	11	15	6
65	*6*					---	4	7	12	10	11	7
40	*7*						---	5	8	8	12	9
95	*8*							---	9	11	13	12
30	*9*								---	10	6	13
45	*10*									---	6	15
60	*11*										---	19

8 Project Scheduling
PERT and CPM

8.1 INTRODUCTION

Scheduling is defined as a timetable for a plan of project indicating time duration of all involved activities. Thus, scheduling cannot be done till project is well planned. This implies that planning of a project precedes project scheduling. Project planning should be well thought out as any later changes in the project would alter the activities and duration of all activities, making changes difficult. Project planning, however, determines and identifies the activities that are required for conduct of project. Workforce allocated depending on their expertise to each activity and the sequence in which they need to be carried out. For instance, a computer assembly operation was broken down into predominantly three activities, namely, monitor assembling (A), CPU assembly (B) and finally testing of complete assembly of monitor and CPU (C). Second, whether each activity has to be performed by an expert, i.e. one who does only a kind of activity or can these be clubbed and done by the same individual. Finally, and most importantly, sequence of activities is determined which indicates an inter-relationship between activities. This relationship is dictated by a logical sequence of operation. For instance, shoe can be worn only after sock has been worn.

Scheduling determines the time of all these related activities. This time includes time required to complete an activity plus waiting time if any. Now, this duration of each activity can be determined in two ways: first, if in addition to computer assembly, the company diversifies into mobile phone assembly that is new to project team then it becomes difficult to ascertain time of each activity with certainty. This happens predominantly in projects that are new for the organization and they have little experience in their conduct. Scheduling of such projects is done by using a network technique known as program evaluation review technique (PERT). On the other hand, in the case of well-established and experience projects, time duration of each activity is known with certainty. Critical path method (CPM) is the network technique used for scheduling purposes in such cases. Thus, the major difference between two methods is that PERT is a probabilistic method, whereas CPM is deterministic in nature. Both methods involve the calculation of following, understanding of which is required for effective project scheduling. These are as follows:

1. Estimation of total elapsed time for each activity.
2. Establishing when the project will start and when it will finish considering activity duration.
3. Based on the established start time, the calculation of earliest start and finish times of each activity.

DOI: 10.1201/9781003212966-8

4. Similarly, calculating the latest start and finish times so that entire project finishes at estimated finish time.
5. Finding of slack, i.e. time between when an activity *can* start or finish and when an activity *should* start or finish. This calculation is possible only after calculations in steps 3 and 4.
6. Finally, identification of critical i.e. most important path. Critical path involves activities that consume maximum resources and any delay in any of them would alter project completion time calculated in step 2.

In addition, this chapter utilized the understanding of PERT and CPM to examine the effect on project cost if time duration of activities lying on critical path is reduced to some feasible level. Conduct of an activity in less time would ask for more resources, thus increasing total cost. This relationship between time and cost is termed as crashing and is a specific application of CPM. In the end, a section is devoted to the concept of resource leveling, which analysed the process of reallocating of resources from over allocated activities to under allocated activities resulting in much smoother network diagram and proper usage of resources.

8.2 NETWORK PLANNING

PERT and CPM are two network planning techniques, which are used for project planning, scheduling and control. Both techniques use network diagram as a pictorial representation to indicate time duration, sequence and inter-relationship between activities. Gantt chart is another used network technique. The benefit of Gantt chart was that it depicted planning and scheduling at the same time on the same chart. However, it does not clearly show the inter-relationship between activities. Furthermore, with complex projects involving numerous activities that are interlinked in a web Gantt chart, it becomes too cumbersome. Network diagrams are helpful in such cases. In these diagrams, the relationship between activities is clearly depicted. Such relationship is termed as precedence–successor relationship that means that each activity is preceded by another activity and some other activity succeeds it. For instance, monitor assembly precedes CPU assembly and final testing succeeds it.

Network diagrams can be drawn in two ways: Activity-on-arrow (A-o-A) method is utilized when an arrow shows activities. Tail of such arrow depicts start and head represents the finish of activity. Each activity is associated with a combination of events; one at the start and another at the end of activity. These events are represented by nodes. An illustration is shown below.

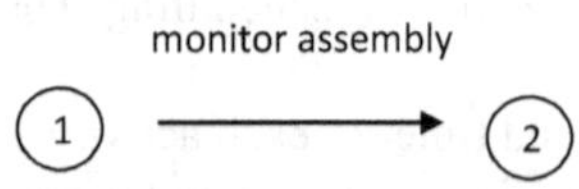

In addition, it is important to understand that whenever term activity is used it means the consumption of resource, whereas an event implies point in time indicating only start or finish of an activity. For instance, resources such as time, money or labor are consumed during assembly of monitor indicated by an activity, whereas events 1 and 2 indicate start and finish of this activity.

On the other hand, in case of Activity-on-Node (A-o-N) method, nodes represent activities, whereas arrow indicates an event. Arrow shows linkage between two activities. For instance, monitor assembly activity is shown by node 1 and CPU assembly by node 2.

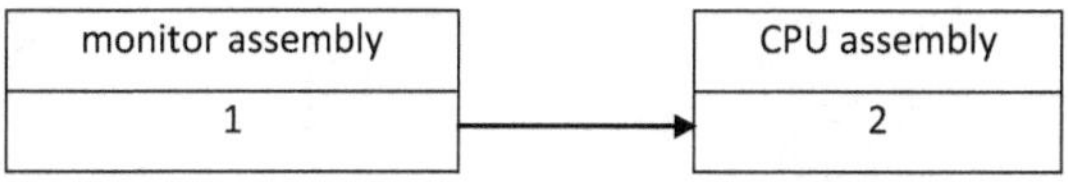

Thus, a network diagram includes activities, events, and inter-relationship between activities that can be of precedence, succession or activities happening simultaneously. This relationship is further explained by the following illustration. This is an A-o-A network diagram. A is preceding activity to C, which is preceding to E. Similarly, B is preceding to D, which is preceding to F. Activities can also be explained in successor terms. E activity will start after finish of C at node 4, i.e. it succeeds C which succeeds A. Similarly, F succeeds D that succeeds B. Events are depicted by circles called as nodes. They indicate the start and finish of an activity. It is important to note that A and B start from the same node, i.e. 1 at the same time, making them concurrent or parallel activities.

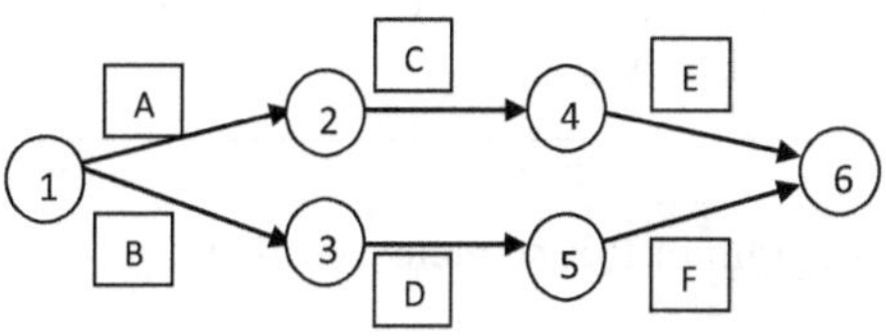

8.2.1 Rules for Construction of Network Diagrams

Rule 1: Each activity is represented only by one arrow.

This also implies that each activity has a unique combination of two events; one of which is predecessor and the other is successor. The following illustration violates this rule.

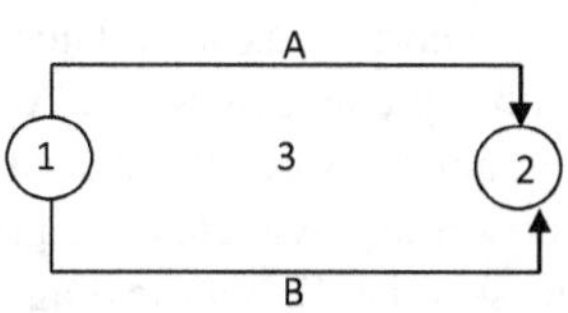

In this case, two activities A and B are occurring during the same events 1 and 2. Multiple activities can happen at the same time interval. For instance, two classes of quantitative techniques being taught by two different teachers might happen during the same time say during 9:00 am to 10:00 am. However, the network diagram drawn for such case is wrong, as activity (1–2) does not imply whether it is referring to activity A or B. There cannot be two activities between the two same events. Thus, each activity should be represented by two unique events.

Such issue can be resolved by using *dummy activities*. Dummy activities are used only to show a logical sequence or a linkage between two events without consuming any resources. A correct version of the above illustration by using a dummy activity is shown below.

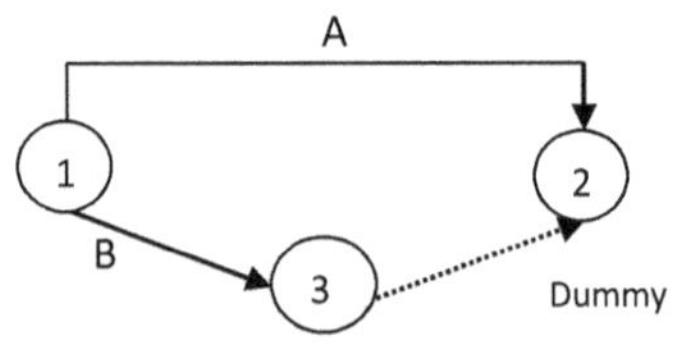

Now both activities follow rule 1 as each of them has a unique predecessor–successor event number combination. Activity A occurs between 1 and 2, whereas B between 1 and 3. Dashed arrow is dummy activity showing linkage of B from 1 to 2 via 3.

Rule 2: Precedence criteria involving preceding, succeeding and any parallel activity to a particular activity must be fulfilled.

This rule is understood by the following illustration.

Activity	A	B	C	D	E	F
Predecessor	---	A	A	B, C	B	D, E

A version of network diagram of this sequence could be:

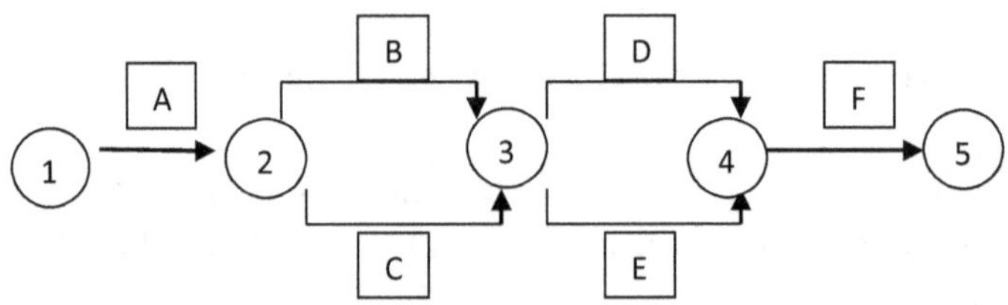

The above network diagram is incorrect as it violates rule 1 on a few occasions. Activities B and C have the same predecessors, i.e. A but both occur between the same events 2 and 3, which is not allowed. So, activities 2 and 3 do not indicate whether it is B or C. This version also violates the precedence relationship of E. Activity E has only one predecessor, i.e. B, but the diagram indicates that it can start only after completion of B and C which is not the case. Finally, to fulfill the precedence relationship of F, activities D and E occur between the same events 3 and 4 that again violates rule 1.

These flaws can be corrected by introducing dummy activities as shown in the following network diagram.

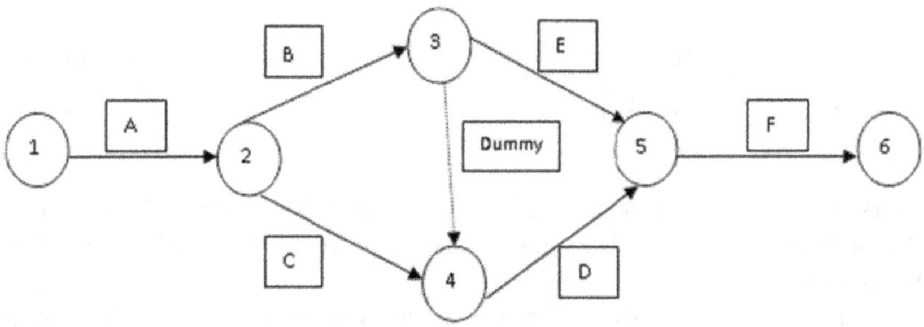

The addition of dummy activity indicated by 3–4 resolves all issues. Both B and C have unique event combinations, D is preceded by B and C, E is preceded only by B and both D and E have different start and finish nodes. Thus, in the above version, both rules 1 and 2 are not violated.

Rule 3: Lopping of activities is not allowed

Looping indicates a path where all activities are repeated continuously. For instance, in the case of computer assembly, three activities A, B and C need to happen in sequence, resulting in the following network diagram.

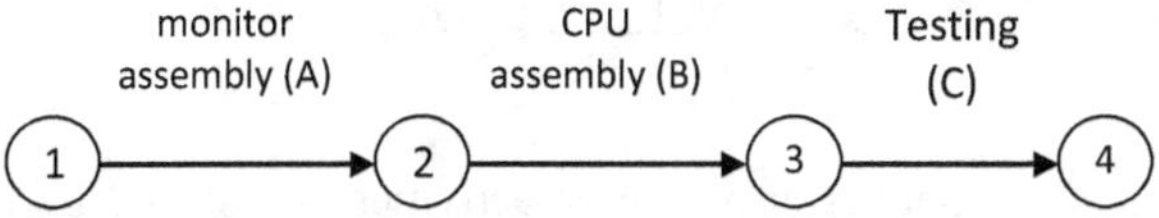

It should not be drawn in loop fashion as shown below. The following network diagram implies activities being continuously repeated. After B, C happens which is succeeded by A, and it continues in never-ending fashion. No doubt activities are repeated, i.e. after assembling of one complete computer involving A, B and C next computer again starts from A and follows the same process. Drawing of such case is shown in Rule 4.

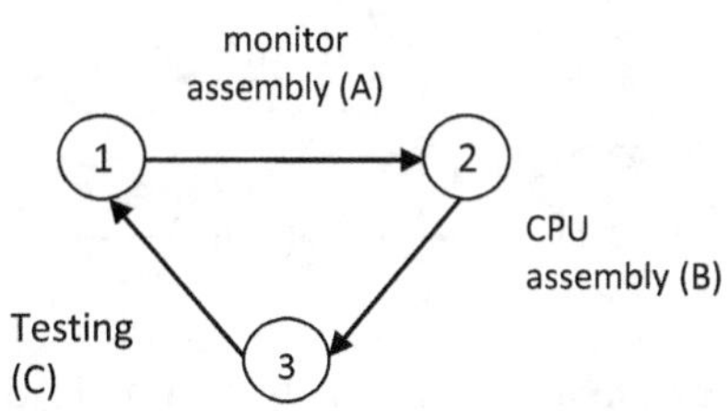

Rule 4: Repetitive activities:

Most manufacturing processes involve activities that are repeated continuously. Computer assembly illustration was discussed earlier for such process. A network diagram by taking an example of process being repeated thrice is shown below (a). A1, B1 and C1 are three activities pertaining to process 1; A2, B2 and C2 for process 2 and similarly for process 3.

(a)

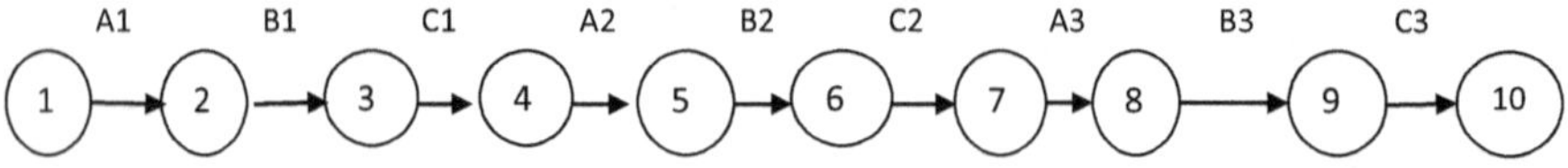

This version means A1 will be repeated again after completion of C1. Worker remains idle for duration of B1 and C1 leading to wastage of resources. This version should not be followed.

The same process can also be drawn in the following manner (b). This version is also unacceptable as activities A1, A2 and A3 are performed simultaneously, which is not possible as only one expert labor is available or capable to carry out this activity of assembling monitors.

(b)

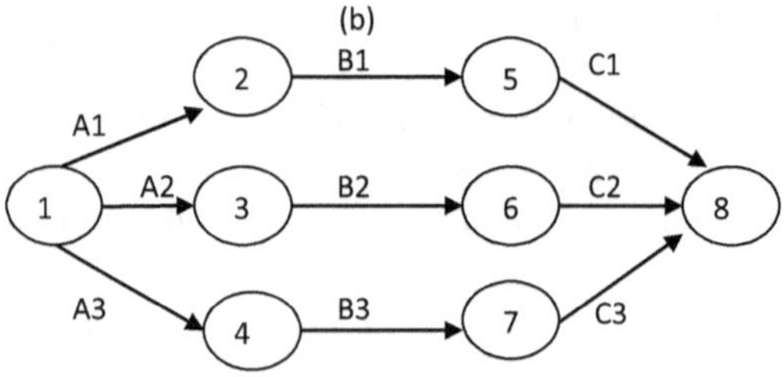

Thus, this problem of conduct of repetitive activities is resolved by the concept of laddering. In this method, each expert after doing job on one product can then move to perform the same process on other product. This is shown in (c).

(c)

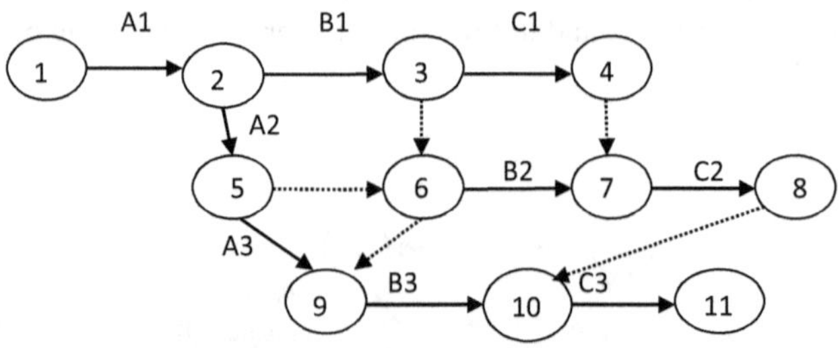

According to network diagram (c), A2 starts after A1 and A3 after A2. B2 can only start after completion of A2 and worker gets free after performing B1. So, if a dummy activity is connected from event 3 to event 5, it would imply in addition to that B2 is preceded by A2 and B1 (which is correct) but also that start of A3 depends on finish of A2 and B1 (which is incorrect). That is why another dummy event 6 is added and dummy activities between 3–6 and 5–6 are incorporated to connect to event 7. The same explanation can be applied for process when repeated third time.

This section explained certain fundamental principles required for the construction of network diagrams. These are important as diagrams are the first step in the application of network techniques such as PERT and CPM. The following sections are devoted to these two methods.

8.3 CRITICAL PATH METHOD

8.3.1 Case: Mattel Inc.

Mattel is the world's largest toy maker with strong memorable brands like Barbie and Hot Wheels. The company after acquiring infant product manufacturing company Fisher-Price increased its market position both in terms of sales and profits. The company designated these brands as core products to be manufactured in-house. However, brands with small shelf life of less than a year and peripheral promotional items such as dresses, compact discs, etc. were categorized as non-core to be manufactured by third-party vendors. With an increase in demand, the company management decided to modify or change its production system. For instance, a Mattel factory manufacturing Barbie dolls in such a scenario needed to increase its manufacturing capacity of plastic legs from 200 units to 2,000 in an 8-hour shift. Operations manager planned to change the production system from small-volume time-consuming job shop to high-volume less time-consuming mass operation system. This system was based on extensive division of labour where an operation is broken down into small tangible activities involving standardized and low skill operations. In the entire process of manufacturing, a Barbie, doll including moulding the manager initiated a change of system in the section involving dressing of dolls. In previous system of job shop, similar operations were arranged in a group carrying out an activity on different products. For instance, a section in dressing of dolls involved putting up shirt on different dolls. However, different dolls require different sizes and colour of shirts, making operator to stop for looking for particular shirt for a specific doll. The system worked well with low demand for variety of dolls. However, with the increase in demand in mass operation system, a line of operations was dedicated to producing only one type of doll. The line was further divided into sections carrying out only specific standardized activity. All activities were arranged in a specific sequence. This resulted in a few bottlenecks resulting in activities being performed quickly and efficiently. The time taken to perform each activity and its relation with other are shown in Table 8.1.

The precedence relationship shows activities that must be completed prior to start of next activity. For instance, as trouser cannot be worn (D) before putting on of undershorts (A) so A has to be completed before D gets started. On the other hand, A, B and C are independent activities with no predecessor and can start at same time. In similar vein, G can start only after finish of E and F, whereas H can start after finish of G. Depending on this precedence relationship created on the basis of logic of activities, the following network diagram is drawn (Figure 8.1). This project network helps a manager to visualize the inter-relationship of activities and forms a basis in solving of PERT/CPM problems.

This technique uses the following notation for representing a node in case of A-o-A method. However, for A-o-N method in case of event number name of activity is shown.

TABLE 8.1
Relationship diagram

Activity	Description	Predecessor Activity	Time Duration (Seconds)
A	Wear undershorts	---	8
B	Wear undershirt	---	10
C	Put on socks	---	12
D	Put on trousers	A	20
E	Put on shoes	C, D	25
F	Wear shirt	B	18
G	Put on belt	E, F	14
H	Put on tie	G	40
I	Put on coat	H	15

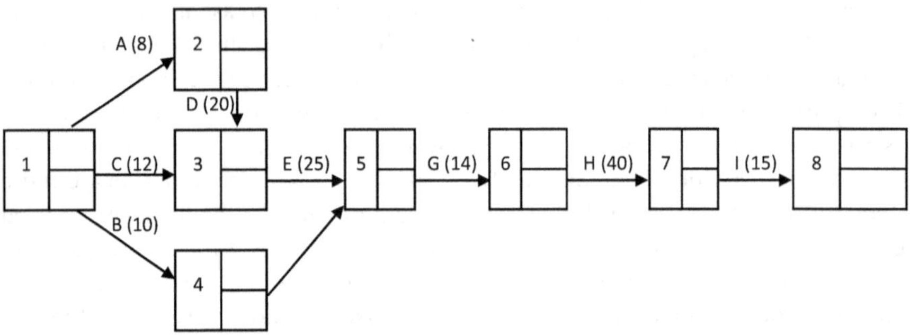

FIGURE 8.1 Activity-on-Arrow diagram.

Event number/Activity name	EST of succeeding activity LFT of preceding activity

By using this notation, the network diagram for the above illustration is drawn below.

Step 1: *Construct network diagram:*

A-o-A diagram (Figure 8.2) consists of nodes representing events from 1 to 8 and arrows, indicating activities from A to I. The bracketed values next to activity name indicate each activity's completion time. The length of arrow is not an indication of time of activity. An arrow only indicates the direction of relationship between activities.

Step 2: *Estimation of project schedule*

After estimation of duration for each activity and representing their interrelationship through a network diagram, a project schedule can be calculated. A project schedule includes the calculation of:

- **Earliest start time (EST)** at which activity can start and earliest finish time (EFT) by which an activity can be completed.

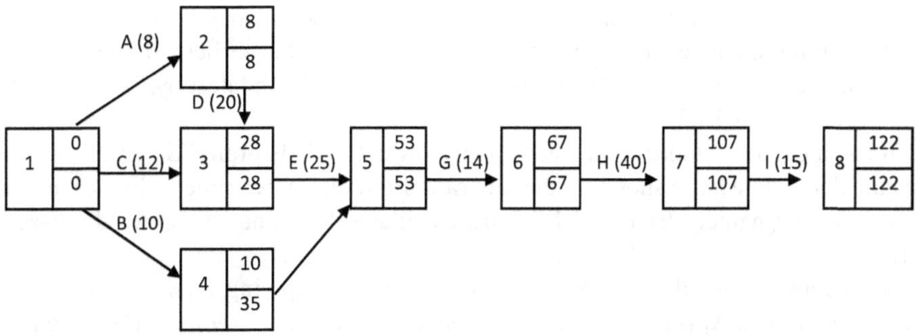

FIGURE 8.2 Activity-on-Arrow diagram showing EST and LFT.

- **Latest start time (LST)** by which an activity should start and latest finish time (LFT) by which it should get completed.

These terms are explained below:

- **EST** of an activity indicates the time at which an activity can start as soon as possible.

  The calculation of EST is governed by *forward path* rule. This rule states that for calculating EST we should move forward from first event to the last event in a network diagram. As an activity cannot start until all preceding activities have been completed, so EST of an activity is equal to the largest finish time of immediate preceding activity.
- **EFT** of an activity indicates the earliest possible time at which an activity can be completed. It is the sum of EST and duration of that activity, i.e.

$$\text{EFT} = \text{EST} + \text{duration of activity}(t)$$

- **LFT** of an activity indicates the finish of an activity as late as possible without having an effect on scheduled finish of entire project.

  The calculation of LFT is governed by *backward path* rule. This rule states that for calculating LFT we should move backwards from the last event to the first event in a network diagram. As purpose is to finish an activity as early as possible, so LFT for an activity is the smallest of LST of immediate succeeding activities.
- **LST** of an activity indicates start of an activity as late as possible without having an effect on scheduled finish of entire project. LST is the difference of LFT and duration of particular activity, i.e.

$$\text{LST} = \text{LFT} - \text{duration of activity}(t)$$

These terms can be understood by a simple example. Suppose a class is scheduled to start at 9:00 am and finish at 10:00 am. Teacher being lenient one allows students to come 5 minutes late. In addition, class of quantitative techniques is quite exhaustive.

Teacher decides to leave the class 5 minutes early so students can take a breath before the beginning of the next class. Thus, in this case, class at the earliest starts at EST = 9:00 am and latest by LST = 9:05 am. Similarly, it can finish latest by LFT = 10:00 am and earliest by EFT = 9:55 am.

It is important to understand that by analysing network diagram EST and LFT can be calculated which are shown in the diagram. From these two time estimates, other two estimates namely EFT and LST are calculated by using the above-discussed formula.

We begin by calculating EST of each activity by applying *forward path rule*. Each of activities A, B and C as has no predecessor so any of these can be taken as first to begin. Thus, EST of each activity would be 0. EFT for any activity would be:

$$\text{EFT} = \text{EST} + \text{time duration}(t)\text{of that activity}$$

This EFT would be EST of succeeding activity. Therefore, EFT for A would be $0+8 = 8$ seconds. After 8 seconds, D starts so its EST would be 8 seconds. Thus, the activity of wearing trousers finishes after 8 seconds, which is succeeded by E. Activity E can start only after completion of C and D. C takes 12 seconds to reach event 3, i.e. start of E, whereas D reaches event 3 after 28 seconds. So, according to forward path rule, an activity will only begin after finish of all preceding activities making start of E at 28 seconds. Similarly, according to network diagram, G can only start after finish of E and F. If we move along paths B and F, then G should be ready to start after 28 seconds, whereas as it is dependent on finish of E that happens only after 53 seconds so EST of G would be 53 seconds. After G, project route involves activities H and I and results in completion of the entire project in 122 seconds.

LFT for each activity is calculated by applying backward path rule that implies beginning from last activity and then moving backwards to first activity. In the case of multiple activities, emanating from a particular activity, the LFT of that activity would be the smallest of LST of all emerging activities. The project was completed after 122 seconds. Activity I take 15 seconds to complete and reach to the last event. So, it has to start latest by $122-15 = 107$ seconds which would be the LFT of the preceding activity, i.e. H. Similarly, LFT for other activities can be calculated.

It is important to note that node of event 1 represents only the start of activity A. So times shown in the nodes are earliest and LSTs and not LFTs.

After calculating LFT for each activity, their LST can be calculated by subtracting it from the duration of each activity. All the times described are shown after calculation in Table 8.2.

Let us explain a node of each format. In A-o-A diagram shown in Figure 8.2 event 5 explains three aspects. First is event number, i.e. it is fifth event; second, EST of an activity (G) emerging from it. As its predecessor is, E that consumes 25 seconds so earliest G can start is 53 seconds. Third, 53 seconds indicate latest time by which preceding activity, i.e. E can finish. Therefore, times can be interpreted as showing EST of G and LFT of E or EFT of E and LST of G.

Step 3: Critical Path

A network diagram shows the start and finish of a project involving a certain number of activities. The finish point can be reached mostly by adopting different paths.

TABLE 8.2
Start and finish times

Activity	Time	EST	LFT	EFT	LST
A	8	0	8	8	0
B	10	0	35	10	25
C	12	0	28	12	16
D	20	8	28	28	8
E	25	28	53	53	28
F	18	10	53	28	35
G	14	53	67	67	53
H	40	67	107	107	67
I	15	107	122	122	107

For instance, according to Figure 8.2, project can adopt A-D-E-G-H-I, C-E-G-H-I or B-F-G-H-I paths from start to end. This does not imply that project can be completed by adopting only one path. It only means that the completion of project depends on finish of the longest path, as it would then also encompass time duration of other paths. In our project path, A-D-E-G-H-I takes 122 seconds (calculated by adding time duration of all activities lying on the path); C-E-G-H-I takes 106 seconds; B-F-G-H-I takes 97 seconds. Now what would be the project completion time? It can only be 122 seconds as all other activities, including that of this path, also get completed in this time. If, for instance, 97 seconds is considered as project completion time, then during this time, other activities cannot be finished. Thus, the longest path in overall network diagram is considered as *critical path*.

A way of determining activities that would form critical path is the determination of *total slack* (TS) of each activity. Slack, also termed as float, indicates the maximum time by which an activity can be delayed without delaying completion of the project. Slack can be zero, negative or positive. A zero slack implies that an activity cannot be delayed; a positive slack means amount of time by which an activity can remain idle without effecting finish of entire project; negative slack of an activity means that activity is running short and needs to be accelerated so as to complete the project on schedule. Let us explain this by example quantitative class schedule. If class begins on time at 09:00 and not at 09:05 so that students does not remain idle for 5 minutes. In addition, it would finish on time at 10:00 am then next class would begin on schedule indicating no delay. Thus slack is zero. If class can begin at 09:00 am and should begin by 09:05 am without effecting succeeding activities and, ultimately, entire project, then this idle time of 5 minutes is positive slack. But suppose class begins at 09:10 am then most likely it will surpass finish time of 10:00 am and might get completed by 10:10 am effecting start of next class by 10 minutes. Therefore, the teacher has to accelerate the class by some means, implying a negative slack of 10 minutes.

From the above illustration, it can be deduced that TS is difference between earliest and LSTs or earliest and LFTs of each activity.

TABLE 8.3
Calculation of Total Slack

Activity	Time	EST	LFT	EFT	LST	TS
A	8	0	8	8	0	0
B	10	0	35	10	25	25
C	12	0	28	12	16	16
D	20	8	28	28	8	0
E	25	28	53	53	28	0
F	18	10	53	28	35	25
G	14	53	67	67	53	0
H	40	67	107	107	67	0
I	15	107	122	122	107	0

$$\text{Thus Total slack}(\text{TS}) = \text{LST} - \text{EST} = \text{LFT} - \text{EFT}$$

These calculations also help in estimating critical path. This path, being most important and consuming maximum resources, should be delayed least or should have least TS. Thus, when activities with least slack are added, it results in giving TS of critical path. TS for all activities is shown in Table 8.3.

Least TS of 0 was found for A-D-E-G-H-I activities indicating it to be the *critical path*. As discussed a network diagram can have many paths from beginning to end. Paths with positive critical path are termed as *non-critical* paths indicating involving activities that can be delayed by slack amount without delaying entire project.

Free slack (FS) is another type of slack that is determined to find activities that can be delayed by certain time without effecting earliest start of immediate succeeding activities. By definition, FS is the relative difference of TS between activities entering into same activity. This makes FS to always be either zero or a positive value. It can never be negative. In addition, as difference can only be calculated if there are at least two activities, so FS will occur only if more than one activity is entering into the same activity. This understanding helps to calculate TS by first identifying the lowest value of TS from all entering activities and then subtracting this from TS of each of these activities. For instance, in our illustration of dressing of doll, C and D are multiple activities entering into event 3 and E succeeds these. Thus, in all likelihood, all these activities would have positive FS. It is calculated by identifying minimum TS of C and D from Table 8.3, which is 0 of D and subtracting it from TS of C and D. This approach would give FS of C as 16 – 0 = 16. Similarly, FS of all other activities is calculated and shown in Table 8.4.

Another way of finding FS is by calculating head even slack (HES) of each activity. In A-o-A diagram, an arrow denotes an activity and if it begins from an event 1 and finishes at event 2, then head event of that activity would be event 2. Its slack, i.e. HES is calculated by subtracting the times shown in event 2. For instance, in Figure 8.2, activity A begins at event 1 and finishes at event 2 that becomes its head event. The times shown in this head event are either EST of E and LFT of A or EFT of A and LST of E. As slack by definition is the difference, between earliest or latest

TABLE 8.4
Calculation of Free Slack

Activity	Time	EST	LFT	EFT	LST	TS	HES	FS
A	8	0	8	8	0	0	0	0
B	10	0	35	10	25	25	25	0
C	12	0	28	12	16	16	0	16
D	20	8	28	28	8	0	0	0
E	25	28	53	53	28	0	0	0
F	18	10	53	28	35	25	0	25
G	14	53	67	67	53	0	0	0
H	40	67	107	107	67	0	0	0
I	15	107	122	122	107	0	0	0

start and finish times so HES of A would be LFT of A minus EFT of A. In this case, it would be 8 – 8 = 0 seconds . Ultimately, FS of A = TS (A) – HES (A) = 0 – 0 = 0 seconds. Similarly, FS of all other activities is calculated and shown in Table 8.4.

This shows that activities C and F can be postponed by 16 and 25 seconds, respectively, without effecting start of their succeeding activities, i.e. E and G.

Independent slack (IS), popularly known as independent float, is another type of slack that is sometimes used in scheduling various activities of a network. It is defined as time by which an activity can be delayed without effecting finish of preceding and start of succeeding activity. Thus, this type of slack is concerned with LFT of preceding and earliest start of the following activity. Independent float is present if an activity starts at LFT of preceding activity but finishes before the following activity's EST. This can also be interpreted as time left even if all predecessors finish are at late finish and all successors are at early start. In scheduling of quantitative technique class suppose independent float of class from 10:00 am to 11:00 am is to be found which is clubbed between classes of 9–10 and 11–12. If LFT of class from 9 to 10 is 10:05 am and earliest start of class from 11 to 12 is 11:00 am and class of interest i.e. from 10 to 11 completes its lesson by starting at the earliest of 10:05 am and finishes at 10:55 am. Therefore, next class will only begin at 11:00 am. Neither the previous nor the following activity is impacted.

One way of calculating IS of an activity is subtracting the latest finish of preceding from early start of following and duration time of interested activity. This gives the formula:

$$\text{IS} = \text{EST of succeeding} - \text{LFT of preceding} - \text{duration of activity}$$

For instance, IS of activity E = EST of G – LFT of D – 25

$$= 53 - 28 - 25 = 0$$

There can be cases where IS comes out to be negative. Negative IS has no meaning. It only indicates that activity with negative IS is starting much earlier than finish of preceding activity which is not possible. So if IS comes to be negative, then it is shown as 0 (zero). Another formula of IS is

$$IS = FS - \text{Tail Event Slack}(TES)$$

FS as shown is calculated by involving slack of head event, i.e. succeeding activity, whereas IS as shown in the above formula involves slack of tail, i.e. preceding activity. Thus, formula becomes quite pertinent for the calculation of IS. Now tail event is represented by the event from where an activity starts. For instance, in Figure 8.2, activity A begins at event 1 and finishes at event 2 so event 1 is tail event. Calculation of TES involves difference of times shown in tail event. Times shown in tail event of activity A would be 0 – 0 = 0 seconds. Similarly, TES of all activities is calculated as shown in Table 8.5.

To summarize, these calculations provide us the information regarding project completion time (in this case it is 122 seconds), scheduled start and finish times for each activity. It was done by showing the earliest and latest start and finish times, critical activities (A, D, E, G, H and I) which must be completed exactly according to the schedule and finally, non-critical activities that can be delayed by their slack amount without effecting scheduled project completion time. Such information is very helpful in planning, scheduling and control of the project. Though entire process becomes quite complex with large projects but is essential in proper scheduling of the project.

The next section is devoted to same purpose of finding schedule of project when activity times are not known with certainty.

Excel Solution

CPM algorithm

Calculations involved in finding out earliest, latest times and TS in the CPM can be summarized as under which are utilized in forming an algorithm required for solution in Excel.

TABLE 8.5
Calculation of Independent Slack

Activity	Time	EST	LFT	EFT	LST	TS	HES	FS	TES	IS
A	8	0	8	8	0	0	0	0	0	0
B	10	0	35	10	25	25	25	0	0	0
C	12	0	28	12	16	16	0	16	0	0
D	20	8	28	28	8	0	0	0	0	0
E	25	28	53	53	28	0	0	0	0	0
F	18	10	53	28	35	25	0	25	25	0
G	14	53	67	67	53	0	0	0	0	0
H	40	67	107	107	67	0	0	0	0	0
I	15	107	122	122	107	0	0	0	0	0

Step 1: To find EST and EFT:

- EST of an activity with no predecessor would be zero i.e. EST = 0.
- EFT of all activities would be = EST + duration of that activity.
- EST of an activity with only one predecessor = EFT of predecessor activity.
- EST of activities with more than one predecessor = Max (EFT of all predecessor activity).
- Finally, find maximum of all EFTs. It would give project completion time

Step 2: Create list of successor activities corresponding to each activity:

Step 3: To find LST and LFT:

- LFT of an activity with no successor = project completion time.
- LST of all activities = LFT – duration of that activity.
- LFT of an activity with only one successor = LST of successor activity
- LFT of activities with more than one successor = Min (LST of all successor activity).

Step 4: Calculate slack for each activity

$$\text{TS} = LST - EST \; or \; LFT - EFT$$

Step 5: Find critical activities and critical path

Activities with zero TS are termed as critical activities and path traversed by them is critical path.

Calculations

The illustration of Mattel is solved by using the above-discussed algorithm in Excel.

The data showing activities, description, predecessor and duration in seconds is inputted in the spreadsheet (Figure 8.3)

In Figure 8.4, formulas are shown, whereas in Figure 8.5, the final solution is shown.

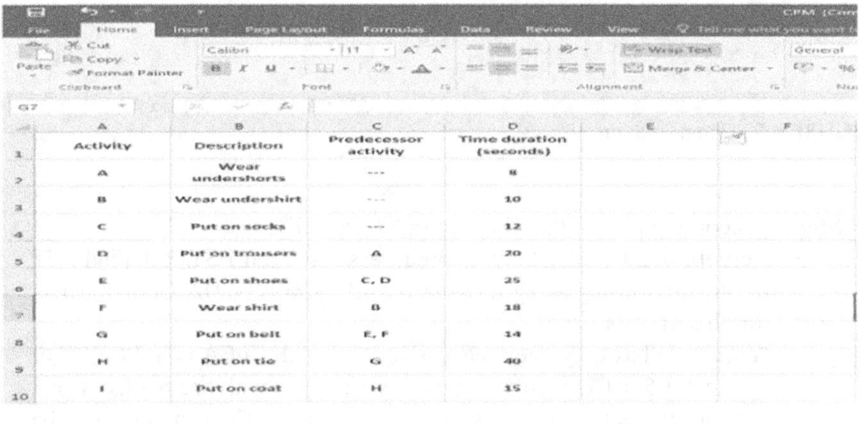

Activity	Description	Predecessor activity	Time duration (seconds)
A	Wear undershorts	---	8
B	Wear undershirt	---	10
C	Put on socks	---	12
D	Put on trousers	A	20
E	Put on shoes	C, D	25
F	Wear shirt	B	18
G	Put on belt	E, F	14
H	Put on tie	G	40
I	Put on coat	H	15

FIGURE 8.3 Input data.

	A	B	C	D	E	F	G	H	I	J	K
1	Activity	Description	Predecessor activity	Time duration (seconds)	EST	EFT	Successor	LFT	LST	total slack	Critical activity
2	A	Wear undershorts	---	8	0	=E2+D2	D	=I5	=H2-D2	=H2-F2	=IF(J2=0,"critical activity", "non-critical activity")
3	B	Wear undershirt	---	10	0	=E3+D3	F	=I7	=H3-D3	=H3-F3	=IF(J3=0,"critical activity", "non-critical activity")
4	C	Put on socks	---	12	0	=E4+D4	E	=I6	=H4-D4	=H4-F4	=IF(J4=0,"critical activity", "non-critical activity")
5	D	Put on trousers	A	20	=F2	=E5+D5	E	=I6	=H5-D5	=H5-F5	=IF(J5=0,"critical activity", "non-critical activity")
6	E	Put on shoes	C, D	25	=MAX(F4,F5)	=E6+D6	G	=I8	=H6-D6	=H6-F6	=IF(J6=0,"critical activity", "non-critical activity")
7	F	Wear shirt	B	18	=F3	=E7+D7	G	=I8	=H7-D7	=H7-F7	=IF(J7=0,"critical activity", "non-critical activity")
8	G	Put on belt	E, F	14	=MAX(F6,F7)	=E8+D8	H	=I9	=H8-D8	=H8-F8	=IF(J8=0,"critical activity", "non-critical activity")
9	H	Put on tie	G	40	=F8	=E9+D9	I	=I10	=H9-D9	=H9-F9	=IF(J9=0,"critical activity", "non-critical activity")
10	I	Put on coat	H	15	=F9	=E10+D10		=F12	=H10-D10	=H10-F10	=IF(J10=0,"critical activity", "non-critical activity")
11											
12				Project completion time		=MAX(F2:F10)					

FIGURE 8.4 Formulas.

	A	B	C	D	E	F	G	H	I	J	K
1	Activity	Description	Predecessor activity	Time duration (seconds)	EST	EFT	Successor	LFT	LST	total slack	Critical activity
2	A	Wear undershorts	---	8	0	8	D	8	0	0	critical activity
3	B	Wear undershirt	---	10	0	10	F	35	25	25	non-critical activity
4	C	Put on socks	---	12	0	12	E	28	16	16	non-critical activity
5	D	Put on trousers	A	20	8	28	E	28	8	0	critical activity
6	E	Put on shoes	C, D	25	28	53	G	53	28	0	critical activity
7	F	Wear shirt	B	18	10	28	G	53	35	25	non-critical activity
8	G	Put on belt	E, F	14	53	67	H	67	53	0	critical activity
9	H	Put on tie	G	40	67	107	I	107	67	0	critical activity
10	I	Put on coat	H	15	107	122		122	107	0	critical activity
11											
12				Project completion time		122					

FIGURE 8.5 Final solution.

Step 1: According to the algorithm, find EST and EFT.

- Activities A, B and C have no predecessor so their EST = 0. Find EFT of all activities by using formula in cell F2 = E2+D2. Copy formula for all activities.
- Activity D has only one predecessor i.e. A. EFT of A = 8. So EST of D in cell E5 = F2. Similarly, for other activities with one predecessor.
- Activity E has two predecessor C and D. In cell E6 formula would be =Max(F4, F5).

Step 2: List successor of every activity. For instance, as predecessor of D is A so activity D becomes successor of A. Similarly, by understanding predecessor data successor activities are written.

Find project completion time which is =Max(F2:F10)

Step 3: Now according to the algorithm, find LFT and LST.

- Activity I has no successor, so its LFT is project completion time given by = F12.
- LST of all activities is calculated in cell I10 as = H10−D10. Copy it to all other cells of LST column.
- LFT of an activity with one successor. H has one successor I so its LFT is given as =I10. Similarly, all LFTs are calculated.

Step 4: Finally, TS is calculated for every activity by using formula in cell J2 as = H2−F2.

Step 5: Critical activities with zero TS are mentioned by using formula = if(J2 = 0, "critical activity", "non-critical activity")

8.4 PROGRAM EVALUATION AND REVIEW TECHNIQUE

This section deals with scheduling of project involving uncertain activity times. As discussed in the introduction section, this scenario happens mostly when a project is new to an organization or employees have no experience in dealing with such project. For instance, Bajaj Scooters ltd had no experience in building motorcycles, so entry into bike market was something very new to the company. Now if they consider entering into car market by producing small cars, then also scheduling of such project will involve the application of PERT. Another instance happened with when Toyota immediately needed to shift to manufacturing of V8 engines instead of V6 to power their cars. Company engineers were not trained and experienced in manufacturing of V8 engines so by considering it a new project PERT was considered to be an appropriate technique for estimating schedule of project. The technique was also important in finding whether project would be able to finish within estimated time or not. It is significant to understand that as PERT deals with uncertain times so outcome of this application is in probability terms indicating quantum of likelihood of project completion within schedule.

8.4.1 Case: East Fork Roofing

East Fork Roofing in Reno of Northern Nevada has been in business of providing roofing facilities to residential and commercial facilities for the last three decades. Roofing projects require meticulous planning and coordination with various other construction works such as electrical, plumbing and adjoining rooms. Improper sequence of operations could result in rework leading to wastages in time, effort and money. Accurate scheduling is the highlight of company as it in advance indicates their customer the sequence of activities and resources required to finish every activity. More importantly, project completion date is projected to close accuracy helping customers to make decisions about using the facility. In one of the roofing projects of a residential facility, a small module of entire project was to install a fibre sheet

TABLE 8.6
Activities and Predecessors

Activity	Description	Predecessor
A	Procure material	---
B	Cut three aluminium channels into specifications	A
C	Place all channels on roof top at proper distance from each other and align them	B
D	Drill holes on side S1 of roof and fix the channel on this side	C
E	Apply adhesive on that side so that fibre sheet can be placed and fixed	D
F	Drill holes on side S2 of roof and fix the channel on this side	C
G	Apply adhesive on that side (S2) so that fibre sheet can be placed and fixed	F
H	Place sheet and balance it properly	E, G
I	Fix S1 side of roof with sheet by drilling and meshing with channel	H
J	Apply silicon adhesive at joints and fix sheet	I
K	Fix S2 side of roof with sheet by drilling and meshing with channel	H
L	Apply silicon adhesive at joints and fix sheet	K

roof on a makeshift kitchen. The kitchen is built in such a manner that roof of fibre sheet when installed can be fixed only from two sides. Let us name these sides as S1 and S2. The project involves procurement of fibre sheet, aluminium channels and various fixtures. Fibre sheet can be installed after aluminium channels are placed for support. Thus, both sheet and channels need to be cut to specifications and then fixed properly. Finally, adhesives are applied for proper fixing of sheet. All these activities in detail and their relationship are shown in Table 8.6.

After the identification of activities, the next step is to build time estimates for each activity. This information is helpful in the calculation of project completion time and scheduling of activities. As labour employed had little experience installing fibre sheet roof, so accurate estimation of times of each activity was considered difficult. Instead, a range of possible times was estimated. The following three time estimates are considered to find the possible range of times for uncertain activities.

- **Optimistic time (t_o)**: is the time taken by an activity when performed under ideal conditions. This is the minimum time taken as conditions are most favourable for its performance.
- **Most likely time (t_m)**: is the time taken by an activity when performed under normal conditions. This is the most possible time under which an activity would be performed as conditions are most realistic.
- **Pessimistic time (t_p)**: is the time taken by an activity when performed under the worst conditions. This is the maximum time taken by an activity as it is performed when significant delays are most probable.

$$\text{So, } t_o < t_m < t_p$$

The average time of an uncertain activity in a new project is calculated by finding average of three probable activity times discussed above. This average time is called as expected time (t_e). Maximum weight age is given to t_m, as this is the most probable time an activity would take to perform. Thus,

$$t_e = (t_o + t_m + t_p)/6$$

Because of uncertain times, the variance in them can be calculated as:

$$\text{Variance} = \left[(t_p - t_o)/6\right]^2$$

Due to uncertainty in times, each activity time can take any value within this range. Thus, these values are considered as random variables with associated probability of occurrence. So, instead of exact values of project completion time, a probability value of it is estimated.

These activity times duration and variance are assumed to follow beta probability distribution (Figure 8.4). This distribution belongs to family of continuous probability distribution that defines the likelihood of occurrence of random variables. As discussed, uncertain times in PERT are random variables that are limited to intervals of finite length bounded in this case by optimistic and pessimistic time. Thus, these are assumed to follow beta distribution. This assumption allows us to safely calculate expected time of each activity with high degree of correctness.

A general form of beta distribution is shown in Figure 8.6 where lower bound of an activity for instance activity A is t_o = 60 minutes; upper bound of A is t_p = 120; t_m = 90 and t_e = 90 minutes. X-axis of graph represents activity time and Y-axis gives probability values. Similarly, expected times and variances of all activities are calculated and shown in Table 8.7.

In light of the above discussion, major purposes of PERT are as follows:

- Estimation of expected time of each activity.
- Estimation of scheduled completion time of project.

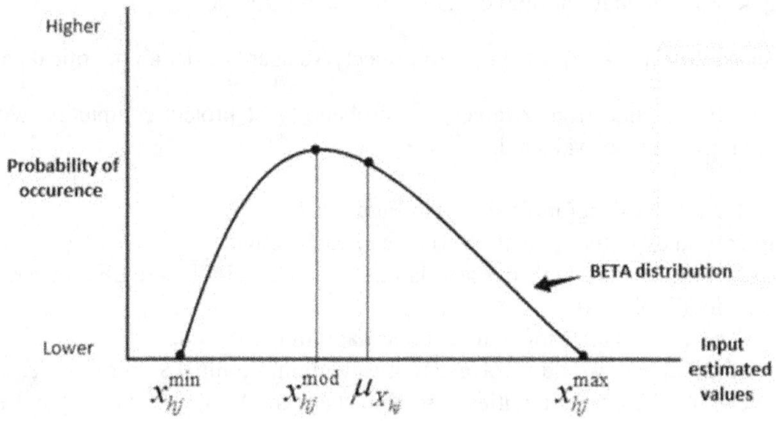

FIGURE 8.6 General form of Beta distribution.

TABLE 8.7
Calculation of Expected Time

Activity	Optimistic Time (Minutes, t_o)	Most Likely Time (Minutes, t_m)	Pessimistic Time (Minutes, t_p)	Expected Time (Minutes, t_e)	Variance
A	60	90	120	90	100.00
B	20	30	34	29	5.44
C	8	10	18	11	2.78
D	12	15	30	17	9.00
E	4	5	6	5	0.11
F	12	21	30	21	9.00
G	4	7	10	7	1.00
H	11	17	29	18	9.00
I	20	24	40	26	11.11
J	6	8	10	8	0.44
K	20	24	34	25	5.44
L	7	8	9	7	0.11

- Finding out critical and non-critical activities.
- Most importantly, PERT is used to find probability that whether project would complete in given scheduled time.

The above-mentioned objectives are achieved by the following steps mentioned below:

Step 1: Draw network diagram of the project.
Step 2: Find expected time and variance of each activity.
Step 3: Find critical path by calculating EST, EFT, LST and LFT of each activity.
Step 4: Find standard normal deviation by using formula.

$$z = (\text{scheduled time} - \text{expected time of project})/\text{standard deviation of critical path}$$

By using this z value from z tables, the probability of project completion within scheduled time can be evaluated.

Step 1: Network Diagram is shown in Figure 8.7.
Step 2: Find expected time and variance of each activity
Step 3: Find critical path by calculating EST, EFT, LST and LFT of each activity (Table 8.8).

These times are shown in network diagram Figure 8.8.

Critical path would involve activities with minimum TS. In this case, as shown in Table 8.8, activities with zero slack are A, B, C, F, G, H, I and J. Adding their expected times would give us project completion time. It would be 210 minutes. Variance of critical path would be addition of variances of activities lying on critical path. Observing variance values from Table 8.7 would give variance of critical path as 138.78.

Step 4: Find standard normal deviation by using formula

$$z = (\text{scheduled time} - \text{expected time of project}) / \text{standard deviation of critical path}$$

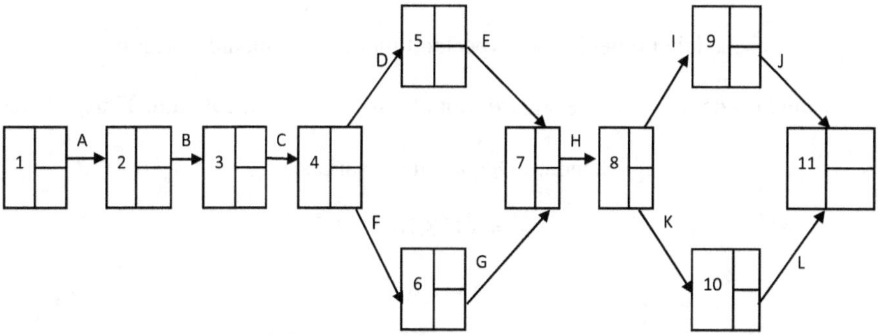

FIGURE 8.7 A-o-A diagram.

TABLE 8.8
Calculation of Total Slack

Activity	Expected Time (Minutes, t_e)	EST	LFT	EFT	LST	Total Slack
A	90	0	90	90	0	0
B	29	90	119	119	90	0
C	11	119	130	130	119	0
D	17	130	153	147	136	6
E	5	147	158	152	153	6
F	17	130	151	151	130	0
G	5	151	158	158	151	0
H	18	158	176	176	158	0
I	26	176	202	202	176	0
J	8	202	210	210	202	0
K	25	176	202	201	177	1
L	7	201	210	209	202	1

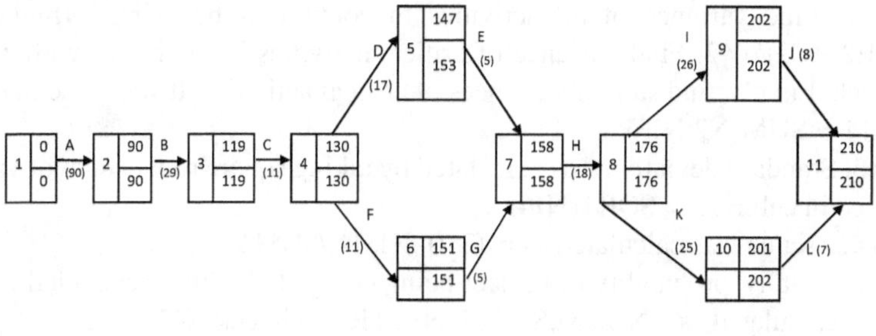

FIGURE 8.8 Activity times.

This formula is used to find probability of completing the project in given scheduled time. For instance, suppose it was estimated that installation of roof would not take more than 4 hours, i.e. 240 minutes, then what is the probability that project will be completed within this scheduled time. Therefore, our aim is to find probability of finishing the project in less than 240 minutes. This can be calculated as:

$$z = \text{scheduled time}(240) - \text{expected time}(210)/\text{standard deviation}$$

Standard deviation is the square root of variance of critical path. Thus,

$$\text{Standard Deviation} = \sqrt{\text{variance}}$$

$$= \sqrt{138.78}$$

$$= 11.78$$

$$\text{Value of z} = (240 - 210)/11.78 = 2.546$$

At 2.75 standard deviations from normal distribution tables, the probability of a value to occur between limits of scheduled and expected time is 0.4917 (Figure 8.9). However, question is to find entire probability of less than scheduled time that would be found by adding 0.5000 of area thus final probability of asked query would be: 0.5000 + 0.4945 = 0.9945 which is very high.

Let us take another illustration. What is the probability that project will cross-scheduled time of 240 minutes? Entire area under a normal curve is equivalent to 1. We have calculated probability of less than 240, that is 0.9945. Therefore, probability of more than 0.9945 would be 1.000 – 0.9945 = 0.0055.

Thus, the above-mentioned detailed approach can be adopted for calculation of project schedule in case of activities with uncertain times.

Excel Solution

Step 1: Find critical path as discussed under the CPM algorithm. It involves calculation of EST, EFT, LFT and LST. Calculations are shown in Figure 8.10. However in case of PERT three times, namely, t_o, t_m and t_p are given. As discussed use this to calculate expected time t_e. So in column F formula entered is = (C2+4*D2+E2)/6.

Step 2: Find variance of all activities in column M by using formula = ((E2 – C2)/6)^2. Find variance of critical activities i.e. activities with zero slack. Finally, find sum of variances of critical activities. It was done in cell N14 = SUM(N2:N13).

Step 3: Standard deviation was calculated by taking square root of sum of variance in cell N15 = SQRT(N14).

Step 4: z value was calculated as = (240-$H$14)/$N$15.

Step 5: Finally, probability of project being completed within scheduled time was calculated as =NORM.S.DIST(N16,TRUE) in cell N17.

PERT Calculations are shown in Figure 8.11.

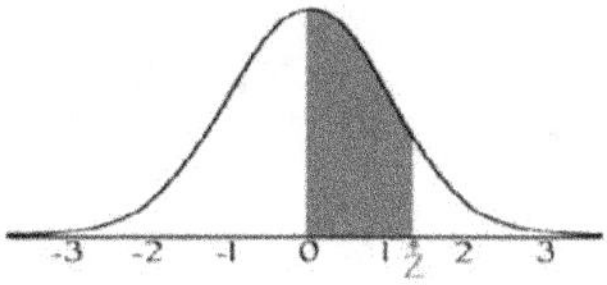

STANDARD NORMAL TABLE (Z)

Entries in the table give the area under the curve between the mean and z standard deviations above the mean. For example, for $z = 1.25$ the area under the curve between the mean (0) and z is 0.3944.

z	0.00	0.01	0.02	0.03	0.04	0.05	0.06	0.07	0.08	0.09
0.0	0.0000	0.0040	0.0080	0.0120	0.0160	0.0190	0.0239	0.0279	0.0319	0.0359
0.1	0.0398	0.0438	0.0478	0.0517	0.0557	0.0596	0.0636	0.0675	0.0714	0.0753
0.2	0.0793	0.0832	0.0871	0.0910	0.0948	0.0987	0.1026	0.1064	0.1103	0.1141
0.3	0.1179	0.1217	0.1255	0.1293	0.1331	0.1368	0.1406	0.1443	0.1480	0.1517
0.4	0.1554	0.1591	0.1628	0.1664	0.1700	0.1736	0.1772	0.1808	0.1844	0.1879
0.5	0.1915	0.1950	0.1985	0.2019	0.2054	0.2088	0.2123	0.2157	0.2190	0.2224
0.6	0.2257	0.2291	0.2324	0.2357	0.2389	0.2422	0.2454	0.2486	0.2517	0.2549
0.7	0.2580	0.2611	0.2642	0.2673	0.2704	0.2734	0.2764	0.2794	0.2823	0.2852
0.8	0.2881	0.2910	0.2939	0.2969	0.2995	0.3023	0.3051	0.3078	0.3106	0.3133
0.9	0.3159	0.3186	0.3212	0.3238	0.3264	0.3289	0.3315	0.3340	0.3365	0.3389
1.0	0.3413	0.3438	0.3461	0.3485	0.3508	0.3513	0.3554	0.3577	0.3529	0.3621
1.1	0.3643	0.3665	0.3686	0.3708	0.3729	0.3749	0.3770	0.3790	0.3810	0.3830
1.2	0.3849	0.3869	0.3888	0.3907	0.3925	0.3944	0.3962	0.3980	0.3997	0.4015
1.3	0.4032	0.4049	0.4066	0.4082	0.4099	0.4115	0.4131	0.4147	0.4162	0.4177
1.4	0.4192	0.4207	0.4222	0.4236	0.4251	0.4265	0.4279	0.4292	0.4306	0.4319
1.5	0.4332	0.4345	0.4357	0.4370	0.4382	0.4394	0.4406	0.4418	0.4429	0.4441
1.6	0.4452	0.4463	0.4474	0.4484	0.4495	0.4505	0.4515	0.4525	0.4535	0.4545
1.7	0.4554	0.4564	0.4573	0.4582	0.4591	0.4599	0.4608	0.4616	0.4625	0.4633
1.8	0.4641	0.4649	0.4656	0.4664	0.4671	0.4678	0.4686	0.4693	0.4699	0.4706
1.9	0.4713	0.4719	0.4726	0.4732	0.4738	0.4744	0.4750	0.4756	0.4761	0.4767
2.0	0.4772	0.4778	0.4783	0.4788	0.4793	0.4798	0.4803	0.4808	0.4812	0.4817
2.1	0.4821	0.4826	0.4830	0.4834	0.4838	0.4842	0.4846	0.4850	0.4854	0.4857
2.2	0.4861	0.4864	0.4868	0.4871	0.4875	0.4878	0.4881	0.4884	0.4887	0.4890
2.3	0.4893	0.4896	0.4898	0.4901	0.4904	0.4906	0.4909	0.4911	0.4913	0.4916
2.4	0.4918	0.4920	0.4922	0.4925	0.4927	0.4929	0.4931	0.4932	0.4934	0.4936
2.5	0.4938	0.4940	0.4941	0.4943	0.4945	0.4946	0.4948	0.4949	0.4951	0.4952
2.6	0.4953	0.4955	0.4956	0.4957	0.4959	0.4960	0.4961	0.4962	0.4963	0.4964
2.7	0.4965	0.4966	0.4967	0.4968	0.4969	0.4970	0.4971	0.4972	0.4973	0.4974
2.8	0.4974	0.4975	0.4976	0.4977	0.4977	0.4978	0.4979	0.4979	0.4980	0.4981
2.9	0.4981	0.4982	0.4982	0.4983	0.4984	0.4984	0.4985	0.4985	0.4986	0.4986
3.0	0.4987	0.4987	0.4987	0.4988	0.4988	0.4989	0.4989	0.4989	0.4990	0.4990
3.1	0.4990	0.4991	0.4991	0.4991	0.4992	0.4992	0.4992	0.4992	0.4993	0.4993
3.2	0.4993	0.4993	0.4994	0.4994	0.4994	0.4994	0.4994	0.4995	0.4995	0.4995
3.3	0.4995	0.4995	0.4995	0.4996	0.4996	0.4996	0.4996	0.4996	0.4996	0.4997
3.4	0.4997	0.4997	0.4997	0.4997	0.4997	0.4997	0.4997	0.4997	0.4997	0.4998

FIGURE 8.9 Standard normal distribution table.

	A	B	C	D	E	F	G	H	I	J	K	L	M	N
1	Activity	Predecessor	Optimistic time (mins., t_o)	Most likely time (mins., t_m)	Pessimistic time (mins., t_p)	Expected time (mins te)	EST	EFT	Successor	LFT	LST	TS		
2	A	----	60	90	120	90	0	90	B	90	0	0		
3	B	A	20	30	34	29	90	119	C	119	90	0		
4	C	B	8	10	18	11	119	130	D,F	130	119	0		
5	D	C	12	15	30	17	130	147	E	153	136	6		
6	E	D	4	5	6	5	147	152	H	158	153	6		
7	F	C	12	21	30	21	130	151	G	151	130	0		
8	G	F	4	7	10	7	151	158	H	158	151	0		
9	H	E, G	11	17	29	18	158	176	I	176	158	0		
10	I	H	20	24	40	26	176	202	J	202	176	0		
11	J	I	6	8	10	8	202	210	****	210	202	0		
12	K	H	20	24	34	25	176	201	L	202	177	1		
13	L	K	7	8	9	8	201	209	****	210	202	1		
14					project completion time			210						

FIGURE 8.10 Critical path.

Activity	Predecessor	Optimistic time (mins., t_o)	Most likely time (mins., t_m)	Pessimistic time (mins., t_p)	Expected time (mins te)	EST	EFT	Successor	LFT	LST	TS	Variance	Variance of Critical ctivities
A	----	60	90	120	90	0	90	B	90	0	0	100.00	100.00
B	A	20	30	34	29	90	119	C	119	90	0	5.44	5.44
C	B	8	10	18	11	119	130	D,F	130	119	0	2.78	2.78
D	C	12	15	30	17	130	147	E	153	136	6	9.00	
E	D	4	5	6	5	147	152	H	158	153	6	0.11	
F	C	12	21	30	21	130	151	G	151	130	0	9.00	9.00
G	F	4	7	10	7	151	158	H	158	151	0	1.00	1.00
H	E, G	11	17	29	18	158	176	I	176	158	0	9.00	9.00
I	H	20	24	40	26	176	202	J	202	176	0	11.11	11.11
J	I	6	8	10	8	202	210	****	210	202	0	0.44	0.44
K	H	20	24	34	25	176	201	L	202	177	1	5.44	
L	K	7	8	9	8	201	209	****	210	202	1	0.11	
					project completion time		210					Total variance	138.78
												Std dev	11.780
												z	2.5460
												prob	0.9946

FIGURE 8.11 PERT calculations.

8.5 CRASHING: TIME – COST TRADE-OFFS

A major application of CPM is the identification of critical activities that decide project completion time. Any delay or expediency in them would alter this time. A project manager would not like to delay the project, but in certain cases, a manager would like to finish the project at an earlier date than estimate completion time. This can happen only by allocating more resources such as manpower, capital or material to these activities. More resources and attention would encourage these activities to be completed at a faster pace. On the other hand, allocation of more resources results in burdening the project with additional cost. Reduction in activity times affect both direct and indirect cost of the project. Direct cost includes cost of material, equipment, permanent labour, etc., whereas indirect cost includes overhead cost, interest cost, etc. relationship between these two types of costs which add up to form total project cost is shown in Figure 8.12.

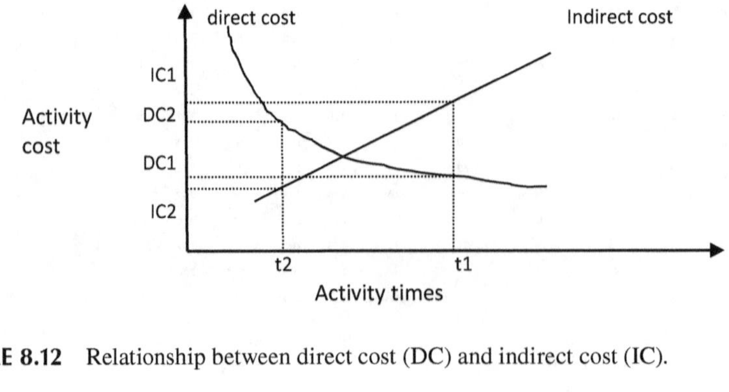

FIGURE 8.12 Relationship between direct cost (DC) and indirect cost (IC).

According to Figure 8.12, if time of an activity is reduced from t1 to t2, then there is an opposite effect on direct and indirect costs incurred in conduct of the activity. It can be clearly seen that reduction in time increases direct cost from DC1 to DC2 and reduces indirect cost from IC1 to IC2. Thus, there is an inverse relationship between two types of costs. As seen in the numerical of crashing, with a reduction in time due to activity occurring in less time, the indirect cost will show downward trend, whereas direct cost will increase. However, project duration cannot be decreased indefinitely, as at one time proportion in increase of direct cost would outweigh reduction in indirect cost. Also practically, reduction in time after some time is infeasible, as it would require inappropriate allocation of resources that no company can afford.

Thus, the purpose is to find optimum reduction in time that would increase total project cost to the minimum. This is done under *crashing*. The following are certain terms that must be understood in application of crashing:

- **Normal time (T_n)**: It is the expected time of completion of an activity.
- **Crash time (T_c)**: It is the reduced time of completion of an activity.
- **Normal cost (C_n)**: is the cost of performance of an activity when performed under normal conditions.
- **Crash cost (C_c)**: the cost of performing an activity in a shorter amount of time.

These times and costs are shown in Figure 8.13 which is derived from graph of total cost.

The following are the steps to take when crashing a project:

Step 1: Draw network diagram

Step 2: Find critical path. As discussed, critical path is found by calculating EST, EFT, LST and LFT. By definition critical path is the longest path in the network diagram.

Step 3: Find cost slope of each activity. Cost slope is the increase in *direct cost* of an activity when its duration is reduced by 1 day.

$$\text{Cost slope} = \text{change in cost / change in time}$$

$$\text{Cost slope} = (C_c - C_n)/(T_n - T_c)$$

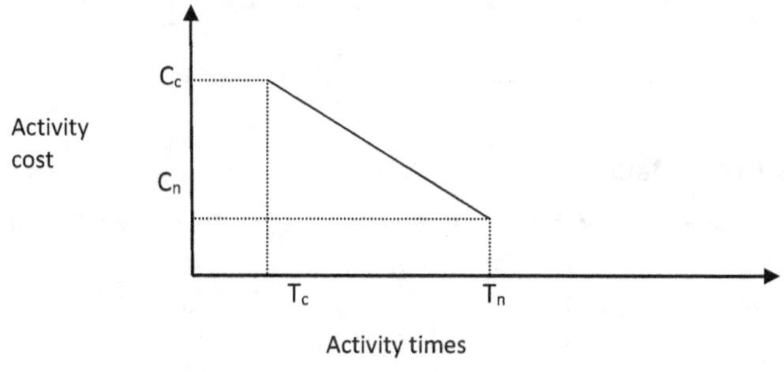

FIGURE 8.13 Time–cost relationship.

Step 4: In the critical path, find an activity with least cost slope because the aim is to reduce project time with least increase in cost.

Step 5: Estimate the time for which selected activity can be crashed. For this a general rule that is followed is that activity should be crashed for so much time that it does not result in change of critical path.

Step 6: Keep on repeating the process till further reduction in time period leads to increase in cost or when all paths in network diagram have become critical.

8.5.1 An Illustration of Crashing

Consider an example of a construction project involving four activities A, B, C and D. The project manager intends to finish the project in 10 months. It was found that deadline cannot be met if project activities are performed under normal times. Therefore, data regarding times by which each activity can be expedited by allocating extra resources incurring extra cost is determined as shown in Table 8.9. The manager intends to use CPM to determine most economical way of crashing the project to meet the deadline given indirect cost of project per month to be $5,000.

Step 1: Draw network diagram

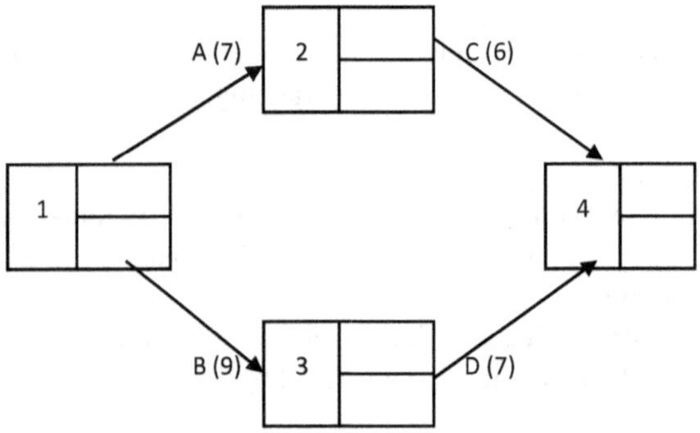

Step 2: Find critical path

Critical path is B --- D = 9 + 7 = 16 months (Table 8.10)

Non-critical path is A --- C = 7 + 6 = 13 months

TABLE 8.9
Normal and Crash Data

Activity	Predecessor	T_n (Months)	T_c (Months)	C_n (in Thousands)	C_c (in Thousands)
A	---	7	6	20	35
B	---	9	7	15	25
C	A	6	5	10	18
D	B	7	5	21	32

TABLE 8.10
Calculation of Total Slack

Activity	EST	LFT	EFT	LST	Total Slack
A	0	9	7	2	2
B	0	8	8	0	***0***
C	7	15	13	7	2
D	8	15	15	8	***0***

Step 3: Find cost slope of each activity

It was clearly explained that only activities lying on the critical path should be crashed as project completion time is dependent only on these activities. Now other purpose is that increase in cost should be minimum. Thus, an activity lying on the critical path with least cost slope should be selected.

- Cost slope of activity A = (35 – 20)/ (7 – 6) = 15
- Cost slope of activity B = (25 – 15)/ (9 – 7) = 5
- Cost slope of activity C = (18 – 10)/ (6 – 5) = 8
- Cost slope of activity D = (32 – 21)/ (7 – 5) = 4.5

Step 4: An iterative procedure of reducing project duration and estimating costs is applied to answer questions such as:

Can the project time be reduced? If yes, then by how much? What would be the effect on direct, indirect and total project cost? These questions are answered in the following steps.

i. Iteration 0:

If project is performed under normal conditions then total project cost would be:

Project completion time = 16 months

Direct cost = 20 + 15 + 10 + 21 = \$66.

Indirect cost = 16 * 5 = \$80

Total project cost = 66 + 80 = \$146.

ii. Iteration 1:

B---D was found to be critical path implying project completion time to be 18 months. Out of B and D, activity B has the least cost slope of 5, indicating that if time is reduced by 1 unit cost will increase by 5 units. Normal time of B is 9 months, whereas crash time is 7 months meaning that activity B will take a minimum of 7 months for completion. So it can be crashed by maximum of 2 months. Crashing it by 2 days would make both paths A --- C and B --- D critical. Thus:

Project completion time = 16 – 2 = 14 months

Direct cost = 66 + 2 * 5 = Rs.76.

Indirect cost = 14 * 5 = Rs.70

Total project cost = $144.

iii. B---D is still critical path. However, B cannot be crashed further so remaining activity, i.e. D would be crashed. According to data, given D can be crashed by 2 months, i.e. by adding more resources, it can be finished in 5 months rather than 7 months. But if that is done, then project completion time would be 14 – 2 = 12 months, making path B --- D non-critical and path A --- C as critical. In such case, even after crashing project will still take 13 months (time duration of A --- C) to finish. Thus, expediting D by adding resources would be a waste. Thus, on path B --- D, activity D should be crashed only by 1 month. This would give:

Project completion time = 14 – 1 = 13 months

Direct cost = 76 + 1 * 4.5 = $80.5

Indirect cost = 13 * 5 = $65

Total project cost = $145.5

iv. Now both paths A --- C and B --- D are critical. Look for activities with least cost slope on both paths. In path A --- C, activity has least cost slope of 8, whereas in path B --- D, only activity D can be crashed. Both activities C and D can be crashed only by 1 month. Thus:

Project completion time = 13 – 1 = 12 months

Direct cost = 80.5 + 1 * 8 + 1 * 4.5 = $88.5

Indirect cost = 12 * 5 = Rs.60

Total project cost = $148.5

v. Finally, A can be crashed by 1 month, but it should not be as it would make other path B --- D critical violating the rule that an activity on critical path should be crashed only by that duration which does not make it non-critical.

The results can be summarized as below:

TABLE 8.11
Summarized Results

Iteration	Project Completion Time (Months)	Direct Cost ($'000)	Indirect Cost ($'000)	Total Cost ($'000)
0	16	66	80	66 + 80 = 146
1	14	76	70	76 + 70 = 144
2	13	80.5	65	80.5 + 65 = 145.5
3	12	88.5	60	88.5 + 60 = 148.5

The summarized results (Table 8.11) clearly indicate the increase in direct cost and decrease in indirect cost with each iteration of reduction in project completion time. However, total project cost has increased as the proportion of increase in direct cost outweighs reduction in indirect cost.

Excel Solution

Step 1: Input the data as shown in Figure 8.14. From the data, find maximum time by which an activity can be crashed by subtracting crash time from normal time. It is shown in column H by using formula = D2 − E2. Similarly, find increase in cost by subtracting normal cost from crash cost. It is shown in column I by using formula = G2 − F2. Finally, slope indicating increase in cost with decrease in time was calculated in column J by using formula = I2/H2. These values for all activities were calculated.

Step 2: In the next table, critical path is calculated by already discussed algorithm (Figure 8.15). However, in crashing before doing so activity duration is calculated shown in cell D8. As because of crashing time to carry an activity will vary from normal time, so activity duration is a function of time by which an activity is crashed. This is shown in cell C8.

Step 3: *Solver solution to find project completion time:* The objective in crashing ise twofold. First, the aim is to estimate *minimum* project completion time after all possible crashing activities and then to find *minimum* possible cost in fulfilling aim of reducing project time. So in first aim objective function is set in cell B13 = max(F9:F12). Decision variables are corresponding

G10

	A	B	C	D	E	F	G	H	I	J	K
1	Activity	Predecessor	Successor	T_n (months)	T_c (months)	C_n (thousands)	C_c (thousands)	Maximum crashing possible (months)	Cc - Cn (thousands)	Slope	
2	A	---	C	7	6	20	35	1	15	15	
3	B	---	D	9	7	15	25	2	10	5	
4	C	A	***	6	5	10	18	1	8	8	
5	D	B	***	7	5	21	32	2	11	5.5	
6											

FIGURE 8.14 Input data.

	Activity	Predecessor	Successor	T_n (months)	T_c (months)	C_n (thousands)	C_c (thousands)	Maximum crashing possible (months)	Cc - Cn (thousands)	Slope	
1											
2	A	---	C	7	6	20	35	1	15	15	
3	B	---	D	9	7	15	25	2	10	5	
4	C	A	***	6	5	10	18	1	8	8	
5	D	B	***	7	5	21	32	2	11	5.5	
6											
7	Calculation of critical path										
8	Activity	Predecessor	Activity crashed by	Activity duration	EST	EFT	Successor	LFT	LST	TS	Indirect cost
9	A	---		7	0	7	C	10	3	3	35
10	B	---		9	0	9	D	9	0	0	45
11	C	A		6	7	13	***	16	10	3	30
12	D	B		7	9	16	***	16	9	0	35
13	Project completion time		16								
14											

FIGURE 8.15 Calculation of critical path.

	A	B	C	D	E	F	G	H
1	Activity	Predecessor	Successor	T_n (months)	T_c (months)	C_n (thousands)	C_c (thousands)	Maximum crashing possible (months)
2	A	---	C	7	6	20	35	1
3	B	---	D	9	7	15	25	2
4	C	A	***	6	5	10	18	1
5	D	B	***	7	5	21	32	2
6								
7	Calculation of critical path							
8	Activity	Predecessor	Activity crashed by	Activity duration	EST	EFT	Successor	LFT
9	A	---		7	0	7	C	10
10	B	---		9	0	9	D	9
11	C	A		6	7	13	***	16
12	D	B		7	9	16	***	16
13	Project completion time		16					
14								

FIGURE 8.16 Solver.

times by which each activity has been crashed shown from C9 to C12. These crashed time values should be less than or equal to time indicated by maximum crashing possible from cell H2 to H5. This forms the constraint. By using this information, go to solver function and set *objective* as cell C13. Objective of minimization is selected. Decision variables shown from C9 to C12 are selected as *by changing variable cells. Constraints* would be C9:C12 ≤ H2:H5. Also to get only integer value of decision variables, another constraint is inputted as C9:C12=int (Figure 8.16).

Solver solution is shown in the figure. Project completion time is reduced from 16 to 12 months. Also time by which each activity is crashed is calculated (Figure 8.17).

Step 4: *Solver solution to find minimum project cost:* The second aim was to find total project cost. It is a combination of two parts: total normal cost and cost of crashing. Total normal cost is the sum of C_n of all activities. Cost of crashing is function of time by which activities have been crashed shown in C9:C12 and slope shown in J2:J5. Thus, total crashing cost is calculated in

	A	B	C	D	E	F	G	H	I	J	K
1	Activity	Predecessor	Successor	T_n (months)	T_c (months)	C_n (thousands)	C_c (thousands)	Maximum crashing possible (months)	Cc - Cn (thousands)	Slope	
2	A	---	C	7	6	20	35	1	15	15	
3	B	---	D	9	7	15	25	2	10	5	
4	C	A	***	6	5	10	18	1	8	8	
5	D	B	***	7	5	21	32	2	11	5.5	
6											
7	Calculation of critical path										
8	Activity	Predecessor	Activity crashed by	Activity duration	EST	EFT	Successor	LFT	LST	TS	Indirect cost
9	A	---	1	6	0	6	C	7	1	1	30
10	B	---	2	7	0	7	D	7	0	0	35
11	C	A	1	5	6	11	***	12	7	1	25
12	D	B	2	5	7	12	***	12	7	0	25
13	Project completion time		12								

FIGURE 8.17 Solution.

	A	B	C	D	E	F	G	H	I	J	K
1	Activity	Predecessor	Successor	T_n (months)	T_c (months)	C_n (thousands)	C_c (thousands)	Maximum crashing possible (months)	Cc - Cn (thousands)	Slope	
2	A	---	C	7	6	20	35	1	15	15	
3	B	---	D	9	7	15	25	2	10	5	
4	C	A	***	6	5	10	18	1	8	8	
5	D	B	***	7	5	21	32	2	11	5.5	
6											
7	Calculation of critical path										
8	Activity	Predecessor	Activity crashed by	Activity duration	EST	EFT	Successor	LFT	LST	TS	
9	A	---	1	6	0	6	C	7	1	1	
10	B	---	2	7	0	7	D	7	0	0	
11	C	A	1	5	6	11	***	12	7	1	
12	D	B	2	5	7	12	***	12	7	0	
13	Project completion time		12								
14	Total Normal cost		66								
15	Total crashing cost		44								
16	Total Project cost		110								
17											

FIGURE 8.18 Total project cost.

cell C15 by using formula = sumproduct(C9:C12,J2:J5). The results of these formulas are shown in Figure 8.18.

The aim is to find whether total project cost of 110 can be reduced further. To do so, solver can be used again with objective being to minimize objective in cell C16 = C14 + C15 by varying decision variables from C9:C12. In addition to earlier constraints, one more is added involving project completion time, which has been found to be 12 months. So, it would be C13 = 12. These are shown in Figure 8.19.

Step 4: Solver solution is shown in Figure 8.20.

Solution shows change in activities being crashed by how much time, for instance now activity has not been crashed and it doe stake normal time for its completion. Also, solution provides minimum possible total project cost by adding normal cost with crashing cost.

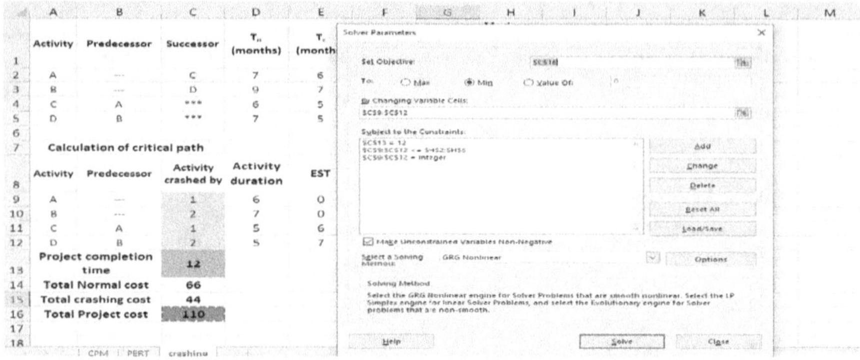

FIGURE 8.19 Solver.

	A	B	C	D	E	F	G	H	I	J
1	Activity	Predecessor	Successor	T_n (months)	T_c (months)	C_n (thousands)	C_c (thousands)	Maximum crashing possible (months)	Cc - Cn (thousands)	Slope
2	A	---	C	7	6	20	35	1	15	15
3	B	---	D	9	7	15	25	2	10	5
4	C	A	***	6	5	10	18	1	8	8
5	D	B	***	7	5	21	32	2	11	5.5
6										
7	Calculation of critical path									
8	Activity	Predecessor	Activity crashed by	Activity duration	EST	EFT	Successor	LFT	LST	TS
9	A	---	0	7	0	7	C	7	0	0
10	B	---	2	7	0	7	D	7	0	0
11	C	A	1	5	7	12	***	12	7	0
12	D	B	2	5	7	12	***	12	7	0
13	Project completion time		12							
14	Total Normal cost		66							
15	Total crashing cost		29							
16	Total Project cost		95							

FIGURE 8.20 Solution.

8.6 RESOURCE PLANNING

Till now, we have discussed methods of project scheduling, namely PERT and CPM, based on time duration of each activity. In these methods, it was assumed that resources required to conduct an activity such as labour, equipment, material, etc. are readily available as and when required. But in reality, the availability of resources is always a constraint. The number of activities involved in project competes for similar resources. Certain activities are performed simultaneously and certain in sequential order. Activities performed in parallel require limited available resources at the same point in time making their performance difficult. In such cases, activities have to be prioritized in terms of resource allocation delaying other activities. This can lead to alteration of project completion time as activities are rescheduled for later date. Which activities needs to be prioritized and how many resources are to be allocated to each of them are some issues that are discussed in resource planning.

Network diagram is again a helpful tool in planning of resources by understanding concepts discussed earlier such as project completion time, start and finish times of each activity and their TSs. Network diagram is particularly useful as it is drawn by considering logical sequence of activities resulting in representing the interrelationship between activities. For instance, in computer assembly project activities of monitor assembly, CPU assembly and testing needs to be done in particular sequence as shown in the following diagram. This sequence cannot be altered as testing cannot be done before assembly.

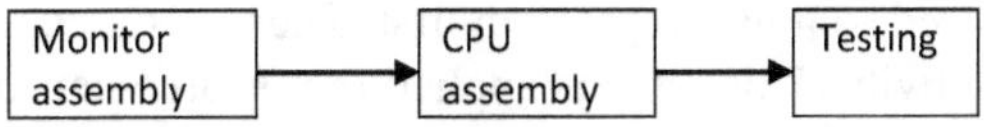

However, some activities need to be performed in parallel. For instance, a class of quantitative techniques for three different sections (QT1, QT2 and QT3) can be arranged between the same time period of 09:00 to 10:00 am, provided three different teachers are available. If it is done so, then the following network diagram is possible.

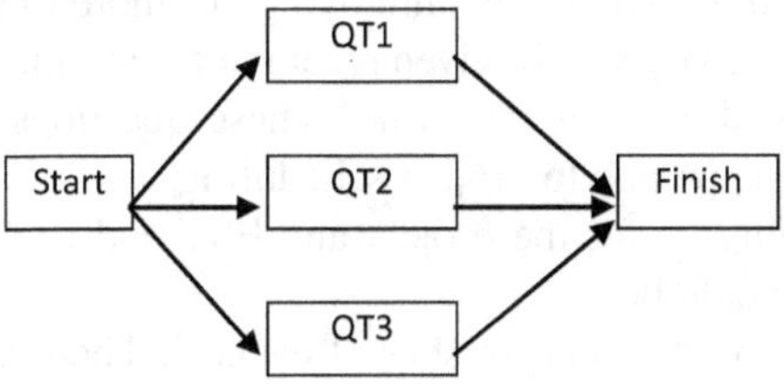

But, in this case, though project of teaching three different classes finishes within scheduled 1 hour, but it requires more resources, i.e. three teachers. On the other hand, if all three sections are taught in sequence rather than simultaneously then network diagram would be like:

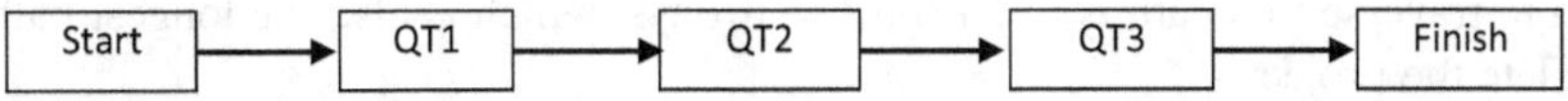

In this case, though project takes 3 hours to finish, but it utilizes only one teacher (resource) rather than three. Thus, there is a trade-off between project completion time and allocation of resources. More resources allocated for the performance of selected activities can pre-pone or complete the project within scheduled time, whereas less resource availability can delay the project. This planning of resources is important and done in the following two ways.

8.6.1 Resource Limited Scheduling

As the name suggests this type of resource planning deals with problems when the performance of each activity is constrained with limited available resources. This limit can delay completion of project from its scheduled completion time.

For instance, due to budgetary constraints, if department cannot hire more teachers to teach quantitative techniques, then project is finished from 09:00 am to 12:00 noon, i.e. in 3 hours rather than in 1 hour.

Projects involving multiple activities are scheduled in resource-constrained environment by understanding the significance of TS of each activity. Network analysis techniques helped us to calculate slack indicating activities that can be delayed (non-critical activities) without effecting the project completion and which cannot be delayed (critical activities). The amount of slack implies priority of each activity. The higher the slack, the lower is priority granted to the activity. An activity with zero slack cannot be delayed so it has to perform first. Thus, resource would also be allocated first to such activity. This logic is applied in resource-constrained scheduling problems. Activities with least or zero slack (high-priority activities) are allocated resources first and then, if any left, resources are allocated to activities with the next least slack (low-priority activities). If, after allocation to high-priority activities, no resources are available, then low-priority activities would be delayed till resources are freed. This delay might cause delay in project completion.

This is illustrated by taking an example. Consider a project with five activities with following precedence, time and labour requirement (Table 8.12). The project manager needs to finish the project by employing not more than eight workers. Is it possible to carry each activity within given resource constraint of eight labours? Can the project complete within scheduled time? These questions can be answered by applying the following approach to project scheduling.

Step 1: *Network diagram* for the project and EST and LFT of each activity are shown in Figure 8.21 would be:

Project starts from event 1 and finishes at event 4. There can be three possible paths from start to end:

Path 1: A---D

Path 2: A---C---E

Path 3: B---E

Step 2: Calculation of TS (Table 8.13)

Critical path is the path with least or zero slack that in this case is A---C---E. Time taken to traverse this path is = 2+6+2 = 10 days, which is also the longest path to complete the project.

Step 3: Now let's draw on a time scale. Draw a resource diagram (Figure 8.22), which shows the amount of labour consumed by each activity on a time scale.

TABLE 8.12
Resource Data

Activity	Precedence	Time Duration (Days)	Labour
A	---	2	8
B	---	3	4
C	A	6	3
D	A	6	2
E	B, C	2	5

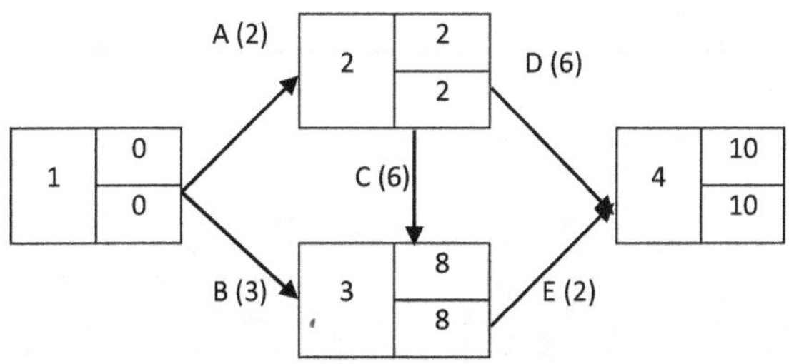

FIGURE 8.21 Network diagram.

TABLE 8.13
Calculation of Total Slack

Activity	Precedence	Time Duration (Days)	Labour	EST	LFT	EFT	LST	TS
A	---	2	8	0	2	2	0	0
B	---	3	4	0	8	3	5	5
C	A	6	3	2	8	8	2	0
D	A	6	2	2	10	8	4	2
E	B, C	2	5	8	10	10	8	0

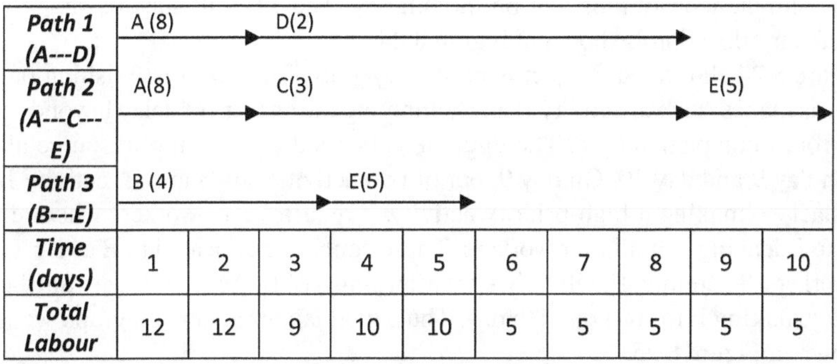

FIGURE 8.22 Resource diagram 1.

The values in brackets of each path as shown in the following figure indicate the number of labour required for a particular activity.

According to Figure 8.22, there is a problem during days 1, 2, 3, 4 and 5 as more workers are required to accomplish several activities than available. Modification of labour allocation as discussed is done based on the priority of activity determined by its slack. On day 1, two activities are initiated, i.e. A and B. Activity A has a slack of 0, whereas B can be delayed by 5 days. Thus, first, A would be allocated with labour as required by it. In this case, it requires 8 which are equal to available. As further, no labour is left so B has to wait till A gets finished, which happens after 2 days delaying start of B by same amount. New labour allocation is shown in Figure 8.23.

Path 1 (A---D)	A (8)		D(2)							
Path 2 (A---C---E)	A(8)		C(3)						E(5)	
Path 3 (B---E)			B (4)			E(5)				
Time (days)	1	2	3	4	5	6	7	8	9	10
Total Labour	8	8	9	9	9	10	10	5	5	5

FIGURE 8.23 Resource diagram 2.

As B has utilized 2 days from TS of 5 days, so now its slack would be 3 days. According to Figure 8.23, the problem of resource allocation persists from day 3 to day 7. Starting from day 3, activities B, C and D are initiating. B has a slack of 3, C has 0 and D has a slack of 2 days. Therefore, C becomes high-priority activity and is allocated first. It requires three labourers, so out of total available eight labourers, five are left which would be allocated to next activity on priority list which is D. After allocating two labourers to D, only 3 are left but next activity i.e. B requires four workers. So B cannot start on day 3 and has to be delayed till labour is freed from other activities. Labour gets free after conduct of C and D on the 8th day. Both activities are simultaneously finishing on the 8th day. Thus, B will start on the 9th day. New labour allocation is shown in Figure 8.24.

Figure 8.24 shows that project is now delayed by 3 days as it is finishing on 13th day. This makes slack of activity B as -3, implying the amount of delay beyond scheduled project completion time. The figure also poses the problem of resource allocation on day 9 and day 10. On day 9, out of two activities of B and E, activity B has least slack so making it high-priority activity. It requires four workers who are allocated to B leaving behind four workers. These cannot be allocated to E as it requires five workers. Therefore, E will only start after finish of B. Activity B gets finished on 11th day making E to start on 12th day. Thus, final labour allocation graph would be as shown in Figure 8.25.

Path 1 (A---D)	A (8)		D(2)										
Path 2 (A---C---E)	A(8)		C(3)						E(5)				
Path 3 (B---E)									B(4)			E(5)	
Time (days)	1	2	3	4	5	6	7	8	9	10	11	12	13
Total Labour	8	8	5	5	5	5	5	5	9	9	4	5	5

FIGURE 8.24 Resource diagram 3.

Path 1 (A---D)	A (8)		D(2)										
Path 2 (A---C---E)	A(8)		C(3)									E(5)	
Path 3 (B---E)									B(4)			E(5)	
Time (days)	1	2	3	4	5	6	7	8	9	10	11	12	13
Total Labour	8	8	5	5	5	5	5	5	9	9	4	5	5

FIGURE 8.25 Resource diagram 4.

Thus, in resource-constrained scheduling, project is delayed, but due to limited resource, no activity consumes more than available resource.

8.6.2 Resource Levelling

In some cases, variable resources (more or less) are available for the performance of each activity in the project but are limited by scheduled completion time of project. In these cases, project cannot be delayed, but the number of resources allocated is not constrained. However, this can lead to other kind of problem in which certain activities being performed simultaneously require a high allocation of resources, resulting in a peak, whereas activities being performed sequentially require a few labour allocation of resources leading to a trough in labour demand. For instance, in roof installation project discussed under the PERT section, fixing of channels or roof on both sides required higher amount of labour as compared to other activities such as procurement of raw material. Thus, the project manager faces the problem of high and low labour requirement during the conduct of project. The manager cannot hire or fire frequently, so instead of paying overtime during peak demand or keeping idle workers during low demand, more uniform or levelled allocation of labour is preferred.

This method of developing a schedule that minimizes fluctuation of labour or other resource requirements is called resource levelling.

Resource levelling does not allow an increase in project completion time, so in the process of levelling of resources, the first focus is on delaying of non-critical activities by the amount of their TS. Delay of such activities beyond their EST but not after their LST would not alter project completion time. Out of multiple non-critical activities, activity with maximum positive slack is first delayed. This delaying of activities would most likely bring down peak demand for labour and shift them to activities with low demand.

The resource levelling process is illustrated by taking the same example shown in Table 8.12. Fluctuations in labour allocation are also represented by the bar chart developed from Figure 8.26.

Resource levelling intends to smoothen the peaks and troughs. For this reason, the method is sometimes also termed as *resource smoothening.* The first step in resource

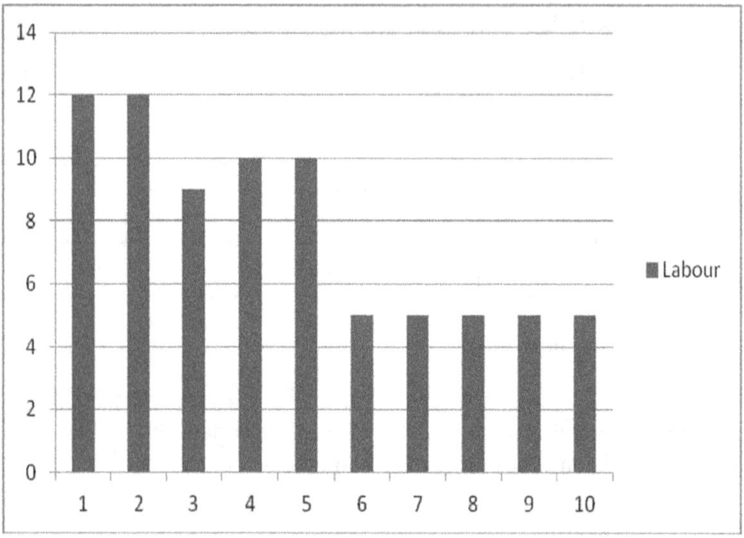

FIGURE 8.26 Bar chart of labour allocation.

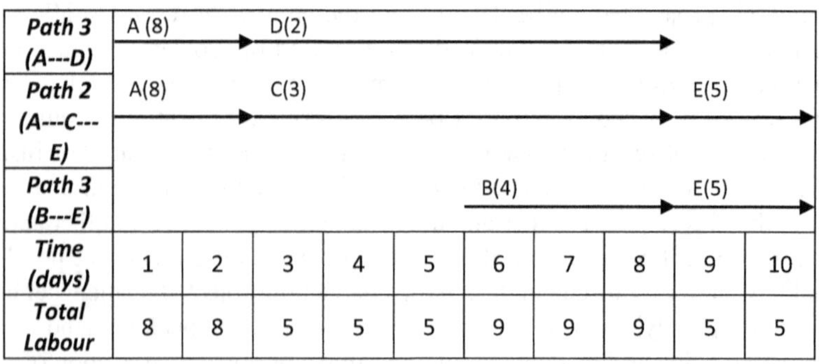

FIGURE 8.27 Labour allocation 1.

levelling is to delay non-critical activity with maximum slack. Activity B has a maximum slack of 5 days that should be delayed by the same amount. Figure 8.27 shows new labour allocation.

Bar chart for such resource allocation is shown in Figure 8.28.

Another alternative could be to delay D as non-critical activity by its TS amount that is 2 days. Figure 8.29 shows resource allocation by doing so.

Bar chart for such resource allocation is shown in Figure 8.30.

The project manager can decide of proper resource allocation among two given alternatives. With the increase in complexity of projects and commensurate increase in project activities process of resource levelling becomes hugely complex and difficult. The usage of various available project management software can be helpful in solving such issues.

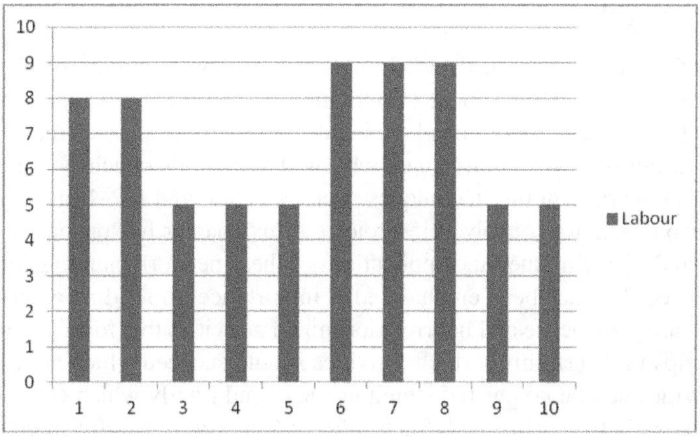

FIGURE 8.28 Bar chart of labour allocation 1.

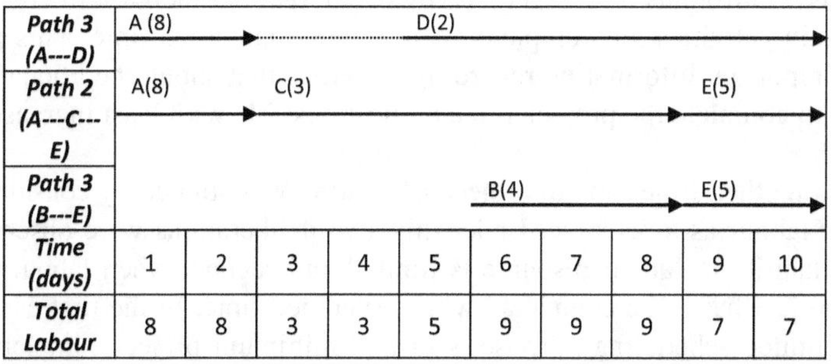

FIGURE 8.29 Labour allocation 2.

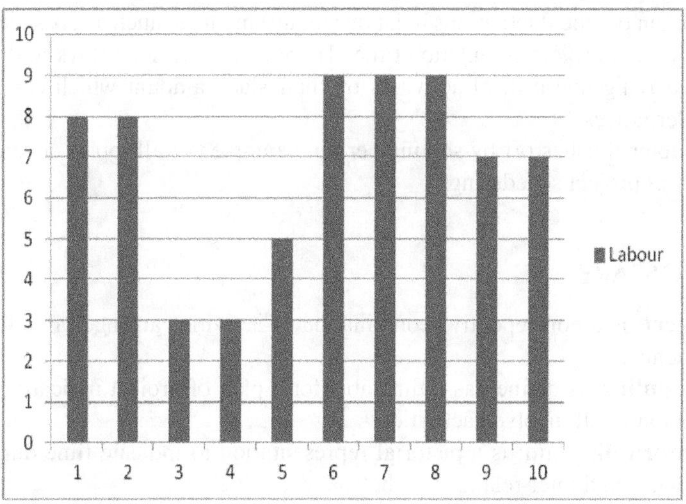

FIGURE 8.30 Bar chart of labour allocation 2.

8.7 SUMMARY

Planning of project is a critical task in successful implementation of a project. It involves a systematic arrangement of activities for the accomplishment of project objective. This critical aspect involves breaking down the project into sub-tasks, assigning responsibility to individuals who will accomplish each of sub-task and then using network planning techniques such as PERT and CPM to schedule and control various activities involved in projects. This chapter has primarily focussed on understanding techniques and applications of these network planning techniques. The first aspect that has been emphasized is importance of building a network diagram illustrating sequence and interrelationship of activities that form a project. This diagram helps in determining which activities should succeed which activity, which of the activities can be conducted simultaneously and finally which cannot be start till previous activity has finished. Such information is very helpful in determining most important and least important path that leads to finish of project. The calculation of various activity times and their floats has also been highlighted.

Schedule control has been discussed by understanding increase in cost of project if certain activities are completed before their scheduled time. This provides manager important information regarding activities that should be allocated extra resources to complete the project at the earliest possible with least increase in cost possible.

Finally, another dimension to project scheduling was studied by considering the number of labour as a resource. Under this, two deliberations were observed: one when availability of labour resource is limited and second, when labour is available but project has to be completed within specified time. In the first case, i.e. of resource limited scheduling purpose is to find minimum project completion time when amount of labour is fixed. It is an iterative method where labour is allocated first to activities with least slack. Most likely in these cases project completion time is exceeded then scheduled time. In the second case, i.e. resource levelling purpose is to smoothen out the fluctuations in labour requirement as much as possible without overshooting the project completion time. There can be various ways to do so as it involves delaying non-critical activities by their slack amount which can result in various alternatives.

The chapter finishes off by solving certain examples of all topics covered under the subject of project scheduling.

8.8 GLOSSARY

- **Project**: is a non-repetitive combination of activities arranged in a logical sequence.
- **Scheduling**: is defined as a timetable for a plan of project indicating time duration of all involved activities.
- **Network diagram**: is a pictorial representation to indicate time duration, sequence and inter-relationship between activities.
- **A-o-A:** method uses arrow to show activities and events are represented by nodes.

- **A-o-N:** method uses nodes to represent activities whereas arrow indicates an event.
- **Dummy activities**: are used only to show a logical sequence or a linkage between two events without consuming any resources.
- **Program Evaluation and Review Technique (PERT)**: is a probabilistic time oriented network technique used to determine probability of completion of project.
- **CPM**: is a deterministic time oriented network technique used to determine critical and non-critical activities.
- **Forward pass**: is a procedure that involves moving forward in network diagram for determining the earliest start and finish times.
- **Backward pass**: is a procedure that involves moving backwards in network diagram for determining the latest start and finish times.
- **Crashing**: is a method of shortening of activity times by adding resources leading to increase in costs.
- **Resource limited scheduling**: This type of resource planning deals with problems when performance of each activity is constrained with limited available resources. This limit can delay completion of project from its scheduled completion time.
- **Resource levelling**: This method involves developing a schedule that minimizes fluctuation of labour or other resource requirements.

8.9 CASE STUDY: POLYPLASTICS INDUSTRIES INDIA PVT. LTD.

Polyplastics Industries India Pvt. Ltd. is a Leading Plastic Automobile Components Manufacturer with a turnover of 70 million USD in year 2018–2019. The Group has a diversified product range to serve multiple industries, with automotive industry being the main industry served. The company is supported by in-house design and development facilities like modern tool room for manufacturing of moulds, dies, jigs and fixtures, and design facility. The Group has a diversified product range to serve multiple industries, with automotive industry being the main industry served. It manufactures Emblems (Electroplated, Painted, Gold Plated & Hot stamped), Automotive Plastic moulded components, Wheel Trims & Wheel Rim Covers, Electroplated Bigger Parts (Radiator Grills. License Plate Garnish, Hood Strips), Decorative Body side moulding, Assemblies Control Brackets, Dash Board Components, Auto Electricals Assemblies, Door Handle, Fuse Box Assemblies, Diesel Water Separator, etc.

Due to an increasing number of products and product variants that have to be offered to the customer and a decrease of the corresponding order quantities, production has to be able to react very quickly. Lots of production time is consumed in mould change over or mould setup time. Since direct production costs are related to the machine performance, an overall equipment effectiveness calculation can easily show the impact of setup reduction on overall machine performance. Setup time needs to be minimized to maximize the capacity available for production.

Shop floor manager of the company faced with problem of reducing change over time of moulds on a 50-ton injection-molding machine. The company used the same machine to manufacture a minimum of three variants of car door handle per company. With an increase in number of consumers (car companies), the variety of door handles increases leading to higher changeover of moulds on injection molding machine. Thus, in order to minimize the cost, improving bottleneck capacities and being flexible to fulfill the customer demand, it is necessary to minimize the change over time.

This purpose was accomplished by applying Single Minute Exchange Die technique of lean manufacturing. The first two steps involved in this technique were data collection regarding set up times of each part for which mould is prepared on injection molding machine and second listing down all activities required in change of mould. It is important to clarify here that changeover of mold would require unloading of mold used to manufacture first part and then setting of new mold for second part and so on. In this case, we have listed down activities only for unloading of mold and obtained times in minutes for each of the activity. The activities applied in unloading of mold pertain to manufacture of three door variants of a car company. Thus, scheduling of one part can be applied to other parts as well. The following is a list of activities with time performed for mould change over.

Activity	Description	Predecessor	Time (Minutes)
A	Collect necessary tools for mold change	---	3
B	Prepare injection unit	A	1
C	Remove pipe from mold	B	2
D	Anti-rust spray application on core and cavity of mold	B	1
E	Lock plate	C	1
F	Fixation of i-bolt	D, E	2
G	Remove clamp from core	F	3
H	Remove clamp from cavity	F	3
I	Remove mold from machine	G, H	1

Consider yourself to be the shop floor manager and prepare a schedule of all these activities to find out which activities are critical and which are non-critical. Activities with certain amount of slack, can be either delayed or combined to reduce project completion time. The identification of such activities is important.

Consider times given in the above table as times obtained when activities are performed under normal conditions. The data was obtained from process owner, but when actual data was collected by visiting the shop floor times showed some variation. The following table grouped these times for both ideal and worst conditions under which listed activities can be performed.

Activity	A	B	C	D	E	F	G	H	I
Time (minutes) under ideal conditions	1	1	2	1	1	1.5	2	2	1
Time (minutes) under worse conditions	4	2	3	2	2	3	3.5	3.5	1.5

Considering yourself as shop floor manager find out the probability of finishing the project 1 minute earlier and 1 minute late than estimated time.

8.10 MODEL QUESTIONS

8.10.1 For the following project involving six activities

i. Draw network diagram by using both A-o-A and A-o-N format. Explain the difference
ii. Also calculate EST, LFT, EFT and LST and determine project completion time.
iii. What is the probability that project will finish in 50 days?

Activity	A	B	C	D	E	F
Predecessor	---	---	---	A	B, D	B
Time (days)	3	10	5	10	12	2

8.10.2 For the following project:

i. Draw network diagram by using A-o-A format.
ii. Calculate EST, LFT, EFT and LST and determine project completion time.
iii. Find total, free and IS of each activity.

Activity	A	B	C	D	E	F	G	H	I
Predecessor	---	A	A	C	B	D, E	F	D, E	G, H
Time (minutes)	4	5	2	6	9	6	2	9	11

8.10.3 For the following project:

i. Draw network diagram by using A-o-A format.
ii. Calculate EST, LFT, EFT and LST.
iii. Find TS, critical path and project completion time.

Activity	A	B	C	D	E	F	G
Predecessor	---	---	A	A	B	B	D, E
Time (minutes)	10	20	5	2	10	5	2

8.10.4 ICS ltd. bagged a project to develop a banking solutions project for an international bank. Activities their inter relationship and performance time in weeks is shown in the following table. Project starts with activities A, B and C initiating simultaneously and finishes with completion of I and J. find out:

- Critical activities
- Activities that can be delayed without delaying the project
- Probability that project will not more than 75 days to complete.

Activity	A	B	C	D	E	F	G	H	I	J
Predecessor	---	---	---	B	A, C, D	B	C, D	E	F, G	H
Time (weeks)	11	13	17	15	18	12	10	14	12	14

8.10.5 Preparation of a birthday party involves various activities such as preparing invitation cards, purchasing decoration material, selecting gifts, deciding and ordering birthday cake, organizing games for kids, cake cutting ceremony, distributing return gifts and finally cleaning the venue. These activities with the precedence relationship and time in hours are shown in the following table.

Activity	A	B	C	D	E	F	G	H
Predecessor	---	---	A	B, C	D	D	F	E, G
t_o (hours)	2	4	1	3	2	1	5	2
t_m (hours)	3	6	2	5	4	4	6	4
t_p (hours)	4	8	3	7	6	4	7	6

For this project

- Find expected time of each activity.
- Draw network diagram if it starts with A and B and finishes when H and G are completed.
- Find TS of each activity, critical path and project completion time.
- What is the probability that entire birthday function can finish in less than 12 hours?
- Find the probability of project taking not more than 2 hours of completion time.

8.10.6 For the following project:

- Find expected time of each activity.
- Find TS of each activity, critical path and project completion time.
- Find the probability of project completing in less than 3 hours of completion time.

Activity	A	B	C	D	E	F	G	H
Predecessor	---	---	A, B	A, B	B	C	D, F	E, G
t_o (hours)	5	3	6	8	4	1	5	6
t_m (hours)	6	5	9	10	5	2	7	7
t_p (hours)	8	7	10	11	9	3	11	9

8.10.7 Time-cost trade-offs are shown in the following table of a project involving activities from A to H. Project starts with A and B taking off simultaneously and finishes with completion of G and H. project manager wants to find the least possible time in which project can be completed with increase in cost being minimal. Indirect cost per week of the project was estimated to be $500.

Activity	Predecessor	Normal Time (Weeks)	Crash Time (Weeks)	Normal cost ($'000)	Crash Cost ($'000)
A	---	4	2	15	25
B	---	2	1	5	15
C	A	3	1	12	20
D	A	5	2	20	38
E	B	4	3	17	25
F	B	6	3	25	43
G	C, E	8	4	20	40
H	D, F	7	5	28	42

8.10.8 The data needed to apply crashing method for a project is given in the following table. Find:

- Critical path by estimating TS
- Normal project completion time
- Least economical way of crashing the project.

Activity	Predecessor	Normal Time (Weeks)	Crash Time (Weeks)	Normal cost ($'000)	Crash Cost ($'000)
A	---	22	18	60	80
B	---	18	15	25	46
C	A	26	21	50	90
D	---	6	3	40	52
E	B	22	17	25	50
F	B	40	37	14	32
G	D	7	5	34	40
H	E, G	10	7	20	32
I	E, G	24	20	44	70
J	C, F	8	6	30	34

8.10.9 The following activities have been identified to carry out a project. Their sequence, time in minutes and labor requirements are also shown.

i. Draw network diagram and find critical and non-critical activities.
ii. Find out the schedule if labor availability is limited to (a) four workers (b) three workers.
iii. Also find the schedule that smoothens labor fluctuations.

Activity	1–2	1–3	3–4	4–5	5–6	2–6	3–5
Time (minutes)	3	3	3	4	5	7	5
Labor	2	3	1	2	2	1	2

8.10.10 The following activities have been identified to carry out a project. Their sequence, time in minutes and labor requirements are also shown.

i. Draw network diagram and find critical and non-critical activities.
ii. Find out the schedule if labor availability is limited to five workers.

Activity	1–2	2–3	2–4	2–5	3–7	4–6	5–6	6–7	7–8
Time (minutes)	2	3	4	3	3	3	3	2	1
Labor	5	2	2	2	2	1	2	1	3

8.10.11 Seven activities numbered A to G were identified in the project of converting seminar hall into computer lab for conduct of a competitive exam. The project begins with simultaneous performance of activities A and B and finishes when both F and G are completed. Precedence relationship and time in minutes of each activity is shown in the following table.

i. Construct network diagram and find critical path by finding TS of each activity.

Activity	A	B	C	D	E	F	G
Predecessor	---	---	A	A	C, B	C, B	D, E
Time (minutes)	30	35	10	50	20	30	40

ii. What would happen to critical path if precedence relationship of activities changes as shown in the following table.

Activity	A	B	C	D	E	F	G
Predecessor	---	---	A, B	A, B	C	C	D, E
Time (minutes)	30	35	10	50	20	30	40

8.10.12 A new training course was developed by a business school to impart analytical skills to doctoral degree aspirants. As the course has been designed for the first time so coordinator has estimated optimistic, most likely and pessimistic time of each activity involved in the conduct of course. The activities, their interrelationship and time in days are shown in the following table. The coordinator wants to estimate:

i. Expected time and variance of each activity.
ii. Critical path and TS.
iii. Probability of completion of course 1 day before expected time.
iv. Probability of completion of project 2 days after expected time.

Activity	Description	Predecessor	Optimistic time (t_o)	Most likely time (t_m)	Pessimistic time (t_p)
A	Formulate training objectives	---	1	2	3
B	Install computers and other infrastructure in training room	---	2	2	3
C	Prepare brochures	A	2	3	4
D	Promote course	B, C	1	2	3
E	Recruit qualified trainers	A, D	1	1	2
F	Identify participants	A, D	1	1	2
G	Schedule classes	E	1	1	1
H	Deliver course	F, G	9	10	12
I	Test and evaluate	H	1	1	2

8.10.13 Consider a four activity (A, B, C & D) project where project starts with A and C simultaneously. A is followed by B and C by D. Project finishes after completion of both B and D. If all activities are performed according to normal time then project takes 14 days to complete. On the other hand, if by expending more resources project duration is reduced to 11 days then total cost increases to \$212,000 from total normal cost of \$160,000. Is this increase in cost is relevant or can project be completed in 11 days by spending lesser amount than \$212,000. The following table shows normal and crash time and cost of each activity.

Activity	T_n (Days)	T_c (Days)	C_n (in Thousands)	C_c (in Thousands)
A	5	3	40	52
B	7	4	70	90
C	8	7	30	40
D	6	4	20	30
Total			\$160,000	\$212,000

8.10.14 The following table provides data regarding a six activity project with their time and labour requirements.

Activity	A	B	C	D	E	F
Predecessor	---	---	---	C	A	B, D
Time (days)	5	4	4	5	4	4
Labour	2	1	3	2	1	3

Find:

i. Critical path and project completion time
ii. Change in project completion time if it is constrained by availability of labour to maximum of three workers.
iii. How much labour requirement can be smoothened if project has to be finished within estimated time?

9 Game Theory

9.1 INTRODUCTION

Game theory facilitates understanding and examining of relationships between at least two parties. The essence of game theory is that parties involved in a relationship have opposing goals or objectives. Such opposition results in conflicting situations wherein one player tends to win and the other player tends to lose. To avoid loss and gain a win, both players have to make a decision regarding selection of a particular strategy from available different strategies. Game theory endeavours to provide a way of selecting most optimum strategy that would allow simultaneous attainment of twofold purpose of maximization of gains and minimization of losses.

A game requires at least two players, so any action by one player causes a reaction from the other player, which results in an outcome. Thus, any outcome is a combination of action and reaction by two players where each player would attempt to win, but ultimately only one would gain and the other would lose. This phenomenon of tackling two purposes simultaneously differs from other decision-making concepts. For instance, in linear programming, the purpose of maximization or minimization is examined individually and not simultaneously.

This indicates that an outcome is strategy-dependent. Players can confront two kinds of strategies pertaining to particular game situations that are discussed in this chapter. One situation deals with a duel kind of situation in which a player has a definite strategy to play that results in an absolute winner or loser. For instance, in a game of chess, tennis or politics involving two players always results in a clear and absolute winner and loser without any intermediate provision. On the other hand, in various situations, such as business and geopolitics, there are no clear-cut winners and losers. Two opposing parties tend to cooperate to reach a solution acceptable to both. For instance, in a vendor and retailer relationship, the ideal situation for a vendor would be to provide low-quality goods at high prices so that it can attain maximum profit. Similarly, the ideal situation for retailers would be to procure the best quality products at the lowest price to maximize the gains. However, both situations are not acceptable. Therefore, vendors, retailers though opposing, and competing parties cooperate to achieve a satisfactory solution. This satisfactory solution might not be a result of a *definite* strategy as was in the first case but a *combination* of alternative strategies.

The first situation is termed as a zero-sum game and the second as a mixed strategy. Both of these situations have been discussed in this chapter. First, the following section is devoted to understanding the characteristics and elements of a game.

DOI: 10.1201/9781003212966-9

9.2 CHARACTERISTICS OF GAME THEORY

1. A game involves *at least two players* where a player can be an individual or a team. For instance, a game of chess has two players, whereas a game of cricket has two teams. In business, two companies compete against each other to capture more customers.
2. Strategies adopted by both players are *opposing and diametrically opposite* to each other. In a game, a player plays a strategy not only to fulfil his/her purpose of maximizing the gain but also to inflict maximum damage in terms of losses to the opponent.
3. Both players are *rational and independent*. Rationality implies that a player can comprehend strategies available and played by the other player. For instance, if a vendor tries to deceive the buyer by providing low-quality products, then vendor can comprehend the reaction of buyer that could be either acceptance or rejection of the order. In addition, selection of a particular strategy requires allocation of limited resources. A player is independent in allocating the amount of resources to various strategies depending on each strategy's effectiveness.
4. Each player has *multiple strategies* at his/her disposal to fulfil the objective of maximum gain or minimum loss. As said, the selection of a particular strategy depends on the availability of resources as well as rules of game. For instance, in a game of cricket, each team cannot have more than 11 players, the game has to be finished within a stipulated time period, etc. Similarly, in a business game, if two are competing for more brand awareness, then each of them has a limited marketing budget to promote their brand by spending on different media. The purpose would be to achieve maximum brand awareness at the expense of competing brands.
5. A strategy's effectiveness is judged by its outcome. As two players are arranged in a duel-like situation, so an outcome is always a combination of the particular strategy of two players. This outcome is termed as *payoff*. Thus, a payoff is directly dependent on the opponent's action. In the case of games played only once, payoff indicates the present value of outcome. For instance, in vendor–retailer relationship, a vendor might be able to deceive a retailer by supplying low quality goods, but it would lead to termination of relationship with no future orders. On the other hand, if a game is played repeatedly, the payoff implies the present value of all current and future outcomes. A vendor to maintain a long-term relationship and indulge in repeat orders should provide (present) and keep on providing (future) good quality products. In this chapter, all explanations pertain to *games played only once*.
6. Every payoff is examined and compared with other payoffs by applying a method of *valuation*. For instance, selection of particular media for increase in brand awareness can be measured in terms of sales or increase in revenue. In politics, the number of votes generated by different media campaigns evaluates the effectiveness of those campaigns. These valuation methods are termed as utility function.

9.3 ELEMENTS OF GAME THEORY

Following payoff matrix is helpful in understanding elements of a game.

		Player B			
Strategies		**B_1**	**B_2**	**----**	**B_m**
Player A	A1	a_{11}	a_{12}	----	a_{1m}
	A2	a_{21}	a_{22}	----	a_{2m}
	----	----	----	----	
	An	a_{n1}	a_{n2}	----	a_{nm}

- As shown, the above game has two players: A and B.
- Each player can play by applying any of the available multiple strategies. For instance, A has A to An and B has B1 to Bm.
- Importantly, each combination of strategies results in a payoff. For instance, if A plays A1 and B plays B1, then outcome would be a_{11}.
- Both players are opponents, so the value of payoff indicates gain for one and loss for the other. Suppose, value of a_{11} is +5 it would indicate that by playing A1 and B1 A has gained five units and B has lost the same. On the other hand, −5 would imply A has lost five units and B has gained the same. This is explained in detail in the next section with an illustration.

9.4 SOLVING GAMES: 5G TECHNOLOGY

Fifth generation technology (5G) is the new technology providing faster Internet speeds with better reach and penetration than presently available 3G and 4G technology. Major telecom companies like AT&T, T-Mobile and Verizon in the US, China Unicom in China and Samsung, the South Koran telecom giant, have taken the lead in developing and providing 5G telecom services. In 2017, more than 34% of the Indian population accessed the Internet, which increased to over 52% by 2020. India, with 688 million Internet users by January 2020, makes it the second largest Internet market in the world only after China. Such exponential growth makes the Indian market very attractive to various telecom companies. Taking the cue, India's two largest players, Reliance Jio Infocomm and Bharti Airtel, have been developing their plans to roll out the 5G technology at the earliest. The companies considered government policy delay and lack of availability of adequate capital to major bottlenecks in efficient development of 5G technology. However, both companies, Reliance Jio (let's call it as A) and Airtel (B) are preparing to launch it on the same low band spectrum which provides great coverage and high speed of Internet.

Looking at socio-economic scenario, companies have initiated the process of creating this new technology at a similar point in time in an attempt to produce it first. They have directed their teams to create the product within 12 months of the start of the product development process. In such a competitive situation, both opponents would like to introduce the product in the marketplace to achieve first

TABLE 9.1
Payoff Matrix

		Company B		
Strategy		**Before 1 Months (B1)**	**Before 3 Months (B2)**	**Before 5 Months (B3)**
Company A	Before 4 months (A1)	3	4	2
	Before 6 months (A2)	2	3	1
	Before 8 months (A3)	4	−2	−1

mover advantage. It would result in a significant advantage in sales. Company A has estimated increase in sales of the new product under three scenarios, i.e. if the product is introduced 4, 6 or 8 months before launch deadline of 12 months. Similarly, company B also estimated its sales under scenarios of 1, 3 and 5 months before the launch deadline. As both companies are engaged in competition of to introduce the product first in the market, so the following payoff table indicates an increase in sales as a combination of two strategies. The strategies in this case are launch dates before the deadline. Managers of both companies want to find the most optimum strategy. Various strategies of each company and payoffs of each combination of strategy are shown in Table 9.1.

9.4.1 Explanation of Payoff Table

The payoff table indicates gain or loss by both players when each of them plays a particular strategy. In this case, values indicate an increase in sales in thousands. So, if A plays A1, i.e. the company is ready to launch the product 4 months before the deadline date and in response, B can launch it only 1 month before, then A would gain a sales of 3 (in thousands) more than B. Similarly A would gain sales of 4 and 2 if B responds by playing B2 and B3, respectively. It is important to note that expediting the process of product development asks for input of extra resources, which might make the entire process financially unviable. In addition, products produced in haste could have quality issues. This can be seen if A decides to launch a product at the earliest, i.e. 8 months before (A3) launch date. B would gain by launching either 3 or 5 months before, i.e. if B plays B2 or B3 it would most likely incur gain two or one units of sales than A. As both players in these companies are opponents, so gain of one is loss of the other. That is why, the gain of B is shown with a minus sign and that of A with a plus sign. The payoff table shows that A can still gain by playing A3 if B is quite late in responding, i.e. if it plays B1. As it has estimated, any flaws in the product can be rectified before B's launch of a similar product. Each strategy selected individually is termed as a pure strategy. Out of three strategies, A1, A2 and A3, company A would be least willing to select A3 as it has chances of incurring losses as compared to A1 and A2. On the other hand, B would most likely play B2 or B3 as B1 would definitely give it a loss, whereas these two strategies might provide him with some gain depending on A's strategy. This choice is done rationally as both companies know how the other will respond. This selection of the most appropriate strategy is solved in the next section.

9.5 GAMES WITH SADDLE POINT

As illustrated in the above example

- There are two players: companies A and B competing with each other to garner maximum sales by launching a new product at the earliest.
- Both A and B have three strategies to select from the most appropriate one. Company A's strategies are represented in different rows and of B's in columns.
- Row represents an evaluation of strategy by company B, which tells them what strategy company A would play to maximize gains. For instance, if company A decides to play A1, i.e. it intends to launch the product 4 months before the deadline, then B has the option to respond by playing either B1, B2 or B3. Positive value of payoffs indicates gain for company A and loss for company B. From all three strategies played by company B, it would always incur a loss. Therefore, to incur minimum loss, company B would adopt strategy B3 that causes it to lose two units. Company A gains the same amount. However, if company A decides to play A2 and launches the product before any of the dates by company B, then it would always gain irrespective of B's strategy. B would opt for a strategy that would incur it minimum loss that is B3 with a loss of 1 unit. Finally, if A expedites the launch to 8 months before the deadline, then it has a chance of incurring losses because of quality problems and company B would take its benefit by playing B2. It results in a gain of 2 and loss of the same amount to A (minus sign indicate loss to A and gain to B as both players are diametrically opposite to each other). Thus, from each row, the minimum value is selected. As these values indicate gain by A from evaluations by B, so A would play with the maximum value, implying maximum gain. Thus, A would play A1 and enjoys a gain of two units. *This principle is called as minimax rule.*
- On the other hand, column indicates the selection of the most appropriate strategy by company B that incurs a minimum loss. For instance, if B decides to play B1, then A would respond by playing either A1, A2 or A3. As its purpose is to maximize the gains, so it would select A3 with a gain of four units. In this case, A would always gain as its launch schedule is in all three strategies much earlier than 1 month before the strategy of B. Company B would like to play B2 because it intends to gain two units, but only if A plays A3. Company A, being a rational player, would, in response, play A1, which gives it a maximum gain of four units. Interestingly, when B launches 5 months before (B3), then A gains maximum even if it launches later than B, i.e. by playing A1, A gains 2, which is more than if it plays A2 or A3. Thus, while selecting a strategy for B, identify the maximum value from each column. However, the purpose of B is to minimize losses, so out

of identified maximum values, select the minimum value. Thus, B would play B3 and incur a minimum loss of two units. *This principle is called as maximin rule.*

The above explanation is summarized as below:

Minimax rule: From each row, select the minimum value. Out of all selected values, choose the maximum value. In the above case, it was strategy A1 with value of 2, indicating gain of 2 to A (Table 9.2).

Maximin rule: From each column, select the maximum value. Out of all selected values, choose the minimum value. In the above case, it was B3 with value of 2, indicating loss of 2 to B.

In this case Maximin = Minimax = 2

This equality of game value of both players is termed as a game having a *saddle point*, also known as equilibrium point. A game value of 2 being positive implies a gain of two units to A by playing strategy A1. This also indicates a loss of the same value to company B. Due to equivalence in gain and loss value, the game is said to have a *zero sum.*

Saddle point is termed as the point of equilibrium because both players does not have a better strategy either to increase its gains or decrease their losses. That is why a game with saddle point is a stable game as selection of another strategy would provoke a reaction from the opponent leading to a worse result. For instance, if instead of A1, company A decides to play strategy A2 as it can increase its gain to three units, then B will respond by sticking to strategy B3 as it would incur a loss of only 1 unit. On the other hand, player A's gain would be reduced from 2 to 1 unit. Suppose A decides to play A3 in the hope of gaining four units, but B, being rational and understanding A's strategy, would respond by playing B2. It would result in a gain of 2 to B and a loss of the same value to A. To avoid this situation, A would revert to A1, forcing B to play the original position, B3.

From company B's perspective, it will never play strategy B1 as it would incur loss in every situation. Suppose B decided to play B2 instead of B3 in the hope that

TABLE 9.2
Minimax and Maximin Rule

		Company B			
Strategy		**Before 1 Months (B1)**	**Before 3 Months (B2)**	**Before 5 Months (B3)**	**Minimum Value from Each Row**
Company A	Before 4 months (A1)	3	4	2	2
	Before 6 months (A2)	2	3	1	1
	Before 8 months (A3)	4	−2	−1	−2
	Maximum value from each column	4	4	2	

it might gain two units, but now A being rational it will shift to A2. It would imply an increase in its gain to three units and inflict the same loss to B. Company B in an attempt to avoid this loss would shift back to playing B3, while company A would play A1 which was the original strategies.

9.5.1 Principle of Dominance

To identify the most appropriate strategy for each player, the payoff table can also be deduced by using the principle of dominance where in inferior strategies are removed in succession until only one choice is left. The principle can be applied for both players. As player in the row (player A) would use a strategy that maximizes its gain, so a strategy is said to be dominated by the second strategy if the payoff values in the second strategy are more or at least equal to those in the first strategy. Player A would always select second strategy, as it would give better gain than the first strategy.

On the other hand, considering player in column (player B) purpose of minimizing losses, a strategy is said to be dominated by the second strategy if payoff values in the second strategy are less or at least equal to corresponding payoff values in the first strategy. Player B would then select the second strategy, as it would result in fewer losses than the first strategy.

These rules are applied on game shown in Table 9.1 that is reproduced as under and renamed as Table 9.3.

For instance, if A1 is compared with A2, then the value of payoff cell a_{11}, i.e. 3 is greater than that of cell a_{21}, i.e. 2. So, A would always select a_{11}. Comparison of a_{12} with a_{22} also results in selection of a_{12} as payoff of 4 is greater than 3. Similarly, payoff in a_{13} of 2 is greater than payoff in a_{23} of 1. This would allow company A to always play A1 in comparison with A2. A1 is called as a dominating strategy and A2 as dominated. This is shown in Table 9.4.

In the case of company B, comparing B1 with B3 would not give a clear dominating strategy. As payoff in cell a_{11} of 3 is less than that of 4 in a_{12}, so B would always select a_{11} indicating strategy B1. However, when the payoff of 4 in a_{31} is compared with that of −2 in a_{32}, B will select a_{32} indicating strategy B2. Comparison of B2 and B3 also gives the same contradicting results. Finally, comparing B1 with B3 provides B3 as a dominating strategy as 2 is less than 3 and −1 is less than 4. So, B would always select B3 when compared with B1. This is shown in Table 9.5.

TABLE 9.3
Payoff Matrix

		Company B		
Strategy		**Before 1 Months (B1)**	**Before 3 Months (B2)**	**Before 5 Months (B3)**
Company A	Before 4 months (A1)	3	4	2
	Before 6 months (A2)	2	3	1
	Before 8 months (A3)	4	−2	−1

TABLE 9.4
Dominating Strategy 1

Strategy		Company B Before 1 Months (B1)	Before 3 Months (B2)	Before 5 Months (B3)
Company A	Before 4 months (A1)	3	4	2
	Before 8 months (A3)	4	−2	−1

TABLE 9.5
Dominating Strategy 2

Strategy		Company B Before 3 Months (B2)	Before 5 Months (B3)
Company A	Before 4 months (A1)	4	2
	Before 8 months (A3)	−2	−1

Repeating the same process for A would show us A1 dominating A3, resulting in Table 9.6.

Repeating for B would result in B3 dominating B2 as B would select strategy that would give minimum loss. This is shown in Table 9.7.

Thus, B would launch new product 5 months before (B3) deadline, whereas A would do so only 4 months (A1). This selection of combination of strategies would provide a gain of two units to company A and loss of same value to company B. as gain value is same as loss value so it results in a zero-sum game.

TABLE 9.6
Dominating Strategy 3

Strategy		Company B Before 3 Months (B2)	Before 5 Months (B3)
Company A	Before 4 months (A1)	4	2

TABLE 9.7
Dominating Strategy 4

Strategy		Company B Before 5 Months (B3)
Company A	Before 4 months (A1)	2

Therefore, in the case of saddle point, the solution is stable as there is no other better alternative for both players. There is no attraction to playing another strategy, as all of them are unviable. These kinds of stable situations with clear and absolute winner and loser are encountered where each player has definite knowledge about other players' choice of strategy. However, in business, most companies confront situations where the best strategy under certain competitive constraints is of cooperation rather than of winning and losing. For instance, in a vendor–retailer relationship, a vendor can opt for various strategies such as low-quality goods at high prices if relationship is one time; reasonable quality products at reasonable prices and so on, i.e. there could be a variety of strategies. Depending on the response from retailers, more than one of these strategies could be appropriate.. *Thus, there can be more than one suitable strategy.* Second, because of huge competition, retailer can procure material from any of the numerous vendors and, similarly, vendor can select any from many retailers. *This makes predicting strategy each player would play difficult.* This was not the case in zero-sum game where each player had one best strategy and was able to predict opponent's strategy. These two differences in availability of suitable alternatives and lack of prediction of the opponent's strategy lead to an explanation of games with mixed strategy that is dealt with in the next section.

9.6 GAMES WITH MIXED STRATEGIES

The following illustration explains the concept of games with mixed strategy. In the above example of two companies fighting to launch a new product at the earliest, suppose if company A decides to introduce the product 4 months before the deadline and company B launches only 1 month before, then the payoff is two units instead of three units. The loss to B is less as launching just 1 month before the deadline would allow it to make product flaw-free. Also, if B decides to expedite the process by trying to introduce the product 5 months before (B3) so as to counter company A's strategy of A1, then A would gain three units and B would lose same. The gain for A is more and loss for B also increases as launching as early as possible can result in the creation of faulty products. This is shown in Table 9.8.

Applying the rule of Maximin and Minimax to find the saddle point would give the result shown in Table 9.9.

TABLE 9.8
Payoff Matrix

		Company B		
Strategy		**Before 1 Months (B1)**	**Before 3 Months (B2)**	**Before 5 Months (B3)**
Company A	Before 4 months (A1)	2	4	3
	Before 6 months (A2)	2	3	1
	Before 8 months (A3)	4	−2	−1

TABLE 9.9
Maximin and Minimax

Strategy		Company B Before 1 Months (B1)	Before 3 Months (B2)	Before 5 Months (B3)	
Company A	Before 4 months (A1)	2	4	3	*2*
	Before 6 months (A2)	2	3	1	1
	Before 8 months (A3)	4	−2	−1	−2
		4	4	*3*	

Maximin = 3 and Minimax = 2 implying A would play A1 to maximize its gain to two units, whereas B would play B3 to minimize its losses to three units. *However, with no saddle point the game is an unstable one.* This is so because both players have better alternative strategies to choose. A would like to play A1 as then it can maximize its gain to four units if B is forced to play B2. However, B would only do so if A plays A3 as then B can gain two units (payoff of −2) incurring a loss of same to A. To avoid this loss A would hope that if it keeps on playing A3, B would be encouraged to play B1 as it would give A gain of four units. However, B would never play B1 as this strategy would always result in loss. Therefore, B would decide to play either B2 or B3. Similarly, suppose B decides to play B3 as it can provide a gain of 1 unit (payoff of −1). Nevertheless, it can happen only if A plays A3. To avoid this loss, A would not like to play A3 and would decide to choose from either A1 or A2. *Thus, there is no one clear and definite strategy for both the players. This is also interpreted as a game having no dominating strategies.*

This lack of definite strategy for both players can also be illustrated by using the principle of dominance. Understanding strategies from Table 9.9 shows that A1 dominates A2, resulting in Table 9.10.

Further reduction of payoff matrix is not possible as no single strategy dominates another for both companies A and B. In such cases, each player selects a strategy randomly by assigning probability to each alternative strategy. So if company A assigns the probability of x to strategy A1, then 1 − x would be the probability of A3 where x would always be positive and lie between 0 and 1. However, company B has three options; thus, probabilities cannot be bifurcated. For instance, if B1 is assigned an

TABLE 9.10
Dominating Strategy 1

Strategy		Company B Before 1 Month (B1)	Before 3 Months (B2)	Before 5 Months (B3)
Company A	Before 4 months (A1)	2	4	3
	Before 8 months (A3)	4	−2	−1

assumed probability of y, then B2 and B3 cannot be assigned $1-y$. So, in the case of 2*m (A has two strategies and B has more than two strategies, i.e. m) or m*2 (A has more than two strategies, i.e. m and B has only two strategies), then game can be resolved by using graphical method which is discussed in the next section. In the case of 2*2, payoff matrix games are solved by assigning probabilities to alternative strategies and the method is termed as algebraic method that will be discussed subsequently.

9.7 GRAPHICAL METHOD

9.7.1 For 2*m Games

After eliminating the dominated strategies, Table 9.10 indicates company A with two strategies A1 and A3 and company B with three strategies B1, B2 and B3. This is a case of mixed strategy. A can select a strategy randomly by assigning a probability of 'x'. Suppose that strategy is A1. Therefore, probability of playing A3 would be $1-x$. To find out the chances of playing a particular strategy, we have to find the value of x. For this purpose, expected payoffs of A are expressed in terms of x for each of its opponent's strategy. These payoffs are plotted on a graph under two conditions:

When A plays strategy A1 i.e. $x = 1$ and $A3 = 1-x = 1-1 = 0$, and

When A does not play strategy A1 i.e. $x = 0$ so it will play A3 i.e. $1-0 = 1$.

To calculate expected payoffs, Table 9.10 has been reproduced here as Table 9.11.

Thus, for each of pure strategies available to company B, the expected payoff of company A would be (Table 9.12):

TABLE 9.11
Assigning Probabilities

		Company B			
Strategy		**Before 1 Month (B1)**	**Before 3 Months (B2)**	**Before 5 Months (B3)**	**Probability**
Company A	Before 4 months (A1)	2	4	3	x
	Before 8 months (A3)	4	−2	−1	1 − x

TABLE 9.12
Expected Payoff

B's Strategy	of Company A
B1	$2x + 4(1-x)$
B2	$4x-2(1-x)$
B3	$3x-1(1-x)$

These lines representing the expected payoff of A with regard to each of pure strategies of B can be plotted on a graph by invoking the above-discussed two conditions. Thus,

i. for B1 : 2x + 4(1 – x)

   When A plays A1, i.e. x = 1, then payoff would be:

$$2x + 4(1 - x) = 2 * 1 + 4(1 - 1) = 2$$

   Similarly, when x = 0, then payoff would be:

$$2x + 4(1 - x) = 2 * 0 + 4(1 - 0) = 4.$$

   Thus, B1 could be represented on the graph by points (2, 4).

ii. for B2 : 4x –2(1–x)

   As stated above, for both conditions, payoff would be:
   When x = 1: 4x –2(1 – x) = 4 and when x = 0: 4x – 2(1 – x) = –2
   Thus, B2 could be represented on the graph by points (4, –2).

iii. for B3 : 3x –1(1 – x)

   When x = 1: 3x – 1(1 – x) = 3 and when x = 0: 3x – 1(1 – x) = –1
   Thus, B3 could be represented (Figure 9.1) on the graph by points (3, –1).

Company A being represented in the row has been discussed as playing by adopting the rule of maximin, i.e. it would always try to maximize its minimum gain. So, in the graph above, identify the highest intersection point in lower bound. It is indicated by entire area below the bold line. The highest point on this line is the maximin point. This point is formed by intersection of two lines, namely, B1 and B3. Thus, out of three alternative strategies by company B graphical method has reduced it to two most appropriate strategies B1 and B3. This leads to the following payoff matrix (Table 9.13).

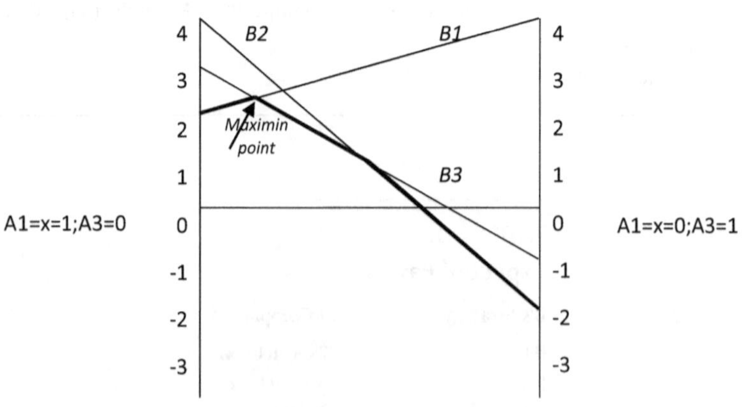

FIGURE 9.1 Graphical representation of 2*m game.

TABLE 9.13
2*2 Payoff Matrix

		Company B		
Strategy		**Before 1 Month (B1)**	**Before 5 Months (B3)**	**Probability**
Company A	Before 4 months (A1)	2	3	x
	Before 8 months (A3)	4	−1	1 − x
	Probability	y	1 − y	

Reduction of 2*m game into 2*2 allows us to assign probability of y to a particular strategy played by B and 1 − y to other strategy. This is shown in Table 9.13. To find most appropriate strategy of both players and value of game, the matrix can further be solved by using the algebraic method.

9.7.2 Algebraic Method for 2*m Games

Assignment of probabilities to random selection of two strategies of both players allows us to adopt the following method.

Expected payoff of A if B plays B1 = 2x + 4(1 − x)
Expected payoff of A if B plays B3 = 3x − 1(1 − x)
Equating both equations would solve for x:

$$2x + 4(1 - x) = 3x - 1(1 - x)$$

x = 5/6 and thus, 1 − x = 1/6

Thus, probability of playing A1 is 5/6 and that of A3 is 1/6.
Similarly,
Expected payoff of B if A plays A1 = 2y + 3(1 − y)
Expected payoff of B if A plays A3 = 4y − 1(1 − y)
Equating both equations would solve for y:

$$2y + 3(1 - y) = 4y - 1(1 - y)$$

y = 2/3 and thus 1 − y = 1/3

Thus, probability of playing B1 is 2/3 and that of B3 is 1/3.

Putting these values of x and y in equations of expected payoffs, we can obtain the value of game as shown under.

Value of game if A plays A1 = 2y + 3(1 − y)

$$= 2 * 2/3 + 3 * 1/3$$

$$= 7/3$$

$$\text{Value of game if A plays A3} = 4y - 1(1 - y)$$

$$= 4 * 2/3 - 1 * 1/3$$

$$= 7/3$$

$$\text{Value of game if B plays B1} = 2x + 4(1 - x)$$

$$= 2 * 5/6 + 4 * 1/6$$

$$= 7/3$$

$$\text{Value of game if B plays B3} = 3x - 1(1 - x)$$

$$= 3 * 5/6 - 1 * 1/6$$

$$= 7/3$$

Thus, A and B by playing any of two available strategies would always win and lose same amount resulting in a zero-sum game. In this case, both players become indifferent as any of the strategies would lead to the same result.

9.7.3 For m*2 Games

Suppose in the same illustration as discussed in Table 9.3, company B inputs more resources to expedite process of new product development. This resulted in reducing its losses from three units to one unit in case product is launched just 1 month before deadline and increase of gain to two units in case of introduction of product in market 5 months before. This leads to the following payoff matrix (Table 9.14).

Applying minimax and maximin rule in order to find saddle point gives indicates its absence. Thus, game involves mixed strategies. By applying the principle of dominance, it can be inferred that B2 is dominated by B3 leading to its elimination. This is shown in Table 9.15.

TABLE 9.14
Payoff Matrix

		Company B		
		Before 1 Month	Before 3 Months	Before 5 Months
Strategy		(B1)	(B2)	(B3)
Company A	Before 4 months (A1)	1	4	2
	Before 6 months (A2)	2	3	1
	Before 8 months (A3)	4	−2	−2

TABLE 9.15
Dominating Strategy

Strategy		Company B Before 1 Month (B1)	Before 5 Months (B3)
Company A	Before 4 months (A1)	1	2
	Before 6 months (A2)	2	1
	Before 8 months (A3)	4	−2
	Probability	y	1 − y

Further reduction of payoff matrix by dominance principle is not possible leading to a payoff matrix with A having more than two strategies and B only 2. Such matrix is termed as m*2 matrix. To solve it graphically, the probability of y and 1-y are assigned to B1 and B3 indicating their chances of selection. For each of the strategies available to A, the expected payoff of B would be (Table 9.16).

To represent A1, A2 and A3, lines on a graph assume first B deciding to play B1 i.e. y = 1 and then B playing B3, i.e. y = 0. This would give the following points for each strategy of A:

i. for A1 : $1y + 2(1-y)$
 When B plays B1, i.e. y = 1, then payoff would be:

$$1y + 2(1-y) = 1*1 + 2(1-1) = 1$$

 Similarly, when y = 0, then payoff would be:

$$1y + 2(1-y) = 1*0 + 2(1-0) = 2$$

 Thus, A1 could be represented on the graph by points (1, 2).

ii. for A2 : $2y + 1(1-y)$
 As stated above, for both conditions, payoff would be:
 When y = 1: $2y + 1(1-y) = 2$ and when y = 0: $2y + 1(1-y) = 1$
 Thus, A2 could be represented on the graph by points (2, 1).

iii. for A3 : $4y - 2(1-y)$
 When y = 1: $4y - 2(1-y) = 4$ and when y = 0: $4y - 2(1-y) = -2$
 Thus, A3 could be represented on the graph (Figure 9.2) by points (4, −2).

TABLE 9.16
Expected Payoff

A's Strategy	of Company B
A1	$1y + 2(1-y)$
A2	$2y + 1(1-y)$
A3	$4y - 2(1-y)$

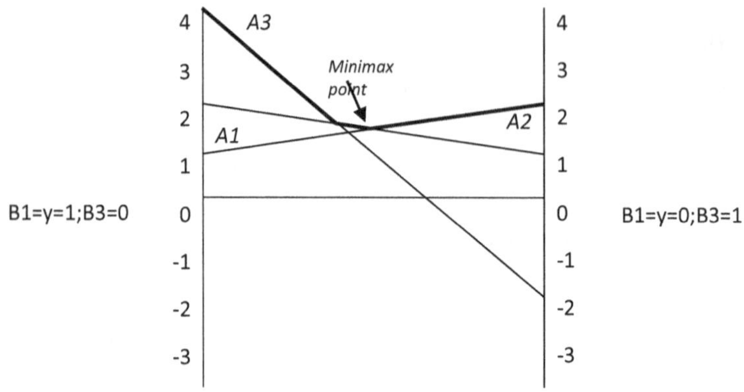

FIGURE 9.2 Graphical representation of m*2 game.

The purpose of company B is to minimize its maximum losses by adopting the rule of minimax. Maximum losses are inferred by identifying the upper bound indicated by all area above bold line. Minimum losses in this uppermost bound is indicated by the lowest intersection value as shown by an arrow. Figure 9.2 shows it to be intersection of lines A1 and A2. This allows us to infer that out of three alternative strategy A would randomly assign probabilities of x and 1 – x to A1 and A2, respectively. Thus, game of 3*2 payoff matrix is reduced to 2*2 as shown in Table 9.17.

This is further solved by using the algebraic method in order to find the most appropriate strategy of both players and value of game.

9.7.4 Algebraic Method for m*2 Games

Assignment of probabilities to random selection of two strategies of both players allows us to adopt the following method.

Expected payoff of A if B plays B1 = 1x + 2(1 – x)

Expected payoff of A if B plays B3 = 2x + 1(1 – x)

Equating both equations would solve for x:

$$1x + 2(1 - x) = 2x + 1(1 - x)$$

TABLE 9.17
2*2 Payoff Matrix

Strategy		Company B: Before 1 Month (B1)	Company B: Before 5 Months (B3)	Probability
Company A	Before 4 months (A1)	1	2	x
	Before 6 months (A2)	2	1	1 – x
	Probability	y	1 – y	

x = 1/2 and thus, 1− x = 1/2

Thus, probability of playing A1 is 1/2 and that of A2 is also 1/2.

Similarly,

Expected payoff of B if A plays A1 = 1y + 2(1 − y)

Expected payoff of B if A plays A2 = 2y + 1(1 − y)

Equating both equations would solve for y:

$$1y + 2(1-y) = 2y + 1(1-y)$$

y = 1/2 and thus, 1 − y = 1/2

Thus, probability of playing B1 is 1/2 and that of B3 is also 1/2.

Putting these values of x and y in equations of expected payoffs, we can obtain the value of game as shown under.

Value of game if A plays A1 = 1y + 2(1 − y)

$$= 1*1/2 + 2(1-1/2)$$

$$= 3/2$$

Value of game if A plays A2 = 2y + 1(1 − y)

$$= 2*1/2 + 1(1-1/2)$$

$$= 3/2$$

Value of game if B plays B1 = 1x + 2(1 − x)

$$= 1*1/2 + 2(1-1/2)$$

$$= 3/2$$

Value of game if B plays B3 = 2x + 1(1 − x)

$$= 2*1/2 + 1(1-1/2)$$

$$= 3/2$$

Thus, A and B by playing any of two available strategies would always win and lose the same amount, resulting in a zero-sum game. In this case, both players become indifferent as any of the strategies would lead to the same result.

9.8 LINEAR PROGRAMMING FORMULATION

The illustration discussed above can be solved by constructing the problem as linear programming model and solving by using Solver application in Excel. The payoff matrix is reproduced in Table 9.18.

TABLE 9.18
Linear Programming Formulation

Strategy		Company B: Before 1 Month (B1)	Before 3 Months (B2)	Before 5 Months (B3)	Probability of A Playing a Particular Strategy
Company A	Before 4 months (A1)	2 (a_{11})	4 (a_{12})	3 (a_{13})	x_1
	Before 6 months (A2)	2 (a_{21})	3 (a_{22})	1 (a_{23})	x_2
	Before 8 months (A3)	4 (a_{31})	−2 (a_{32})	−1 (a_{33})	x_3
Probability of B playing a particular strategy		y_1	y_2	y_3	

Objective of company A is to maximize its gains. Therefore, for A, the objective function would be of maximization. By assigning probability of playing strategy A1, A2 and A3 as y_1, y_2 and y_3, respectively, the following equations can be derived.

$$A1 = 2y_1 + 4y_2 + 3y_3$$

$$A2 = 2y_1 + 3y_2 + 1y_3$$

$$A3 = 4y_1 - 2y_2 - 1y_3$$

Payoff values indicate coefficients of decision variables. They are represented by a_{ij}.
This is generalized as:

$$\textbf{Maximize } Z = \max\left(\sum a_{1i}y_i \sum a_{2i}y_i \sum a_{3i}y_i\right)$$

Constraints:

Probability of A and B playing a particular strategy is indicated by x_i and y_j. Now sum of probabilities will always be equal to one making this as constraint. Mathematically, these are represented as:

$y_1 + y_2 + y_3 = 1$ i.e.

$$\sum y_i = 1$$

$0 \le y_i \le 1$ where i = 1,2….n.

Similarly, *objective* of company B is to minimize its losses. So, for B, the objective function indicated by Z would be of minimization.

The equations for each strategy of company B could be written as:

$$B1 = 2x_1 + 2x_2 + 4x_3$$

$$B2 = 4x_1 + 3x_2 - 2x_3$$

$$B3 = 3x_1 + 1x_2 - 1x_3$$

Payoff values indicate coefficients of decision variables. They are represented by a_{ij}.

This is generalized as:

$$\textbf{Minimize } Z = \min\left(\sum a_{j1}x_j \sum a_{j2}x_j \sum a_{j3}x_j\right)$$

Constraints:

Sum of probabilities of A playing a particular strategy will always be equal to one. Mathematically, it is represented as:

$x_1 + x_2 + x_3 = 1$, i.e.

$$\sum x_j = 1$$

$0 \leq x_j <= 1$ where i = 1,2….m.

Excel Solution

Understanding the problem in linear programming formulation could be used to solve it by using the Solver application of Excel.

Step 1: Input the payoff matrix in Excel spreadsheet.

Step 2: To solve for optimal strategy of company A, probability of B playing a particular strategy indicated as y_1, y_2 and y_3 are shown as decision variables in cells C10:E10. Constraint is sum of all these probabilities shown in cell C12 =sum(C10:E10). Lastly objective function is the minimum of value of game for each strategy. It is calculated as =sumproduct(C3:E3,$C$10:$E$10) in cell C15 if A plays A1. For other two strategies, the same formula is copied in a way that coefficients change. If A plays A2, then in cell C16 =sumproduct(C4:E4,$C$10:$E$10), and if A plays A3, then in cell C17 = sumproduct(C5:E5,$C$10:$E$10). The objective function Z of maximization is =max(C15:C17) shown in cell C19.

Step 3: In Solver application, set objective in cell C19 to maximization with changing variables as C10:E10. Add constraints as C12 = 1. Click OK and then Solve. Solution as shown in Figure 9.3 would be provided.

Step 4: For optimal strategy of company B after creating formulas for objective function and constraints, apply Solver application as discussed. Solution is again shown in Figure 9.3.

9.9 SUMMARY

Game theory provides a conceptual framework to resolve the decision-making problems in a competitive environment. This chapter particularly deals with problems involving two persons, teams or firms indulging in opposite actions. In such opposing situations, both players are rational and play a strategy, which will either maximize the profit or minimize the loss. To reach to this decision, this chapter has focused on two categories of game theory. First, a game where both opponents have

Strategy		Company B: Before 1 month (B1)	Company B: Before 3 months (B2)	Company B: Before 5 months (B3)	Probability of A playing a particular strategy
Company A	Before 4 months (A1)	2	4	3	x_1
	Before 6 months (A2)	2	3	1	x_2
	Before 8 months (A3)	4	-2	-1	x_3
Probability of B playing a particular strategy		y_1	y_2	y_3	

Optimal strategy for A			
Decision variables	y_1	y_2	y_3
	0.3333	0.3333	0.3333
Constraint	1.0000		
Objective function			
if A plays A1	3.0000		
if A plays A2	2.0000		
if A plays A3	0.3333		
Maximize Z	3.0000		

Optimal strategy for B		
Decision variables	x_1	1
	x_2	0
	x_3	0
Constraint		1
Objective function		
if B plays B1	2	
if B plays B2	4	
if B plays B3	3	
Minimize Z	2	

FIGURE 9.3 Game theory solution in excel.

a clear and definite strategy wherein gain by one player would be equivalent to loss by other or vice versa leading to the formation of two person zero-sum game. Such kind of games possess an equilibrium or saddle point, resulting in stable situation. On the other hand, mostly in business and geopolitics, the strategies are not clearly defined, and both parties tend to reach for cooperation wherein both players would win, though not the maximum. Such games are termed as mixed strategy games in which randomness in selection of strategy is done by applying probabilities to various available strategies.

Finally, this chapter has discussed game theory by limiting a game to be played by only two parties, whereas in reality, like in a game of cards or a highly competitive marketplace, there can be multiple players, each with a different set of strategies. N player games require advanced studies on game theory that this chapter has left out.

9.10 GLOSSARY

- **Game**: involves *at least two players* where a player can be an individual or a team
- **Payoff**: is an outcome that is always a combination of particular strategy of two players.
- **Minimax rule**: Selecting minimum of maximum value from each row.
- **Maximin rule**: Selecting maximum of minimum value from each column.
- **Saddle point**: also known as equilibrium point is reached when minimax = maximin

- **Mixed strategy**: When there is no one clear and definite strategy for both the players indicated by no saddle point, implying that the game is an unstable.
- **Zero-sum game**: When A and B play any of two available strategies, they would always win and lose the same amount, resulting in a zero-sum game. In this case, both players become indifferent as any of the strategies would lead to the same result.

9.11 CASE STUDY: VENDOR–RETAILER RELATIONSHIP

Retailers rely heavily on reliable supplies from vendors to provide satisfactory products to customers. This requires a high level of coordination between two parties, i.e. retailer and vendor. However, the relationship is never cooperative. The lure of achieving higher profits at the expense of other party also creates scenarios of conflict. For instance, in a one-time relationship of selling and buying, the vendor might resort to compromise on the quality of product or delay the delivery and tend to achieve higher profit. On the other hand, retailers might cancel the order at the last moment on the pretext of buying similar quality products from other vendors at lower prices. Delaying payments is a common practice among retailers that somehow forces vendors not to cooperate. This is termed as achieving short-term gains at the expense of others by not cooperating or defecting from the agreement. This scenario of relationship involving cooperation or defection between vendor and retailer has been discussed in four following cases.

First, both vendor and retailer indulge in an agreement where the vendor decides to provide good quality products at a reasonable price. In addition, the retailer agrees to fulfil the contract and would clear the dues on time. In addition to cooperating and fulfilling the contract, both parties have an option to defect. Payoffs of both parties are represented on a scale of 1–4 where 4 is the maximum and 1 is the minimum payoff. Under such situation, if both parties cooperate, i.e. good quality products are provided at not the highest price, then payoff of retailer is indicated by a gain of 2. On the other hand, if one party defect and other sticks to cooperation then it gains maximum, i.e. 4. Finally, no one gains anything if both of them cheat or defect from the agreement. In this case, the vendor tries to cheat on the quality of products and retailer cheats by cancelling the order.

Second, in another case, suppose vendor decides to defect from the agreement. Then a retailer has an option to keep cooperating or it can also decide to cheat. In the case of short-term relationships, if a retailer keeps on cooperating, then it would lose 2 which would be gain of vendor. However, if retailer also decides to defect, then it shows a gain of 3 as a retailer can always opt to obtain products from a number of other vendors available in the market. Whereas vendor might get blacklisted impacting its long-term sustainability. Due to high expectation of loss in business, if the vendor switches back to cooperation but retailer decides to defect by delaying the payment, then it results in gain of 1 to retailer. If both keep on cooperating, then the retailer earns a gain of 2 as was in the previous case.

Third, retailers, instead of playing the game in terms of either cooperating i.e. buy or defecting i.e. not to buy, decides to add another strategy of procuring only 50% of order and testing it before finalizing the entire order. This results in following payoff matrix. Interestingly, retailer loses heavily if it intends to defect and vendor sticks to cooperating. This could be due to high-quality goods provided by a reliable vendor and not fulfilling the order retailer would lose out.

		Vendor	
		Cooperate	**Defect**
Retailer	Cooperate (buy)	2	−4
	Defect (don't buy)	−3	0
	Buy 50% of product	1	−1

Finally, suppose now vendor has three strategies to play to counter two strategies of cooperation or defection of retailer. Vendor in addition to aforementioned strategies of cooperating, i.e. providing good quality products at reasonable prices and defecting i.e. providing bad quality products at high prices decides to have another option of proving reasonable quality products at reasonable prices. Payoff matrix of such strategies is shown below.

		Vendor		
		Cooperate (Good Quality at Reasonable Prices)	**Defect (Bad Quality High Prices)**	**Reasonable Quality at Reasonable Prices**
Retailer	Cooperate (buy)	2	−4	−1
	Defect (don't buy)	−3	0	1

a. Create payoff matrix for first two cases and determine the most appropriate strategy for retailer and vendor.
b. For third and fourth cases with given payoff matrix determine the most appropriate strategy for both players.

9.12 MODEL QUESTIONS

9.12.1 Find saddle point of following games:

i.

		Player B		
		B1	**B2**	**B3**
Player A	A1	3	2	1
	A2	3	4	5
	A3	−6	3	−2

ii.

		Player B			
		B1	B2	B3	B4
Player A	A1	−3	1	0	7
	A2	4	4	3	5
	A3	2	−1	−2	−3

iii.

		Player B		
		B1	B2	B3
Player A	A1	5	7	4
	A2	3	9	1

9.12.2 Find saddle point of following games. If games do not have saddle point, find appropriate strategy of both players and value of game by algebraic method.

i.

		Player B	
		B1	B2
Player A	A1	4	0
	A2	2	3

ii.

		Player B	
		B1	B2
Player A	A1	1	3
	A2	6	4

iii.

		Player B	
		B1	B2
Player A	A1	−2	5
	A2	1	−1

9.12.3 Reduce following games by using dominance principle and then find the most optimal strategy of both players:

i.

		Player B			
		B1	**B2**	**B3**	**B4**
Player A	A1	4	−1	2	0
	A2	1	3	1	4
	A3	2	0	−1	3

ii.

		Player B		
		B1	**B2**	**B3**
Player A	A1	3	1	−2
	A2	−1	−4	−3
	A3	1	−5	−1

iii.

		Player B		
		B1	**B2**	**B3**
Player A	A1	3	2	0
	A2	−1	0	1

iv.

		Player B		
		B1	**B2**	**B3**
Player A	A1	0	5	1
	A2	4	1	3
	A3	2	0	3

v.

		Player B			
		B1	B2	B3	B4
Player A	A1	2	1	3	−1
	A2	1	2	−1	3
	A3	4	4	3	−1
	A4	0	3	0	5

9.12.4 For the following payoff matrices:

a. Use minimax or maiximin criterion to determine most appropriate strategy for both players.
b. Illustrate the principle of dominance in order to eliminate dominated strategies for the same purpose on the same payoff matrices.

i.

		Player B		
		B1	B2	B3
Player A	A1	−2	0	3
	A2	0	2	2
	A3	0	−1	−3

ii.

		Player B			
		B1	B2	B3	B4
Player A	A1	3	−2	−2	0
	A2	−2	0	−3	2
	A3	−1	3	−2	2

9.12.5 Two firms were engaged in a duel of increasing sales of their competing product. For this they decided to promote the product via three mediums namely TV, online and print. Selection of particular medium by a firm would be retaliated by its opponent by selecting same or other available mediums. Appropriateness of a medium is decided by amount of increase in sales it can generate. Following payoff matrix shows increase in sales of a firm when both firms have adopted a particular medium. For instance, if both firms decide to promote through TV then value of 1 in the matrix imply firm A sold 1 unit (in millions) more than B. Find the most appropriate strategy of both firms.

		Firm B		
		TV	**Online**	**Print**
Firm A	TV	1	3	−1
	Online	3	1	−2
	Print	2	−1	2

9.12.6 Leaders of two political parties engage in a bitter campaign to garner more votes in order to win a state election. Both parties have short-listed three major issues on which they decide to target their opponent. Selection of a particular issue depends on its effectiveness in terms of capturing opponents' votes. It is assumed that there are fixed number of voters and all will vote. One party can only gain by garnering more votes than its competitor. Also selection of one issue is countered by selecting an issue by other party. So, the following payoff matrix shows the expected increase in votes of a party with regard to combination of issues of respective parties. Apply game theory principle to find most appropriate issue of each party.

		Party B		
		Issue 1	**Issue 2**	**Issue 3**
Party A	Issue 1	5	2	−2
	Issue 2	0	1	−1
	Issue 3	−2	3	1

9.12.7 Solve the game of Problem 9.12.6 if payoff matrix gets modified to following where instead of gain of three units to party A if it selects to campaign with issue 3 and party B adopts issue 2, it loses 1 unit which is gained by party B. Also party A now loses one and two units on issue 1 and issue 3, respectively, if party B selects issue 3.

		Party B		
		Issue 1	**Issue 2**	**Issue 3**
Party A	Issue 1	5	2	−1
	Issue 2	0	1	−2
	Issue 3	−2	−1	1

9.12.8 Two FMCG companies A and B are engaged in a price war. Both companies can either offer the product at a high price or a low price. A product offered by one company at a low price forces other company to follow the price cut. This price cutting, however, results in loss in profits though it increases sales, but a company has to follow price cutting strategy of its competitor. On the other hand, if one company maintains high price, then so does other company. Hence, no action is undertaken unilaterally. Thus, a company's strategy depends on its rival strategy and a change in strategy would depend on opponents' action. Under such scenario, suppose both companies maintain a high price strategy of selling the product and thus sharing the market by recording a sale of 100 units each. Company B initiates price war by deciding to cut the price of its product while B still sells the product at a high price, resulting in an increase of sales to 130 units. Sale of company B drops to 70 units. Similarly, if company A initiates a price cut and company B keeps on following high price strategy, then A gains with sales of 150 units while sales of B drops to 50 units. Finally, if both companies indulge in similar price cut then both experience in drop of sales to 80 units each.

a. Create a payoff matrix indicating strategies and payoff of both companies.
b. Does the game have a dominant strategy? If yes, which? Explain its implication.
c. Suppose company A follows low price strategy and company B high price. Instead of earning a gain due to price cut sales of company A falls to 40 units and that of B remains same at 50 units. How does this change (a) and (b)?

9.12.9 Two players A and B are playing game of rock, paper and scissor. In case of tie, i.e. for instance both players play rock then none of them gains or losses anything and game has to be played again. On the other hand, as rock supersedes scissor so in such mix of strategies a player with rock would gain three units and one with scissor would lose the same. In duel of paper and scissor player opting to play scissor would win two points and player with paper would lose same. Finally, in case of rock and paper duel player with paper would win one point and other would lose the same.

a. Create a payoff matrix indicating strategies and payoff of both players.
b. Does the game have a dominant strategy? Explain by emphasizing on the concept of stability.

Index

Activity-on-arrow 316
Activity-on-Node 317
additivity 22
Adidas AG 5
advertising 25
algebraic method 375
allocation of limited resources 20
allocation problem 296
allowable range 132
alternative diets 35
alternative evaluation 4
alternative solutions 114
application of algorithms 8
arcs 179
artificial cell 206
artificial variable 86, 87, 88, 94, 111
assignee 233
assignment 233
assignment of workforce 39
assumptions 236

backward path rule 323
balanced problem 238
basic variable 70, 78, 138
basis variables 69, 70, 86
Big M method 87
blending problems 37
branch and bound method 286
branch network 287

certainty 22
change in coefficient 134, 138
choices 4
Clark Wright 298
closed loop 193
coefficients 6, 7
common area 41, 43
competitive 127
competitive forces 23
complementary 127
constrained 9
constrained optimization 20
constraint(s) 6, 7, 8, 42
constraint functions 127
cost slope 340
crash cost 339
crash time 339
crashing 338
crew assignment 260
crew pairing 262
criteria 4, 233
critical path 325
critical path method 315, 321
current tour 298
customer perspective 23
cyclic 282

decision-making 2, 3
decision variables 6
degeneracy 107, 206
degenerate 107, 206
demand 181
destinations 179
Dijkstra's algorithm 303
divisibility 22
dual 144, 145
dual variables 145
duality 126
dummy activities 318
dummy column 244
dummy demand centre 198
dummy row 244
dummy supply centre 198

earliest finish time 322, 323
earliest start time 322
estimates 125, 128, 144
excel solution 45, 47, 75, 96, 141, 196, 201, 205, 225, 243, 249, 252, 257, 262, 270, 294, 304, 328, 336, 343, 381
expected payoff 373, 375, 378
expected time 333

feasibility 159, 161, 163, 182, 238
feasibility of dual 150
feasibility of optimal solution 21
feasible region 22
feasible solution 41
financial instruments 27
formulate the problem 3
forward path rule 323
free slack 326

game theory 363
Gantt chart 316
George Dantzig 67
graphical method 41, 374
greater than equal to constraint 101

head even slack 326
heuristic method 296
Hungarian algorithm 238

incoming cell 191
incoming variable 72
independent slack 327
inequalities 7
infeasibility 111
infeasible 22, 44
initial solution 186
insensitive parameters 136
integer value 182
intersection points 41

latest finish time 323
latest start time 323
least cost method 186
linear programming 19
linearity 20
looping 319

make-or-buy decisions 33
managing processes 2
marketing research 23
mathematical representation 7
maximin 368, 374
maximization 10, 41, 67, 209
maximization problem 253
media channels 25
minimax 367, 368
minimization 10, 49, 82
mixed strategy 363, 371
model(s) 5
model solution 8
MODI method 191
modified model 156
most likely time 332
multi criteria problems 35
Musashi Auto Parts 182

net evaluation row 71, 72, 79
network flow 179
network planning techniques 316
nodes 179
non-basic variable 134
non-critical path 326
non-negative coefficients 131
non-negativity criteria 70, 148
non-standard form 155
normal cost 339
normal time 339
North-West Corner method 188
number of allocations 186

objective function 6, 7, 44
operations 1
operations management 2
operations research 2
optimal solution 8, 44, 241
optimality 159, 161, 163, 240
optimistic time 332
optimization 8
optimum 4, 6, 10
original model 131
origins 179
outcome 363
outgoing cell 191, 193
outgoing variable 72

payoff 364
pessimistic time 332
players 363
portfolio selection 27
post-optimality analysis 125, 126
precedence relationship 321
predecessor 317
primal 144, 145
principle of dominance 369
problem 3
processes 2
product-mix problems 29
program evaluation review technique 315, 331
programming 20
project planning 315
project schedule 322
proportionality 21

range of optimality 140
repetitive activities 319
resource levelling 351
resource limited scheduling 347
resource planning 346
resource smoothening 351
revised model 131
right hand side values in constraint functions 127, 129
routing problem 296

saddle point 368
scheduling 315
sensitive parameters 125, 140
sensitivity analysis 125, 126
shadow price 129, 131
shortest path problem 302
shortest route 180
simplex method 67, 68, 78, 84, 97, 102
slack variable 46, 54, 86, 147
square matrix 282
stepping stone method 193
strategy 363
sub-optimality 154
sub-tours 285
successor 317
supply 181
surplus variable 54, 86, 147

tail event slack 328
task 233

Taylor, F.W. 1
total slack 325
transhipment 221
transportation 179, 180
travelling salesman problem 180, 281
true values 125

unacceptable assignment 257
unacceptable routes 216
unbalanced assignment 244
unbalanced 110, 198
unconstrained 9, 11, 155, 157
unconstrained optimization 11
unoccupied cells 206

valuation 364
value of game 375, 376, 379
vehicle routing problem 180, 295
vendor selection 30
Vogel's approximation method 189

what-if analysis 125

zero-sum 363, 368

www.ingramcontent.com/pod-product-compliance
Lightning Source LLC
LaVergne TN
LVHW011107300926
849160LV00008B/300

9781032081076